Scripturalist Islam

Islamic Philosophy, Theology and Science

Texts and Studies

Edited by

H. Daiber

VOLUME 72

Scripturalist Islam

The History and Doctrines of the Akhbārī Shīʿī School

by

Robert Gleave

BRILL

LEIDEN • BOSTON
2007

This book is printed on acid-free paper.

ISSN 9004-15728X
ISBN 978 90 04 15728 6

PRINTED IN THE NETHERLANDS

For My Parents

CONTENTS

LIST OF FIGURES

ACKNOWLEDGEMENTS

If I forget to mention the names of the numerous people who have helped me in the course of writing this book, I ask their forgiveness. This book is the result of ten years of research, off and on, which began under the tutorship of Dr Paul Luft and the late Dr Norman Calder in Manchester. I had planned to write something similar to the present volume for my doctoral dissertation, but Norman and Paul, wisely, persuaded me to write a thesis on one Akhbārī thinker (Yūsuf al-Baḥrānī) rather than tackle the whole Akhbārī school. I owe an immense debt of gratitude to them both for this piece of wise counsel. If I had insisted on my original plan, I would probably have finished my doctorate a decade later than I did. This more general study was mothballed again, as I completely re-wrote some of my doctoral dissertation to focus on al-Baḥrānī's *uṣūl al-fiqh*, comparing it with that of the Uṣūlī Muḥammad Bāqir al-Bihbihānī. The result was my previous work, *Inevitable Doubt*, on the Akhbārī-Uṣūlī dispute. Norman's ideas and writings on the development of Shīʿī legal thought (and Islamic jurisprudence more generally) continue to influence me, even if my position diverges from him on specific points. Similarly Paul's insistence on interpreting the Akhbārī school in the context of Iranian history is also evident in my research, even though my focus here concerns the ideas of Akhbarism. Students are always indebted to their teachers, but teachers should not be held responsible for their students' mistakes.

During visits to Iran and elsewhere, various scholars have generously given me their time. Mansur Sefatgol and Susan Asili of the University of Tehran have found and copied manuscripts, provided delicious meals and allowed me to interrogate them until the early hours about the arcane elements of the Akhbārī-Uṣūlī dispute. Ustad Abdol Husayn Ha'iri of the Majlis Library, together with his friendly assistants, made me most welcome in their manuscript library whilst in Tehran. Drs Mousavinezhad, Isfandiar and Shariatmadari of the Markaz-e Matala'at va Tahqiqat-e Adyan va Mazaheb in Qum allowed me to test out my ideas in an extended discussion seminar at the Markaz. The staff of the Marʿashi library in Qum and the Astan-e

Quds library in Mashhad were gracious in their patience as I got to grips with their catalogue and ordered manuscript copies.

My research also benefited from two periods of intensive research as visiting scholar, first at the University of Michigan and then at the University of Washington, Seattle. At Ann Arbor, Professor Juan Cole showed me great hospitality, providing me with sources and acting as a soundboard for ideas. This trip was made possible through the support of the British Academy. In Seattle, Professor Michael Williams and the staff at NELC made my stay in that wonderful city comfortable and productive. This was made possible through a grant from the Worldwide Universities Network.

In Britain, my colleagues in the Department of Theology and Religious Studies at the University of Bristol, and more recently at the Institute for Arab and Islamic Studies at the University of Exeter have provided productive research environments in which to work. In particular Dr Ayman Shihadeh who read a chapter and gave useful comments, and Dr Sajjad Rizvi, who provided me with a pre-publication copy of his work on Mullā Ṣadrā. The Arts and Humanities Research Council provided matching study leave in 2005, without which research on this book would have been stalled for an even longer period of time. The British Institute of Persian Studies have provided me with facilities and support throughout the last decade. Of course, my wife, Sarah and son, Joshua, have had to suffer the trials of a book which has been "nearing completion" for too long.

Finally, the anonymous readers of parts of this book, both from Brill and elsewhere, gave excellent advice on areas for improvement at various stages. I may not have followed up on all their lines of enquiry, but I have benefited from their insights. I think I know who they are, but etiquette demands that I thank them anonymously.

NOTE ON TRANSLITERATION, DATES AND REFERENCES

Transliteration in this volume follows the now standard Arabic conventions, even when Persian words are transliterated. This was thought preferable to running two transliteration systems simultaneously. Hence the Arabic *ḍ* or *dh*, though found in a Persian passage (and hence pronounced, and according to some transliteration systems written, *z*) are written as *ḍ* and *dh* respectively. A few allowances have been made to Persian transliteration, however. The *tāʾ marbūṭa* is left as an *-a* in Arabic passages (except in an *iḍāfa* when it is written *-at* before a *ḥamzat al-waṣl*). However, when a word with a *tāʾ marbūṭa* is used in a Persian passage, it is written *-ah* (unless it has been converted to a *tāʾ* in the course of its transplantation into Persian, when it is written *-at*). For consistency, the Persian *-eh*/*-ah* ending is written *-ah*. The Persian *ezāfa* is written as *-i* or as *-yi* as appropriate.

When two dates are given, the format is Hijrī Qamarī/Common Era. When a single date is given, unqualified by subsequent abbreviations, it is Hijrī Qamarī. I give only Common Era dates when absolutely necessary, and hence most dates in this book are Hijrī Qamarī. Common Era dates are followed by "CE" and Hijrī Shamsī dates by "Sh". In the References section, dates are given in the form supplied by the publisher (Hijrī Qamarī or CE, without qualification), though "Sh" is added to Shamsī dates to avoid Qamarī/Shamsī confusion.

References are given in abbreviated form, and publication details can be found in the References section at the end of the book. References to manuscripts take the form f.1a.1 (folio 1, verso, line 1). When a manuscript has numbered pages (usually added by a later hand), this is given in parenthesis (p. 1.1, meaning page 1, line 1) rather than using "a" and "b".

PREFACE

Shīʿī Muslims consider the political and religious leadership of the Muslim community to be the right of the descendents of the Prophet Muḥammad. Early Muslim historical writings concur on the fact that Muḥammad's only descendents were through the marriage of his daughter to the Prophet's cousin, ʿAlī b. Abī Ṭālib (d. 40/661). According to Shīʿī Muslims the Prophet had designated ʿAlī as his successor before his death in 9/632, and reserved future leadership of the Muslim community to ʿAlī's descendents, collectively known as the "People of the House" (*Ahl al-Bayt*). The existence of a leader, designated by the Prophet himself, certainly stunted the development of a distinct Shīʿī jurisprudence. This is not to say there was no Shīʿī law. The various descendents of the Prophet who received Shīʿī devotion (the Imams) were asked questions about right conduct and proper compliance with the Sharīʿa. They gave answers with which their followers were ordered to comply. However, when an Imam was present there was no need for an overarching jurisprudence. Since the Imam could answer all legal enquiries, there was no need to create a framework into which the Imams' rulings collectively might fit. The doctrine of the Imamate, then, reduced the need for legal theory. This at least partially explains why the Shīʿa were tardy in producing works of *uṣūl al-fiqh* (legal theory) when compared to those developed by Sunnis. However, as the Shīʿa began to fissure internally, the doctrinal paths taken by Shīʿī groupings led to varying levels of theoretical need. The Zaydī Shīʿīs, for example, proposed an Imam with reduced charismatic authority, whose rulings were perhaps the most authoritative; however, they left space for individual jurists to propose their own legal opinions. In order to do this, the jurists had to base their opinions on theoretical principles—whimsical adherence to a particular ruling because it appealed to a jurist was not considered sufficient grounds upon which to argue for its superiority over the opinions of other jurists. Hence, theoretical writings amongst the Zaydiyya began to be written in the early Fifth/Eleventh Century at the latest.[1] There were those who argued for continued devotion

[1] An early (perhaps the earliest) Zaydī work of *uṣūl al-fiqh* is *al-Mujzī fī uṣūl al-fiqh* by Abū Ṭālib Yaḥyā b. al-Ḥusayn al-Nāṭiq biʾl-Ḥaqq (d. 424/1033, see Ziriklī, *al-Aʿlām*, v. 8, p. 141).

to the Imam, since he was available and able to give answers. The Ismāʿīlī Shīʿī tradition—both under the Fatimids and later amongst the Nizārī Ismāʿīlīs—presents the most dedicated adherence to this doctrine. There may have been a need for the law to be collected, as it was in al-Qāḍī al-Nuʿmān's (d. 303/974) *Daʿāʾim al-Islām*, but there was little need for a general theory when the Imam was present.[2] This lack of interest in legal theory eventually resulted in a lack of interest in the law generally, and the Ismāʿīlī Shīʿī tradition after Qāḍī Nuʿmān produced few significant legal works.

The Imāmī Shīʿa developed an interest in legal theory following the disappearance of their Twelfth Imam in 260/874. The doctrine of the Imam's occultation (*ghaybat al-imām*) was explained by later Shīʿīs as being necessitated by the oppression of the Imāmī Shīʿī community by the Sunnis. Whether or not this explanation is credible, the removal of the Imam certainly enabled Imāmī Shīʿī scholars to think more creatively about the law. Deciding whether the doctrine of the *ghayba* was a result of scholars' efforts (to free up intellectual space for their activities), or emerged on the Imam's own initiative (as Shīʿī tradition portrays it) is outside the scope of this study. What is clear, though, is that theoretical reflection on the legal coherence of the Imams' rulings emerged tentatively prior to the *ghayba*, and flourished thereafter. Scholars could now speculate on what the Imams' meant when they made a particular legal pronouncement, how such pronouncements might be married with other sources of revelation (such as the Qurʾān) and whether their statements reflected the true law or the Imams' own dissimulation in an attempt to protect the Shīʿa from suffering at the hands of the Sunnis. Within a hundred years of the Twelfth Imam's "greater occultation" in 329/941 (after it became clear his promised return was not imminent), works of *uṣūl al-fiqh* were being composed by Imāmī (or Twelver) Shīʿī scholars and the basis for a tradition of juristic scholarship was being laid. Since this book is only concerned with Imāmī jurisprudence, from hereon, reference to Shīʿī, Shīʿa and Shiʿism refers to the Imāmī or *Ithnāʿashariyya* (Twelver) expression of Shiʿism.

---

[2] Al-Qāḍī al-Nuʿmān's *Ikthilāf Uṣūl al-Madhāhib* shows the author's great interest in legal theory, but is ultimately a rejection of the need for a discipline of *uṣūl al-fiqh*. For a debate on this, see "Alta Discussion", pp. 419–420.

The early development of Imāmī Shīʿī law has been outlined by other writers.[3] The increasing rationalisation of the law and the search for coherence amongst the Imams' sayings (*akhbār* or *ḥadīths*) did not happen in isolation. One sees the influence of Sunni legal theory in the earliest extant works of Shīʿī *uṣūl al-fiqh*, and the subsequent tradition was, in part at least, an attempt to establish Shīʿī legal theory as a viable scholastic discipline in competition with that of Sunni jurists. Stewart has argued for a close connection between Shāfiʿī jurisprudence and the developing Shīʿī tradition.[4] Biographical records point towards Shīʿī scholars studying with Shāfiʿī jurists, and later, Shāfiʿī works of *uṣūl al-fiqh* became important elements in the training of Shīʿī jurists. Central Sunni *uṣūl* concepts, such as *ijtihād* and the division between "religious matters known with certainty" (*ʿilm, qaṭʿ*) and those which are open to informed dispute (*ẓann*), were incorporated into Shīʿī *uṣūl* from at least the time of the great Shīʿī jurist al-ʿAllāma al-Ḥillī (d. 726/1325). The *akhbār*, collected in the famous "Four Books" of Shīʿī *ḥadīths*,[5] were considered insufficient in themselves to provide answers to all the elements of a comprehensive legal system. There was a need to test the *akhbār*, to distinguish the historically accurate from the dubious, to resolve apparently contradictory reports and to discern what the Imams' rulings might have been on issues never put to them. These are basic requirements of a workable legal system, and require procedures. Or alternatively, when the law was already known, a fundamental legal theory was necessary to justify it. Through such a theory, individual rules—particularly those which appeared to be anomalous—could be justified or modified. Some of the Imams' pronouncements explicitly forbade certain Sunni exegetical procedures (such as *qiyās*) and Shīʿī scholars were unable to interpret such reports away. When the Imams condemned *ijtihād*, it was interpreted as referring simply to *qiyās*, and not to the broader set of hermeneutic procedures found in the works of al-ʿAllāma and those who came after him.[6] In this way, Shīʿī *uṣūl*

[3] See Modarressi, *Introduction to Shīʿī Law*, pp. 1–57; Calder, "The Structure of Authority"; Stewart, *Islamic Legal Orthodoxy*.

[4] Stewart, *Islamic Legal Orthodoxy*, pp. 61–110.

[5] Namely, Muḥammad b. Yaʿqūb al-Kulaynī's *al-Kāfī*, al-Shaykh al-Ṣadūq Ibn Bābūya's *Man lā Yaḥḍuruhu al-Faqīh* and the *Tahdhīb al-Aḥkām* and *al-Istibṣār* of Muḥammad b. al-Ḥasan al-Ṭūsī. On the composition and structure of these texts, see Gleave, "Between *ḥadīth* and *Fiqh*".

[6] See Gleave, "Qiyās".

works began to resemble those composed by Sunnis, most obviously in structure, but in large part in conclusions also.

Resistance to the creeping rationalisation of the law (and the accusation of Sunni influence on Shīʿī legal theory) was patchy. There is little textual evidence of any immediate Shīʿī opposition to al-ʿAllāma's theory. Instead, subsequent writers of Shīʿī *uṣūl al-fiqh* endorsed his epistemological dichotomy between certainty and opinion. Even if his views on specific areas of the law (such as his view on the probative force of the "isolated opinion", *khabar al-wāḥid*) were subjected to criticism and refinement, the general framework he instituted became the mainstay of classical Shīʿī scholarship. The fact that al-ʿAllāma's theory enhanced the authority of the opinions of a *mujtahid* (the one qualified to carry out *ijtihād*) certainly contributed to the popularity of his views amongst later Shīʿī *uṣūl* writers.[7] His work inspired a series of textbooks of legal theory up to and including the famous *Maʿālim al-Dīn* of Ḥasan b. Shahīd II (d. 1011/1602). These were used to train Shīʿī scholars, and were the subject of extensive commentaries and super-commentaries.

Opposition to al-ʿAllāma's theory of *ijtihād* and the division of knowledge into *ʿilm* and *ẓann* eventually emerged, three hundred years later, in the Akhbārī movement. This, at least, is one of the theses argued for in this book. Some scholars have argued that the Akhbārī movement pre-dates its rise to scholarly prominence. In Chapter 1, I argue that whilst there certainly was Shīʿī opposition to legal rationalisation, after al-ʿAllāma it was effectively extinguished amongst the Shīʿī scholarly elite. A rigorous critique of al-ʿAllāma's "innovations" was first delineated in the work of Muḥammad Amīn al-Astarābādī (d. 1036/1626–7), and later through those who claimed to continue his Akhbārī approach. The occasional use of the term Akhbārī in texts before Astarābādī is not, I argue, consistent. It cannot be used as evidence of a defined intellectual trend with a known set of doctrines. It is not clear from the sporadic deployment of the term in classical Shīʿī literature (or Sunni literature about the Imāmiyya) that there was a definable group called the Akhbāriyya. No scholars are ever named as being members of the early Akhbāriyya until much later (and such references are clearly back projections), and no distinctive doctrines are explicitly assigned to the group. In short, I argue

[7] See Calder, "Doubt and Prerogative".

that the Akhbārīs are best seen as starting with Muḥammad Amīn al-Astarābādī. After him, the term has a stability of reference which enables us to speak of an Akhbārī trend (perhaps even an Akhbārī "school") within Shīʿī jurisprudence. The Akhbārīs themselves, of course, wished to trace their history back to the earliest Shīʿī scholars in order to claim an historical pedigree. This should not persuade us that the early Shīʿī scholars were either called "Akhbārīs" at the time (there is little evidence that they were, or that they claimed the name for themselves), or that the later Akhbārīs shared all the juristic methodology of the early scholars. The interplay of reason and revelation in early and classical Shīʿī theology raises a set of interesting questions. However, I argue these questions are not immediately germane to an analysis of the later Akhbārī movement inspired by Astarābādī. The later Akhbārī movement was concerned primarily with questions of legal theory, and how the system established by al-ʿAllāma might be reformed (or more radically, rejected).

Having established Astarābādī's work as marking the beginning of the Akhbārī school, Chapter 2 is an examination of what is known of his life and works. This requires an excursus on the nature of the sources usually employed in such historical accounts. Outside of the biographical and legal literature of the period, the Akhbārī movement is rarely mentioned in either Safavid or Qajar historical sources. Safavid court histories, for example, make extensive reference to the *ʿulamāʾ* (both individually and as a group), and to their conflict with other groups claiming religious authority (in particular the Sufis).[8] However, they are largely silent on the dispute between the Akhbārīs, and their opponents, the Uṣūlīs (or *mujtahids*, who sought to maintain al-ʿAllāma's juristic system). This in itself indicates that tensions between Akhbārīs and Uṣūlīs were, in the main, internal to the scholarly community, rarely expressing themselves in ways which troubled the authors of historical chronicles. Scholars identified as Akhbārīs within the Shīʿī scholarly tradition are mentioned in the historical chronicles: some of them had close relations with the Safavid court and held official positions in the Safavid state. However, reference to their allegiance to the Akhbārī school of jurisprudence is rare in these historical sources. That other (unpublished) documentary sources from the Safavid and Qajar periods might throw further

[8] See in this regard, the detailed work of Babayan (*Monarchs, Mystics and Messiahs*).

light on the Akhbārī movement is inevitable. Sefatgol has intimated that the growth in the number of *waqf* establishing seminaries for the study of *ḥadīth* could be one indication of increased Akhbārī influence.[9] Even then, the dispute surrounding Akhbārī jurisprudence seems confined to the scholarly elite. The view argued for in this work is that the Akhbārī-Uṣūlī dispute was primarily concerned with technical questions of Shīʿī jurisprudence. The dispute had, then, limited societal effect compared to the disputes concerning the role of Sufism (or even philosophy) in Safavid and Qajar society (and the wider Shīʿī world).

Astarābādī rejected *ijtihād* and the epistemological *ʿilm/ẓann* distinction implied by it. In its place, he proposed a legal methodology which attempted to ensure that legal rulings could be derived with certainty from the sources. In order to do this, he had to establish that the sources themselves were inviolable. Two elements of his theory (namely the establishment of the authenticity of the sources and the manner in which they could be interpreted with certainty as to their intended meaning) were to become the central doctrines of subsequent Akhbārī writings. Astarābādī's ideas are analysed in detail in Chapter 3, on the basis of his extant writings on legal theory and in particular, his famous *al-Fawāʾid al-Madaniyya*. He argued that the *akhbār* of the Imams are historically accurate, and that they provide sufficient guidance to enable an individual believer's complete adherence to the law of God. Since the Imams are, by common Imāmī Shīʿī agreement, sinless, the Imams' words are reflections of God's will. In this sense the books of *akhbār* are, for Akhbārīs, "scripture", and Astarābādī's emphasis upon them as a source means that his methodology can be characterised as "scripturalist". How this scripturalist emphasis was developed by subsequent Akhbārīs is described in later chapters. Astarābādī does not consider it necessary for the believer to seek any alternative source of knowledge in order to comply with God's will. "Scripture" provides sufficient guidance, both in terms of substance and in terms of the correct way to proceed when scripture appears inadequate. Scripture makes perfect compliance with the law possible, even if, on a particular point of detail, scripture is ambiguous or silent. Other possible sources of knowledge (reason, philosophy, experience, inspiration) are systematically excluded from acting within the area

[9] Sefatgol, "Safavid Administration", p. 408.

of the law. Akhbārīs agreed on this, though they clearly disagreed on whether these alternative potential sources might usefully inform theological (as opposed to legal) debate. Astarābādī's theological writings are analysed in Chapter 4, and his views on the canonical questions of *kalām* (dialectical theology) are found to be only tangentially influenced by his Akhbārī jurisprudence—confirming that Akhbarism was, at least for its founder, primarily a jurisprudential movement.

The considerable popularity of Astarābādī's ideas amongst the *ʿulamāʾ* led to the establishment of new Akhbārī teaching circles in the Gulf, Iran, Iraq, Jabal ʿĀmil and eventually India. Evidence for the spread of Akhbarism can be found in the sudden appearance of self-avowed Akhbārī works of jurisprudence, in which Astarābādī is cited and his views propounded (and at times developed and modified). The influence of Astarābādī's ideas can also be traced through the network of teacher-pupil relationships, beginning with his teaching circle in Mecca, and spreading to other parts of the Shīʿī community. This network is examined in Chapter 5, and a tentative history of the spread of Akhbarism in the two hundred years following Astarābādī's death is mapped. The establishment of teaching institutions, the construction of a scholarly tradition of Akhbārī writing and the study of a distinctive Akhbārī curriculum are all indicators of the emergence of an Akhbārī "school". The common Islamic term for such a school of thought, *madhhab*, was applied to the Akhbāriyya around a century after Astarābādī, and remained popular thereafter. The Akhbārī *madhhab*, if this is the correct term, is contrasted with the *madhhab* of the Uṣūliyya or *mujtahidūn*. Lists of differences between the two "schools" are composed in great number, sharpening the lines of dispute between Akhbārīs and Uṣūlīs. When composed by Akhbārīs, these lists facilitate an examination of Akhbārī "self-definition". Such an account is carried out in Chapter 6, and the evidence rallied there supports the dating of the emergence of the Akhbāriyya as a "school", in the proper sense of the term, to around 80 years after Astarābādī's death.

Chapters 7, 8 and 9 present an analysis of the major themes within Akhbārī works of jurisprudence between the death of Astarābādī and the work of Mīrzā Muḥammad al-Akhbārī, the last significant Akhbārī scholar in Iraq and Iran. The analysis demonstrates that intra-Akhbārī debate, in addition to the Akhbārī-Uṣūlī dispute, contributed to the formation and consolidation of the Akhbārī school. In Chapter

7, the debate around the extent to which the interpretation of the Qur'ān was dependent upon the Imams' own interpretation (*tafsīr*) in the *akhbār* is analysed. In Chapter 8, the Akhbārī conception of *Sunna*, and the identification of *Sunna* solely with the *akhbār* of the Imams is examined. The different means whereby Akhbārī jurists demonstrated the historical authenticity (and hence the probative force—*ḥujjiyya*) of the *akhbār* exemplify the variation within the Akhbāriyya over how to maintain this fundamental, perhaps defining, Akhbārī doctrine. Akhbārī doctrines such as these also enable us to position some scholars whose Akhbārī allegiance is not unequivocally declared in the sources. In Chapter 9, the manner in which interpretable scripture (whether it comprised Qur'ān and *akhbār*, or *akhbār* alone) might be understood, and legal rulings deduced, is examined. Here, once again, one detects similarities across the Akhbārī school, but also tensions and debates. In all three chapters, it becomes clear that the major debates within the Akhbārī school all related to the extent to which established (Uṣūlī) methodology might be employed in the defence of Akhbārī doctrine. Distinctive Akhbārī "camps" did not emerge in any definable form and it is not possible to talk of coherent "schools within the Akhbārī school". However, it is possible to discern different attitudes towards other Shīʿī trends, in particular, the Uṣūliyya. Questions concerning whether the Uṣūlīs should be considered fellow Shīʿa or heretics meld with issues relating to whether or not established (Uṣūlī) models of legal enquiry should control both the presentation and content of Akhbārī jurisprudence. This indecision about Akhbarism's relationship to Usulism is, I argue in the Conclusion, one of the factors which contributed to its demise in Iran, Iraq, Jabal ʿĀmil and the Arabian peninsula.

The current popularity amongst some Imāmī Shīʿa of the "Akhbārī" scripturalist method is, in part, due to the increased availability of the sources. In particular, the publication of *akhbār* collections in both print and electronic media have encouraged some modern Shiʿites to by-pass the scholarly training of the seminary (*ḥawza*) and the supposedly hidebound juristic methodology of the Uṣūlīs. However, as I hope to demonstrate in the course of this book, Akhbārī jurisprudence was far from straightforward, and Akhbārī jurists showed as much intellectual ingenuity as their Uṣūlī counterparts as they attempted to construct a coherent juristic system. They were not simple "literalists" as they are sometimes portrayed in both Shīʿī and secondary literature,[10] but rather scholars, with a vehement commitment to the maintenance of scholarly authority and the tradition of Shīʿī learning. Modern developments, rising levels of literacy in the Shīʿī world and the increasing number of translations of the sources into languages other than Arabic have all encouraged a perception that the Akhbārīs represented a sort of Shīʿī protestantism in which expert scholars were redundant. The Akhbārīs, it is believed, let the texts "speak for themselves". I hope that my analysis here of the historical Akhbārī school demonstrates that this is a rather partial view of an innovative and intellectually complex juristic tradition.

[10] Stewart makes the common comparison between the Akhbāriyya and the "literalist" Ẓāhiriyya (*Islamic Legal Orthodoxy*, p. 184). Cole glosses the Akhbārīs as "literalist" (Cole, "Sources of Authority", p. 84). Whilst the two groups share a refutation of *qiyās*, their replacement hermeneutics (analysed below in Chapters 7 and 9) are quite distinct. Reflex equations of the Akhbārīs and Ẓāhirīs can also be found in Abtaḥī, "Naqsh-i Sunnat", Asʿadī, "Qurʾān-i 'Mubīn' ", Jābirī, *al-Fikr al-Salafī* and Asʿadī, "Fahm-i Qurʾān".

CHAPTER ONE

THE AKHBĀRĪ-UṢŪLĪ DISPUTE AND THE EARLY "AKHBĀRĪ" SCHOOL

For most Muslim writers, both in the contemporary period and in the past, God's revelation to humanity comprises two principal elements: the Qur'ān and the *Sunna*. However, the form and content of both of these elements continues to be matters of dispute and discussion. The text of the Qur'ān ("the Book", *al-Kitāb*) was subject to variant readings (*qirā'āt*), which at times indicate differences in the possible meaning of a particular passage.[1] Furthermore, some Muslim scholars conceived of verses which were absent from the text of the Qur'ān as we have it today, though present in the recitation revealed to the Prophet Muḥammad.[2] The *Sunna*, on the other hand, was a somewhat theoretical concept which a Muslim could come to know through an examination of the extant reports of the Prophet's actions and an assessment of what is indicated by the words and actions of the Prophet found in the literary genre known as *ḥadīth* (pl. *aḥādīth*). The *aḥādīth* were only potential indicators of the Prophet's example. Other possible indicators included the words and actions of the Prophet's companions and those of his successors who were inspired by his example. The "recited revelation" (*waḥy matlū*, that is, the Qur'ān) was used during the performance of Muslim rituals, and therefore had a ritual function which the unrecited revelation (*waḥy ghayr al-matlū*, that is, the *Sunna*) did not have. This, however, was not always an indication of the Qur'ān's primacy over the *Sunna*. Nor was the relative stability of the Qur'anic text, compared with the amorphous concept of *Sunna*, always sufficient to guarantee the greater authority of the former. Some scholars felt the relationship between Qur'ān and

[1] For a view that the two *qirā'as* (and by implication, most of the other variant readings) in current usage do not affect the sense of variant verses, see Brockett, "The Value of the Ḥafṣ and Warsh Transmissions". That most *qirā'as* are exegetical asides rather than true variants, see Burton, *Collection*, pp. 29–45 and Wansbrough, *Qur'ānic Studies*, pp. 205–206.

[2] The classic example of this is the "Stoning Verse", on which see Burton, *Collection*, pp. 72–85 and his *Sources of Islamic Law*, pp. 122–164.

Sunna to be one akin to text and commentary; others felt them to be two separate (and equal, albeit complementary) sources of knowledge of God's intended message.[3] Occasionally, one finds groups which accept the Qur'ān but reject the *Sunna* entirely.[4]

Amongst the disputed indicators of the Prophet's *Sunna* were both the actions of his companions and those who came after him. In the absence of an explicit (and historically reliable) statement of the Prophet, is it permitted to turn to a statement made by one of his companions?[5] What of those who came after the Prophet but may have been preserving Prophetic practices which were not recorded in *ḥadīth*?[6] Can these reports act as indicators of the *Sunna*? Answers to these questions were much disputed by Muslim scholars, and continue to be important for Muslims into the modern period. What is to be included in the concept of revelation has elicited a variety of responses from Muslim scholars (the *ʿulamāʾ*). Hence "scripture", the record of God's communication with humankind is, at the same time, both centrally important *and* disputed. Beyond these disputes concerning the content of scripture, there were disputes concerning interpretation. What interpretive processes were legitimate, and which were to be rejected? These were formally separate from the question of content, though discussion concerning what scripture meant inevitably impacted on what counted as scripture.[7] At times, a textual variant accorded with doctrine (perhaps generated independently from scripture), and was therefore given preference over other variants. At other times, a segment of possible scripture became so resistant to an acceptable interpretation that it became easier for Muslim writers to argue for its abrogation (or more radically, its exclusion from the canon).[8] Most classical Muslim authors can be described as

[3] On the relationship between Qur'ān and Sunna, see Burton, *Sources of Islamic Law*, pp. 18–31.

[4] See Hawting, "The significance of the slogan *Lā Ḥukm illā lillāh*", and in the modern context, Rippin, *Muslims*, pp. 218–245.

[5] On the Imāmī rejection of the companion reports as sources of information for Prophetic practice, see Kohlberg, "The attitude of Imāmī Shīʿites".

[6] The most pertinent example of this being the practice of the "people of Madina" being a source in Mālikī jurisprudence. See Dutton, *The Origins*, pp. 32–52.

[7] On the effects of disputes about canonicity on the methods of exegesis, see Wansbrough, *Qur'ānic Studies*, pp. 207–227. He refers to the "related but distinct processes of hermeneutical derivation [i.e. exegesis] and textual adjustment [i.e. revelatory content]" (p. 148).

[8] See the examples in Burton, *Sources of Islamic Law*, pp. 56–80.

scripturalist in the sense that doctrines are presented as having been derived from the texts which record God's message. However, what was encompassed within scripture (or more broadly, revelation) has been the subject of extensive dispute.

Further debates around possible sources of knowledge *beyond* scripture enhanced the sophistication of Muslim thought concerning scripture and revelation. Whether reason—operating as a source of knowledge independent from revelation—brought knowledge was, of course, a well-known topic of early Muslim debate.[9] Some placed knowledge attained through reason at a level higher than that provided by scripture;[10] others wished to subordinate reason to scripture. Was it possible, also, to gain knowledge through a personal encounter with God, or (more commonly) through a personal encounter with a savant (who has, in turn, had a personal encounter with God)? The manner in which such questions were answered, and the disputes which arose between different Muslim intellectual traditions should form the principal elements of any analysis of the scholarly accomplishments of Muslim authors.

This book is an account of the Akhbāriyya, a "school" within Twelver Shīʿī Islam which answered the questions posed in the above paragraphs in distinctive ways. The Akhbārīs flourished between the Seventeenth and Nineteenth Centuries (CE) amongst Twelver Shīʿī scholars of Iran, Iraq, Eastern Arabia and India. Some Akhbārī views on the above questions were innovative and had little or no precedent in either Twelver Shiʿism or Muslim thought more generally. Many of their opinions, however, followed well-worn lines of argument. The extent to which Akhbarism was original, and the extent to which it was merely a derivative Shiʿite expression of previously established Sunni positions is one area discussed in the following chapters.

Shiʿism and Akhbārī Shiʿism

The Shīʿī movement within Islam was born out of the disputes which ensued following the Prophet's death in 632 CE. These disputes

[9] Van Ess, *Theologie und Gesellschaft*, v. 2, pp. 223–342.

[10] Famous amongst these was the philosopher Ibn Rushd (see Leaman, *Averroes*, pp. 144–160).

concerned not only *who* should lead the Muslim community, but also the *personal characteristics* of a leader of the Muslims.[11] Whether the former dispute preceded the latter or vice versa is not a matter which concerns us here.[12] Within two centuries of Muḥammad's death, distinct views over the nature of leadership had coalesced into various scholarly (and non-scholarly) opinions. The principal two views (those of the Shīʿīs and the Sunnis) encompassed both history and theology. In terms of history, the Shīʿa claimed that ʿAlī and his descendents had been designated by the Prophet as his rightful successors and leaders of the Muslims. In terms of theology, the greater part of the Shīʿa proposed that this leader, drawn from ʿAlī's descendents, had knowledge of God's purpose for the Muslims which exceeded that gained by the rest of the community. In terms of history, the Sunnis claimed that the Prophet made no such designation, and the question of leadership was left open to community decision. In terms of theology, the Sunnis generally argued for a pious and learned leader of the Muslims from the Quraysh tribe who had no special charismatic powers. The terms *Imām* and *Khalīfa* (caliph) came to be associated with these two conceptions of leadership.

Amongst those Shīʿa who claimed that the Imam had privileged knowledge, there were differences concerning the line of descent from which the Imam should be drawn. The Twelver Shīʿa (or the Imāmiyya as they are also known) argued for a line of twelve descendents from the Prophet, beginning with ʿAlī and ending with Muḥammad "al-Mahdī". Muḥammad al-Mahdī disappeared in 260/874, communicating with his followers through representatives (*sufarāʾ*) for around 70 years. In 329/941 he finally disappeared completely, leaving the Shīʿa with promises of his return (*rajʿa*) at some future time to establish a rule of justice and the beginning of a messianic age. Twelver Shīʿīs still await his return. In his absence, community leadership has generally been held by the scholars (*ʿulamāʾ*), since (they argue)

[11] Good introductions to Shīʿī Islam include those by Momen (*An Introduction to Shiʾi Islam*) and Halm (*Shīʿism* and *Shiʾa Islam*).

[12] That the conception of a leader as a charismatic figure was generally accepted, and the principal dispute was over the leader's identity is one of the interesting conclusions of Crone and Hinds (*God's Caliph*, p. 108 and *passim*).

they were delegated by the Imams (when present) to adopt this role.[13] This assumption of leadership has not gone unchallenged by other charismatic figures in Shīʿa history. At times, individuals have claimed to be the *mahdī* returned. At other times, a rival intelligentsia has argued for a more popularist notion of leadership based on political skills and diplomatic acumen in the absence of the Imam.

Even though Twelvers have generally located religious community leadership in the *ʿulamā'*, this has not prevented disputes about the role of scholars in the life of the community. For the Imāmiyya, knowledge of the law could be gained directly from the Imam when present, and hence scholarship (such as writing works of *uṣūl al-fiqh*, collecting reports of the Imams' words and deeds and speculating about theological matters) was not a pressing concern. Only after the Imam's disappearance did Twelver Shīʿī scholars begin in earnest to develop distinctive Twelver genres of religious literature to rival those of the Sunni theological and legal schools. When Shīʿī literature did emerge, one theological and legal point which distinguished them from the Sunnis was an assertion that certain (that is, indubitable) religious knowledge was attainable. Some argued that this knowledge could be gained through the Imam, who though hidden, had left the Shīʿa records of his words and deeds (*akhbār*). Others supplemented this textual source with a (Muʿtazilī inspired) doctrine which affirmed the ability of every rational person to gain knowledge of theological (and even some legal) truths through reasoned speculation. Ibn Bābūya (d. 381/991) is associated with the former view and al-Shaykh al-Mufīd (d. 413/1022) is associated with the latter. Both Imāmī positions (and also those in between the two views) asserted an epistemology in which certain religious knowledge was attainable, and the Sunni distinction between matters which were certain and matters which were subject to fallible (though learned) human judgement (*al-masā'il al-ẓanniyya* or *al-masā'il al-ijtihādiyya*) was rejected. When this distinction *was* introduced into Shīʿī legal theory by al-ʿAllāma al-Ḥillī (d. 726/1325), it does not seem to have provoked much reaction. Instead, within generations of al-ʿAllāma's work, the distinction was accepted as authentically Shīʿī. Reports of the Imams were reinterpreted as supporting the distinction and a full blown Imāmī theory of *ijtihād* (see below) was developed. It was as a reaction to the hegemony of this

[13] See generally Madelung, "Authority in Twelver Shi'ism".

ijtihādī (also called *mujtahid* or Uṣūlī) position that the Akhbāriyya emerged in the Seventeenth Century CE.

The origins of Akhbarism[14] have been subject to differing assessments in the secondary literature. This, in part, is due to the conflicting evidence of Shīʿī (and non-Shīʿī) sources. The dominant thesis used to be as follows: the first to claim to be an Akhbārī was the Iranian scholar, Muḥammad Amīn al-Astarābādī (d. 1033/1623–4 or 1036/1626).[15] Astarābādī studied in the shrine cities of southern Iraq, and in Iran, before relocating to Arabia where he studied, wrote and taught in the cities of Mecca and Madīna. The work in which he is supposed to have outlined the broad principles of Akhbārī thought was his *al-Fawāʾid al-Madaniyya* in which he criticised the views of his contemporary Shīʿī jurists and proposed an alternative methodology for deriving religious knowledge. Shīʿī jurists contemporary with Astarābādī argued for an interpretive method which they termed *ijtihād*. *Ijtihād* was defined as the "exhaustion of the jurist's effort in order to gain an opinion regarding a legal ruling".[16] This definition, and indeed the whole concept of *ijtihād*, had developed in Shīʿī jurisprudence under the influence of Sunni legal theory. Sunni writers of *uṣūl al-fiqh* had recognised that whilst elements of God's law could be known with absolute certainty, many elements (particularly the details, or *furūʿ*, of the law) were less than certain as the texts were at times ambiguous. The resulting view was that a trained jurist was required to exert effort, and "exhaust himself" in discovering a ruling which was not considered to be "plain" or "obvious" in the texts. The resultant ruling remained the jurist's own opinion (*ẓann*), and other (suitably qualified) jurists might produce different rulings resulting from their own "effort" (*ijtihād*) in interpreting the texts.

---

[14] I use this term to refer to the doctrines of the Akhbāriyya.

[15] An analytical biography of Astarābādī is given in the next chapter, see below, pp. 40–60.

[16] *Istifrāgh al-faqīh wusʿahu fī taḥṣīl al-ẓann bi-ḥukm sharʿī* (Ḥasan, *Maʿālim*, p. 238). Variants in wording of this definition are found in works of Shīʿī (and Sunni) *uṣūl al-fiqh*, though the basic elements are retained. Al-Muḥaqqiq al-Ḥillī defines *ijtihād* as "the exertion of effort to extract (*istikhrāj*) the legal ruling" from the texts (Muḥaqqiq, *Maʿārij*, p. 179). ʿAllāma defines it as "exhausting effort in speculating on those legal questions which permit opinion such that no more [effort] can be made" (ʿAllāma, *Mabādiʾ*, 240). The development of *ijtihād*, and its relationship to the epistemology of juristic knowledge in Imāmī Shiʿism is expertly analysed in Calder, "Doubt and Prerogative".

Which interpretive techniques were legitimate, the schemes of classification of these techniques, the variable assessments of a text's authenticity—all of these were matters of dispute with Sunni authors. In general though, the epistemology and legitimacy of *ijtihād* was affirmed and given a scholarly pedigree which could be traced back to the Prophet. Shīʿī authors of *uṣūl al-fiqh* had originally rejected *ijtihād*, since (they argued) certainty as to both the general principles and the specific details of the law could be attained. This confidence in one's ability to attain certainty was, of course, linked to the general Shīʿī idea that knowledge of God's will is available after the Prophet's death in the form of an inspired individual—the Imam. Under pressure from various factors (the prolonged disappearance of the Imam, societal changes which were not envisaged by the texts and the general establishment of an Imāmī tradition of scholarship), the Shīʿī jurists felt the need to move from this anti-*ijtihād* position. The natural model to choose was that established by the Sunni jurists. This move is particularly associated with al-ʿAllāma al-Ḥillī, though some trace its beginning to the more positive assessment of *ijtihād* by al-ʿAllāma's teacher al-Muḥaqqiq al-Ḥillī (d. 676/1277). Al-ʿAllāma's work in *uṣūl al-fiqh* established the need for Shīʿī *mujtahids* (those who carry out *ijtihād*), and the epistemological division between certainty (*ʿilm, qaṭʿ*) and opinion (*ẓann*) in Shīʿī jurisprudence. By the time of Astarābādī, nearly all Shīʿī jurists argued for the legitimacy of *ijtihād*, though there remained different conceptions of its remit and scope, just as there were in Sunni works of *uṣūl al-fiqh*. Astarābādī, and the Akhbārīs who followed him, argued for a return to the earlier Shīʿī attitude of a rejection of *ijtihād* on the grounds that legal certainty was available. The claim of the early Shīʿī jurists was, then, an element in Astarābādī's attempt to establish the precedence of the rejection of *ijtihād* over the "innovation" of more recent Shīʿī jurists.

Astarābādī's position proved popular with other Shīʿī jurists disillusioned with (what they perceived as) the stagnation of Imāmī scholarship. Various prominent scholars in Safavid Iran either identified themselves as Akhbārīs, or were termed so by subsequent authors. Akhbarism also gained ground outside of Iran. It was supposedly so popular in the shrine cities of Iraq that those who still held fast to the doctrine of *ijtihād* were afraid to admit their opinions in public. An Indian Akhbārī school also developed, since Indian Shīʿism was linked to Iran and southern Iraq through migration and intellectual exchange.[17] In the Eighteenth Century CE, the popularity of Akhbarism in Iran and Iraq began to wane, principally through the efforts of Muḥammad Bāqir al-Bihbihānī (d. 1205/1790–1) and his disciples. It survived a little longer in India, but by the mid-Nineteenth Century CE, there were few Shīʿī Akhbārī scholars of note remaining. This has remained the case until today. Akhbārī communities are said to survive in parts of southern Iran, the Gulf and the Indian subcontinent, though they are not intellectually active to any great degree.[18] *Ijtihād* is accepted as a legitimate enterprise by nearly all major Shīʿī authorities today, and this has been the case for over a century.[19]

The above account has, however, been criticised recently, first by Madelung, and later by Newman, Kohlberg and Stewart.[20] The common element in their reformulations is that Akhbarism can be said to exist before Astarābādī, and that Astarābādī expressed, in his *al-Fawāʾid al-Madaniyya*, an Akhbārī position which predated him. This reduces his originality as a thinker and as the founder of a "school" of thought. Unfortunately, no Akhbārī texts from this earlier period have survived to prove this thesis conclusively. However, there are a number of references to the *akhbāriyya* and *akhbāriyyūn* (or *akhbāriyān* in Persian) in pre-Astarābādī texts. Furthermore, the *akhbāriyya* mentioned in these earlier texts held opinions on a range of subjects beyond the rejection of *ijtihād*. They were associated with opinions on broader theological issues (such as the legitimacy of reason as

---

[17] See Cole, *Roots of North Indian Shīʿism*, pp. 124–168.

[18] Madelung ("Akhbāriyya", p. 56) and Kohlberg ("Aḵbārīya", pp. 716–718) both make mention of these communities, though they make no reference to any source for this information.

[19] Variants of this standard account are given in Browne, *Literary history of Persia*, v. 4, p. 374; see also Scarcia, "Intorno alle controversie" and Falaturi, "Die Zwölfer-Schia" in this regard, and also the account of Gharāwī, *Maṣādir al-Istinbāṭ*. Arjomand (*The Shadow of God*, pp. 145–146) rather optimistically links the rise of Akhbarism to ethnic conflicts between Persian notables and incoming Arab jurists in early Safavid Iran. His theory is discussed further below, pp. 173–174.

[20] The relevant references are Madelung, "Imamism and Mu'tazilite Theology", pp. 20–21; Newman, "Development and Political Significance" and his "The Akhbārī-Uṣūlī Dispute in late Safavid Iran", parts 1 and 2; Kohlberg, "Aḵbārīya"; Stewart, *Islamic Legal Orthodoxy*, pp. 182–183. Stewart's "The Genesis" also argues for an earlier form of Akhbārism, though he also recognises the originality of Astarābādī. Al-Jābirī's's view (*al-Fikr al-Salafī*, pp. 277–321) could be said to argue for a longer history of Akhbarism that predates Astarābādī, though he opts for the more general terminology of *salafī* Twelver thought. In his work, though, he argues for a continuous "traditionalist" current in Imāmī thought from the earliest times to the present day.

a source of knowledge and the nature and role of the Imam). Their rejection of *ijtihād* in jurisprudence also entailed a set of distinctive legal opinions. These included the rejection of the legitimacy of government during the occultation, the validity of Friday prayer and the distribution of community taxes. Newman, in particular, has argued that these legal views can be characterised as Akhbārī before the time of Muḥammad Amīn al-Astarābādī.[21]

We begin, then, with the conjoined problems of origins and definition. The Akhbārīs themselves trace their beginning to the earliest stages of Twelver Shīʿī literary scholarship, to the disappearance of the Twelfth Imam and the *ḥadīth* collections of al-Kulaynī (d. 329/941) and Ibn Bābūya. The Uṣūlīs view Akhbarism as a relatively recent phenomenon, emerging through the work of Muḥammad Amīn al-Astarābādī. Both of these claims are, in truth, elements of the polemic between the two groups; both are claims to historical precedence which should not sway our analysis here. Each group associated their opponents' views with Sunni opinions. This may indicate Sunni historical influence on the formation of the two groups, though it is best understood as a technique of vilification commonly found in Twelver juristic discourse.

Secondary scholarship on the dispute between Akhbārīs and Uṣūlīs displays a similar divergence of opinion, though here the division is associated with different conceptions of Akhbarism. For some (notably Madelung, Newman, Qaysari,[22] Mashayekh[23] and Stewart), Akhbarism encompasses not only views on legal methodology (*uṣūl al-fiqh*), but also issues of substantive law (*furūʿ al-fiqh*, such as the role of the jurist and the legitimacy of the state during the occultation) and theological issues (such as the validity of speculative reasoning in theology and philosophy). For others (Browne, Scarcia, Falaturi, and more recently Calder, Momen and Abisaab),[24] Akhbarism is best seen as a reaction to the introduction of certain processes of legal

[21] Newman, "Development and Political Significance", pp. 1–56. Āl ʿUṣfūr ("ʿAqlgirī") also links the "early" Akhbārīs and Uṣūlīs to early Shīʿī juristic attitudes towards the state.

[22] See Qayṣarī, "Akhbāriyān", pp. 160–163.

[23] See Mashāyikh, "Akhbāriyyah", pp. 7–13.

[24] The relevant references here are, Calder, "Doubt and Prerogative", p. 68, n. 31 (see also his "Structure of Authority", p. 231, n. 18); Momen, *Introduction to Shiʿi Islam*, pp. 222–225; Abisaab, *Converting Persia*, pp. 105–112. See also, Bayat, *Mysticism and Dissent*, p. 21.

reasoning in Twelver works of *uṣūl al-fiqh* (primarily *ijtihād*, but encompassing a more general suspicion of an epistemology which allows opinion in place of certainty). This "reaction" could only occur after the incorporation and acceptance of these processes, and hence this reaction only occurred (and could only occur) after the pioneering work of al-ʿAllāma al-Ḥillī. How one defines Akhbarism inevitably influences one's views as to its origins.[25]

How might one decide between these various positions? In the subsequent chapters of this book, I outline the development of the Akhbārī "school", from Astarābādī until Mīrzā Muḥammad al-Akhbārī (d. 1233/1818). In the course of this presentation, I examine evidence for these different opinions (both within the tradition and outside of it). In the remainder of this chapter, I examine the evidence for a pre-Astarābādī school of Akhbarism, arguing that whilst the early use of the term *akhbārī* (or its derivatives) is undeniable, its employment is rare and ill-defined and is best understood as an *ad hoc* description of certain opinions and not a "name" which designates a "school" as such. Akhbarism, then, is still best understood as starting with Astarābādī.

Akhbarism before Astarābādī

The later Akhbarism of Astarābādī and his followers was legally scripturalist, in the sense that scripture (defined as both the Qurʾān and *sunna*) was seen as providing sufficient legal guidance for the Shīʿa. However, later Akhbārīs did not seem to have a unanimous position on whether or not this scripturalism applied to other areas of religious knowledge. There were Akhbārīs who used reason (*al-ʿaql*) as a proof of the basic elements of the Shīʿī creed, writing *kalām* works in the process.[26] There were Akhbārīs who promoted direct religious experience as a means of gaining religious knowledge out-

[25] More cautious, indeed almost non-committal, on this question are Modarressi, "Rationalism and Traditionalism", pp. 141–158; Cole, *Roots of North Indian Shiʿism*, p. 11 (where he speaks of Astarābādī "*reformulating* conservative jurisprudence", my emphasis); Moussavi, *Religious Authority*, p. 92.

[26] Astarābādī himself presented his theology in *kalām*-style dialectic, and used Muʿtazilī-derived presumptions in his reasoning. See below, pp. 111–112.

side of the law.[27] There were Akhbārīs who simply collected reports of the Imams concerning non-legal matters (in particular theological doctrines current in *kalām* works), and offered no personal reasoning as to how these might be brought together into a coherent doctrine.[28] There were even those who chiselled out a limited role for reason in their legal theory, but based this allowance exclusively on scriptural grounds (in that the Imams had, themselves, give explicit permission for reason to be used in restricted circumstances).[29] It was, I argue, the legal scripturalism of these scholars which defined later Akhbarism, not any particular array of theological doctrines, nor even a particular mode of exegesis.[30] If I am right in this characterisation of the defining doctrines of later Akhbarism, then the relationship between these later Akhbārīs and earlier Imāmī sects described in the texts as *akhbāriyya* (or *akhbāriyyūn* or *akhbāriyān* in Persian texts) requires examination. It should be made clear at the outset that I am primarily interested in whether or not later Akhbārī doctrine had any relationship with the doctrines ascribed to the earlier "Akhbārīs". It is the use of the term *akhbārī* in these earlier texts which has prompted some scholars to assert a continuous line of Akhbārī scholarship from early Imāmī traditionalism to Astarābādī and his successor Akhbārīs. Some Akhbārīs have also been quite willing to use these earlier references since it gave their "school" a pedigree. As is outlined below, my conclusion here is that most pre-Astarābādī references to the *akhbāriyya* have little in common with Astarābādī and the later Akhbārī school.

---

[27] See, for example, my analysis of the theology of Muḥsin Fayḍ and its relationship to Akhbarism in Gleave, "Scripturalist Sufism".

[28] The best examples of this are various collections of theological *ḥadīth* by al-Ḥurr al-ʿĀmilī (such as his *Ithbāt al-Hudā* on miracles and their proofs, and *al-Fuṣūl al-muhimma* on theology more generally).

[29] Yūsuf al-Baḥrānī argued for such a position (see Gleave, *Inevitable Doubt*, pp. 188–204), and Muẓaffar Riḍā, the great twentieth century Shīʿī scholar, describes Akhbārīs generally as allowing rational legal proofs under restricted circumstances (see Riḍā, *Uṣūl al-fiqh*, v. 1, pt. 2, pp. 213–224 and summarized in Gleave, *Inevitable Doubt*, pp. 184–188).

[30] As is seen below (pp. 218–232), some Akhbārīs did restrict *tafsīr* of the Qur'ān to the uncomplicated citation of relevant *ḥadīths* (of course it was they who often decided which *ḥadīths* were relevant). Others, however, allowed direct theological interpretation of the Qur'anic text. Furthermore, Akhbārī writers do not seem to have felt a reticence to comment on *ḥadīth* themselves, and they often argued in quite conventional (ie not-scripturalist) ways (see below, pp. 270–275).

This conclusion does not, however, imply that Astarābādī's critique had no historical precursors, and these are mentioned, *inter alia*, in the course of this book. It is clear that some of the structural features of later Akhbarism are taken both from trends within Shīʿī history, and from wider Muslim debates over jurisprudence. For example, early Imāmī or Ẓāhirī epistemological rigorism (that is, the rejection of *ẓann*) may have been sidelined within Imamism by al-ʿAllāma, but was picked up again by the Akhbārīs.[31] Similarly, the emphasis on *akhbār* as the only source of law can find some precursors in al-Shaykh al-Ṭūsī's promotion of *khabar al-wāḥid*, and Ibn Idrīs's rejection of it.[32] Al-ʿAllāma's criticism of those who accept *ḥadīth* without examining *isnāds* may also be evidence of a recalcitrant traditionalist faction which refused to accept that a workable legal system needed a method of critiquing sources and the interpretive space which such a critique provides.[33] However, the group adhering to these legal doctrines is never described with the name *akhbārī*.[34]

In theological texts, precursors to the Akhbārīs are, perhaps, more obvious. The usual account of the development of early Imāmī thought is that it began as traditionalist (and anti-rationalist), and was radically rationalised, first by the Banū Nawbakht and later by al-Shaykh al-Mufīd under the influence of the (Sunni) Muʿtazilīs.[35] Earlier Imāmī devotion to the words of the Imams gave way to rationalising theology from the wider Muslim community, and this in turn began to affect legal theory. The effect on jurisprudence was slightly delayed (what al-Shaykh al-Mufīd did for Imāmī *kalām*, al-ʿAllāma al-Ḥillī did for *uṣūl al-fiqh*), but was, nonetheless, inevitable. Early Imāmī traditionalism was eclipsed by rationalised theology and jurisprudence, and it is in this earlier traditionalism that one supposedly finds the roots of Astarābādī's Akhbarism. As a corrective to this account, Sander has argued that the reports in the early collections of al-Barqī, al-Ṣaffār al-Qummī and al-Kulaynī show the incorporation of Muʿtazilī-inspired theology into Imāmī theology much earlier

[31] See below, p. 108.

[32] See Ibn Idrīs, *al-Sarāʾir*, v. 1, p. 47.

[33] See, for example, al-ʿAllāma, who criticises scholars for following reports with weak *isnāds* (*ḍaʿīf al-sanad*), in his *Mukhtalaf*.

[34] The only example of the use of the term *akhbāriyyūn* in a pre-Astarābādī Shīʿī *legal* text is in al-ʿAllāma al-Ḥillī's *Nihāyat al-wuṣūl*, analysed below, p. 27.

[35] See, for example, Madelung, "Imamism and Muʾtazilim".

than al-Shaykh al-Mufīd.[36] Whether the reports reflect the Imams' words or later Imāmī theological concerns, the established account of a move from traditionalism to rationalism needs reformulation. Furthermore, these collections of *ḥadīths* are amongst the earliest surviving works which can be classified as recognisably Imāmī. Some pre-Mufīd (and even pre-*ghayba*) Imāmī Shīʿī scholars obviously laid great emphasis on the doctrines laid out in reports such as these, and probably considered them sources of doctrine, legal or otherwise. In this sense they are traditionalist, and therefore share the later Akhbārīs' attitude towards revelatory sources of knowledge. Melchert has argued that the Imāmīs were not traditionalist as such, but semi-rationalist.[37] Much of Melchert's evidence comprises the titles of works ascribed to early Imāmī theologians, taking these to imply theological commitment. Alternatively, it could be argued that many early Imams were "traditionalist", in the sense that they relied upon the sayings of the Imams as the prime source of religious knowledge and rejected reason as a possible alternative source of knowledge. However, devotion to these reports led to adherence to doctrines which the usual traditionalists (namely the *ahl al-ḥadīth* and the Ḥanbalīs) found reprehensible. There is a similarity between some Imāmīs and the Sunni traditionalists in method, though not in doctrine. More extensive evidence for a rationalist—or semi-rationalist—Imāmī camp is adduced by Bayhom-Daou. Her examination of pre-*ghayba* Imāmī theological trends shows variety both in terms of method (that is, the use of what Melchert might call pure rationalist and semi-rationalist arguments, along with traditionalism) and conclusions. Of particular interest to us here is Faḍl b. Shādhān's (d. 260) "scripturalist" legalism in which the Imams are portrayed as infallible transmitters of the Prophet's *sunna*. The *sunna* is, in turn, merely knowledge of the true meaning (or true interpretation—*tafsīr*) of the Qur'ān.[38] Such a position, if it can be accurately ascribed to al-Faḍl b. Shādhān and others in the pre-*ghayba* Imāmī community, has some similarities with later Akhbārī discussions of the Qurʾān as an independent source of legal rulings. However, the doctrine of a single source of law (the Qur'ān) mediated through the Imams is

[36] See Sander, *Charisma und Ratio*, passim.
[37] Melchert, "The Imāmīs".
[38] See Bayhom-Daou, "The Imam's knowledge".

too restrictive for later Akhbārīs. On this doctrine they almost unanimously upheld a theory of dual sources of law: the law was revealed through both the Qur'ān and the Sunna of the Prophet, but can only be understood through the *akhbār* of the Imams.

These early Imāmī currents, coupled with the Imams' (alleged) rejection of *ẓann* already outlined, indicate that the later Akhbārīs did not emerge with doctrines which had no precedent in Imāmī history. A more thorough examination of early Imāmī juristic trends may throw up further similarities. This, however, is not my point in the following analysis. My aim here is to demonstrate that the term *akhbārī* in these earlier texts has little to do with later Akhbārī legal scripturalism. References to the *akhbāriyya* (or some other locution) in pre-Astarābādī texts do not suggest that Astarābādī and his followers were reviving a pre-existent "Akhbārī" tradition, even if they themselves might wish to commandeer these references in their polemic against the Uṣūlīs.[39]

The use of the term *akhbārī* in pre-Astarābādī texts is sporadic, and refers to a sub-group of the Imāmiyya called "*al-akhbāriyya*" (or in Persian texts *akhbārīyān*).[40] The earliest of these is probably in the *Kitāb al-Milal wa'l-Niḥal* of Muḥammad b. ʿAbd al-Karīm al-Shahrastānī (d. 548/1153). This work, composed in 521, is a compendium of the different sects of different religions, including Islam. The passage concerning the *akhbāriyya* reads:

> Hence, the Imāmīs became adherents of justice (*ʿadāla*) in *uṣūl*[41] and anthropomorphism with respect to attributes [of God—the *ṣifāt*], being confused and errant. Between the *akhbāriyya* amongst them and the

---

[39] See, for example, Astarābādī's own use of al-Shahrasānī's *Kitāb al-Milal* (Astarābādī, *al-Fawā'id*, p. 97). Nūr al-Dīn al-ʿĀmilī, commenting on and refuting this passage in his *al-Shawāhid al-Makkiyya* (found as a gloss on Astarābādī's *al-Fawā'id*, pp. 97–98), argues that al-Shahrastānī's *akhbāriyya* are Shīʿī, but not *Twelver* Shīʿī. Fatḥ ʿAlī Zand also wishes to use the *Kitāb al-Milal* as evidence of early Akhbarism (see his *al-Fawā'id al-Shīrāziyya*, f.3a3).

[40] I am excluding the use of the term *akhbārī* to mean historian or transmitter of *akhbār* (that is, the reports of historical figures, rather than the Imams specifically)—which is a much more frequent use of the term and not relevant to my analysis here.

[41] Meaning that they adhered to the Muʿtazilī doctrine of God's justice in the principles of theology (*uṣūl al-dīn*), whilst at the same time adhering to the anthropomorphic interpretation (*tashbīh/mushabbih*) of God's qualities (*ṣifāt*).

> *kalāmiyya* there is violence,[42] and similarly between the *tafṣīliyya* and the *waʿīdiyya*[43] there is fighting and accusations of error.[44]

Al-Shahrastānī tells us nothing of the doctrines of the *akhbāriyya* here,[45] merely that they are distinct from the Twelver Shīʿī "*kalāmiyya*". The latter is probably a reference to Twelver scholars who presented and argued for doctrine in the manner of the theologians (*mutakallimūn*, the proponents of *ʿilm al-kalām*). An implication which could be drawn, then, is that the *akhbāriyya* named here were those who rejected *kalām*, or at least did not use it in their presentation of doctrine. There is clearly a series of contrasting dichotomies here: *al-ʿadāla/al-mushabbih, al-kalāmiyya/al-akhbāriyya, al-waʿīdiyya/al-tafṣīliyya*. In each of these, the former term refers to the doctrines of the Muʿtazila adopted by Imāmī theologians, and the latter refers to doctrines of the Sunni traditionists (*ahl al-ḥadīth*), or even the Ashʿariyya. It may be that al-Shahrastānī is referring simply to two groups of Imāmīs, traditionalists and rationalists. The former group is not named as such, but is described by the doctrines it holds. The Muʿtazilī-influenced Imāmīs hold the doctrines of *ʿadāla* and *waʿīd*, and argue using the methods of *kalām*. The traditionalists hold the doctrines of *mushabbih* (that is, *tashbīh*) and *tafṣīl* and assert the primacy of reports over reason. That the latter group is given the distinctive and exclusive name *akhbāriyya* is far from clear. That they hold views which accord with the doctrines found in later (post-Astarabādī) Akhbārī theology is clearly not the case.[46] There is a second reference to *akhbāriyya* in al-Shahrastānī's *Kitāb al-Milal* where the passage reads:

---

[42] Literally "a sword" (*sayf*).

[43] I.e. *uṣūl al-dīn*: the principle of religion/theology. Those who "distinguished" between God's attributes (*tafṣīl*) in an anthropomorphic manner are opposed by those who adhered to the Muʿtazilī doctrine of God's threat and punishment (*waʿīd*).

[44] Shahrastānī, *al-Milal*, p. 172.

[45] It is probably the editor of al-Shahrastānī's *al-Milal* who vocalises this word *ikhbāriyya* (and not *akhbāriyya*). It is not clear whether this vocalisation might imply a different emphasis. It is possible that it designates a more neutral description of a group of "historians" or transmitters of historical material about the Prophet and the Imams, rather than those who argued that such material was the only source of religious knowledge.

[46] See my analysis of Astarabādī's theology, below (pp. 102–139), and the theological views of later Akhbārīs (in my "Scripturalist Sufism"). Later Akhbārīs explicitly reject the theological doctrines of *tashbīh* and *tafṣīl*.

> They [the Imāmīs] followed their Imams in *uṣūl* [*al-dīn*] at first, but when the reports from their Imams differed and time passed, each group [within the Imāmīs] chose their own path. Some of the Imāmīs became Muʿtaziliyya, either *wadaʿiyya* or *tafṣīliyya*, and others became *akhbāriyya*, either *mushabbaha* or *salafiyya*.[47]

Once again, the reference here is to theological doctrines. The reference to *salafiyya* as a sub-group of the Imāmī *akhbāriyya* is not glossed and remains unclear. It could refer to Imāmīs who claimed to abide by the teachings of the early followers of the Imams, rather than the teachings of the Imams themselves. What is clear is that the doctrines of al-Shahrastānī's *akhbāriyya* are theologically traditionalist, but, once again, have little in common with the distinctive legal doctrines of the later Akhbārīs.[48]

The best known references to the *akhbāriyya* are to be found in the Persian work, *Kitāb al-Naqḍ* by ʿAbd al-Jalīl al-Qazwīnī (d. Sixth Century AH). This work is roughly contemporaneous with al-Shahrastānī's *Kitāb al-Milal wa'l-Niḥal*. It was on the basis of this text that Madelung, in an *obiter* footnote, suggested that the Akhbārī-Uṣūlī dispute did not start with Astarābādī as had been previously believed.[49] It existed before his time in the form of two schools of Imāmī Shiʿism, the Akhbāriyya and the Uṣūliyya, who had made negative and positive assessments respectively, concerning the employment of dialectical reasoning in the service of theological and legal argument. These comments prompted Newman and Stewart to develop Madelung's suggestion that there was a "pre-Astarābādī" form of Akhbarism. Newman traced the history of these two "schools" of Imāmī Shiʿism from their beginning in the Third/Ninth Century to the time of Astarābādī, thereby providing a history of Akhbārī thought before Astarābādī.[50] His argument, as mentioned above, depends upon a broad definition of Akhbarism, concerned with juristic authority and not merely the legitimacy of particular interpretive techniques.[51]

---

[47] Shahrastānī, *al-Milal*, v. 1, p. 165.

[48] The reference to *akhbāriyya* by al-Sayyid al-Sharīf al-Jurjānī (d. 482/1413) in his *Sharḥ al-Mawāqif* (v. 8, p. 392) is entirely based on these passages from al-Shahrastānī's *Kitāb al-Milal*, and therefore does not constitute a distinct reference to pre-Astarābādī *akhbārīs*.

[49] Madelung, "Imamism and Mu'tazilite Theology", p. 21, n. 1.

[50] Newman, "Development and Political Significance", passim.

[51] Newman developed his views further in his edition and commentary upon al-Samāhījī's famous forty points of conflict between Akhbārīs and Uṣūlīs. See Newman, "Akhbārī/Uṣūlī Dispute pt. 2", pp. 250–253.

Kohlberg and Qayṣarī also refer to the *Kitāb al-Naqḍ* as evidence of the existence of the opposing camps called Akhbārī and Uṣūlī in the period before Astarabādī.[52] Stewart also makes brief reference to al-Qazwīnī's work in his discussion of the evidence for the existence of the Akhbāriyya before Astarābādī.[53] It is clear, then, that the *Kitāb al-Naqḍ* is believed to be a crucial piece of evidence in arguments for the existence of a pre-Astarābādī Akhbārī school. It therefore deserves to be examined in some detail.

The *Kitāb al-Naqḍ* is a fascinating account of both popular and scholarly Shīʿī belief in the Sixth/Twelfth Century.[54] The work takes the form of a refutation of a previous work by a Sunni author, identified by some as Shihāb al-Dīn al-Tawārikhī al-Shāfiʿī al-Rāzī[55] and by others as Abū al-Ḥusayn (or al-Ḥasan) al-Baṣrī.[56] Whichever is the case, the author had recently converted from Shiʿism to Sunni Islam, and therefore claimed to have in-depth knowledge of Shīʿī heresies. This author's work, entitled *Baʿḍ Faḍāʾiḥ al-Rawāfiḍ*, is unfortunately lost, though al-Qazwīnī quotes a significant portion (possibly all) of it in the course of his refutation. Al-Qazwīnī gives the impression that he is not at all worried about the *Baʿḍ Faḍāʾiḥ al-Rawāfiḍ* itself. It contains no accomplished arguments and is full of falsehoods and lies (*durūgh*). What worried him was the exposure the work was receiving as it was read in public, particularly in his former home town of Qazvin.[57] The author of *Baʿḍ Faḍāʾiḥ al-Rawāfiḍ* clearly did not consider the Shīʿa part of the Muslim community as he sets about exposing elements of Shīʿī belief and

---

[52] Kohlberg, "Akbarīya", p. 717 and Qayṣarī, "Akhbāriyān", p. 160.

[53] Stewart, *Islamic Legal Orthodoxy*, pp. 202–207.

[54] See, for example, the use made of the text by Calmard in his "Le Chiisme Imamite en Iran".

[55] See Ṭihrānī, *al-Dharīʿa*, v. 24, p. 283.

[56] Al-Afandī states that it is clear to him (*al-ẓāhir ʿindī*) that ʿAbd al-Jalīl al-Qazwīnī is identical to ʿAbd al-Jalīl b. ʿĪsā al-Rāzī and ʿAbd al-Jalīl b. Abī al-Fatḥ al-Rāzī, all of whom are mentioned in Muntajab al-Dīn's *Fihrist* (pp. 76–77). If this is the case, then ʿAbd al-Jalīl's opponent was one Abū al-Ḥusayn (or perhaps al-Ḥasan) al-Baṣrī since ʿAbd al-Jalīl b. Abī al-Fatḥ is recorded as having written a work titled *Naqḍ al-taṣaffuḥ li-Abī al-Ḥusayn al-Baṣrī* (which, al-Afandī argues, is this *Kitāb al-Naqḍ*). See Afandī, *Riyāḍ*, v. 3, p. 74. He does, however, give these scholars separate biographical entries. For al-Qazwīnī, see Afandī, *Riyāḍ*, v. 3, pp. 71–73. Al-Ḥurr, on the other hand, sees ʿAbd al-Jalīl al-Qazwīnī as a different scholar to the other two ʿAbd al-Jalīls (see Ḥurr, *Amal*, v. 2, p. 143 #418), though he does consider the other two to be probably the same person—*yaqrab ittiḥād al-rajulayn*; see Ḥurr, *Amal*, v. 2, p. 145).

[57] Qazwīnī, *Naqḍ*, p. 2. There were three copies of the work, one of which was sent to Qazvin and "anybody who read it fell into heresy and unbelief".

practice which, in his view, demonstrate that the Shīʿa are outside the bounds of orthodoxy. Al-Qazwīnī was asked to write a refutation of *Baʿḍ Faḍāʾiḥ al-Rawāfiḍ*, and a copy of the work was sent to him. The *Kitāb al-Naqḍ* ("The Book of Refutation"), or to give it its full title, *Baʿḍ Mathālib al-Nawāṣib fī Naqḍ Baʿḍ Faḍāʾiḥ al-Rawāfiḍ*, was his response.

Al-Qazwīnī's aim in the *Kitāb al-Naqḍ* is to refute, passage by passage (and at times line by line), the *Baʿḍ Faḍāʾiḥ al-Rawāfiḍ*. A good proportion of the *Kitāb al-Naqḍ* is, inevitably, made up of citations from the *Baʿḍ Faḍāʾiḥ al-Rawāfiḍ*, and the structure of the earlier work dictates that of al-Qazwīnī's refutation. The criticisms expressed in *Baʿḍ Faḍāʾiḥ al-Rawāfiḍ* are of three types, he states. Firstly, there are scurrilous attacks on the Shīʿa with no justification. Secondly, there are attacks on heretical Shīʿī groups, but not on the true Shīʿa. Thirdly, there are accurate depictions of Shīʿī belief which can be defended by recourse to historical evidence and dialectical reasoning. It is only in passages dealing with the second type of criticism that the term *akhbāriyya* (or *akhbārī*, or *akhbāriyān*) appears. Al-Qazwīnī's general point is that the author of *Baʿḍ Faḍāʾiḥ al-Rawāfiḍ* has described accurately some of the beliefs of some groups (three or four different groups are mentioned) who identify themselves as Shīʿī. However, these groups are not Shīʿī; they are not even Muslim. Hence, the Sunni opponent's criticisms are irrelevant. The true believers are given a number of different names in the course of the *Kitāb al-Naqḍ* amongst which are Shīʿa (used to distinguish between the true Shīʿa and heretics), Imāmiyya, Ithnāʿashariyya and Shīʿah-yi Uṣūliyya. The heretical groups "who count themselves as Shīʿa" (*khʷīshtan-rā Shīʿah khʷānand*[58] and *īshān-rā az hisāb-i Shīʿah shumurand*)[59] are given the titles *ghulāt* ("extremists"?), *ḥashwiyya* ("literalists"?), *Zaydiyya*,[60] *Kaysāniyya*,[61] *Faṭaḥiyya*[62] and

[58] Qazwīnī, *Naqḍ*, p. 618.

[59] Qazwīnī, *Naqḍ*, p. 492.

[60] The well known Shiʿite sect, followers of Ḥasan's grandson Zayd b. ʿAlī (d. 122/740). Qazwīnī, *Naqḍ*, p. 492.

[61] The well known "extremist" group who supported the Imamate of Muḥammad b. Ḥanafiyya, led by Mukhtār and after the latter's death by Abū Amrah Kaysān. Qazwīnī, *Naqḍ*, p. 492.

[62] Namely, the group who considered ʿAbd Allāh al-Afṭaḥ, eldest son of the sixth Imam (Jaʿfar al-Ṣādiq), to have been the seventh Imam. There is little evidence of them surviving beyond al-Afṭaḥ's death, since he is commonly believed to have left no descendents. See Qazwīnī, *Naqḍ*, p. 492. A brief description of them is given in Ashʿarī, *Maqālāt*, p. 28.

akhbāriyya.[63] It is, of course, the last of these which interests us here. The fact that al-Qazwīnī identifies himself as a member of the Shīʿah-yi Uṣūlīyya, and opposes the *akhbāriyya* (amongst other so-called Shīʿa groups) prompted Madelung's suggestion that the Akhbārī-Uṣūlī dispute can be traced to an earlier period of Shiʿism.

There are, in all, eight references to the *akhbāriyya, akhbārī* and *akhbārīyān* in the *Kitāb al-Naqḍ*. This, it could be argued, is rather slight evidence for the existence of an important sub-group of the Imāmiyya. The position is confused further by the fact that, according to Muḥaddith's notes, on four of these eight occasions, the reading is not unequivocally *akhbāriyya/akhbārī/akhbāriyān*. On four occasions, some manuscripts record *ijbāriyya* (not *akhbāriyya*).[64] In one other of the remaining four references, the passage concerned is missing in some manuscripts and could be an interpolation.[65] All this, then, leads to the possible conclusion that there are only three references to the *akhbāriyya* in the *Kitāb al-Naqḍ*,[66] reducing the evidence even further.

Presuming that the editor (Muḥaddith) is correct in recording *akhbāriyya*, *akhbārī* and *akhbāriyān* on all these occasions, and that al-Qazwīnī is referring to the same group on each occasion, one is afforded an insight into some of the beliefs al-Qazwīnī attributed to the *akhbāriyya*. The first use of the term *akhbāriyya* occurs in the introductory passage of the *Kitāb al-Naqḍ*, and is one element of a list of heretical Shīʿī groups which the author of *Baʿḍ Faḍāʾiḥ al-Rawāfiḍ* has simply mistaken for the beliefs of the Shīʿah-yi Uṣūliyya. The *akhbāriyya* are, then, linked with *ghulāt* and *ḥashwiyya*. There

---

[63] Another group mentioned alongside these is the Daysāniyya. See Qazwīnī, *Naqḍ*, p. 301, though there is a variant reading of *naṣariyya* (Christians) in place of *daysāniyya* (see Qazwīnī, *Naqḍ*, p. 301, n. 3). Whilst strictly a reference to the supporters of the dualist Bardesanes (d. 201 CE), it is probably a general reference to dualists.

[64] These are found at Qazwīnī, *Naqḍ*, p. 2, p. 256, p. 301 and p. 688. The second of these references is to a possible reading of *ijbārī* for *akhbārī*. The last of these references refers to a possible *ijbāriyān* reading for *akhbāriyān*. If, on these occasions, the reference is to *ijbāriyya* or *ijbārī* or *ijbāriyān*, then it is likely that the reference is to those who support the doctrine of *ijbār* (i.e. the *mujbira*), who hold that human agency is impossible and we are all "compelled" to perform actions by God. *Ijbār* is the term used by Imāmīs attracted to Muʿtazilī thought (amongst whom we can certainly include al-Qazwīnī) to describe the doctrine of the Ashʿariyya (see, for example, Ibn Nadīm, *Fihrist*, p. 180). It is not impossible that al-Qazwīnī is referring to a Shīʿī group which he considered to hold the doctrine of *ijbār*.

[65] Qazwīnī, *Naqḍ*, p. 304.

[66] These would be Qazwīnī, *Naqḍ*, p. 492, p. 571 and p. 618.

is no further indication of what the *akhbāriyya* might believe at this point.[67] The other seven occasions, however, provide more (though still limited) information concerning particular doctrines of the *akhbāriyya*:

1. In the "First disgrace" of the *rawāfiḍ* (that is, the Shīʿa), the author of *Baʿḍ Faḍāʾiḥ al-Rawāfiḍ* explains how the Shīʿa insult the companions of the Prophet Muḥammad, his wives and the *salaf* (early generations). According to the Shīʿa (he says), 33,000 pious Muslims are in hell, including companions, successors, Qurʾān commentators and reciters and all the Zaydiyya (who are accused of not recognising the special powers of the Imams). Al-Qazwīnī's response is that this is slander and sin (*ithm*) on the part of the author and that it is certainly not the belief of the "Imāmiyān-i Uṣūlī". "If, however, an *akhbārī*, or a *ḥashwī* or a *ghālī* has made [such] a statement, and an incorrect report [*naqlī nā-durust*] has come about, then that [person] has, for the Uṣūliyya, gone beyond the limits of belief and is not a Muslim."[68] The impression gained from this passage is that this is a belief an *akhbārī, ḥashwī* or *ghālī* may hold, but the conditional sentence could indicate that these groups are not necessarily associated with these beliefs.

2. The author of *Baʿḍ Faḍāʾiḥ al-Rawāfiḍ* records a story in which Zurāra b. Aʿyan, the companion of Imam Jaʿfar al-Ṣādiq, reports that the Imam was asked to interpret (*taʾwīl*) the verses "On that day, no one will inflict a punishment like his [that is, God's], and no one will bind as he binds" (Q89(al-Fajr).25–26). The Imam is supposed to have answered that this verse refers to Abū Bakr, who will receive a more severe punishment than all others, for he stood on the minbar of the Prophet even though he had no right (*nā-ḥaqq*) to do so. Al-Qazwīnī replies that this report is not to be found in the books of the Uṣūliyya. It is an *akhbārī, ghālī* and *ḥashwī* lie.[69]

3. A little further on from the above reference, the author of *Baʿḍ Faḍāʾiḥ al-Rawāfiḍ* accuses the Shīʿa of ignoring the exegesis of a list of early Qurʾanic interpreters (*mufassirān*) whose views are normally treated with respect.[70] Al-Qazwīnī answers that the Shīʿa respect the exegesis

[67] Qazwīnī, *Naqḍ*, p. 2.

[68] Qazwīnī, *Naqḍ*, p. 256. The last phrase could be literally translated as "he has entered into the extremities of unbelief and 'non-muslim-ness'" (*ān-rā ... bāstān-i ghāyat-i bī-diyānatī va-nā-musalmānī bāshad*; both *bī-diyānatī* and *bī-amānatī* are recorded as possible readings by Muḥaddith, see Qazwīnī, *Naqḍ*, p. 256, n. 10).

[69] The report is also associated with the Daysāniyya (or in some manuscripts, the Christians). See above, n. 63.

[70] The list comprises of Ibn ʿAbbās, Daḥḥāk, Suddī, Muqātil, Jubayr, Ḥakīm, Qalansī, Hishām, Mujāhid and Kalbī. See Qazwīnī, *Naqḍ*, p. 303.

> of Imams Bāqir, Jaʿfar and Ḥasan ʿAskarī, who were sinless. After that they respect the exegesis of al-Ṭūsī, Muḥammad Fattāl, al-Ṭabarsī and Abū al-Futūḥ al-Rāzī.[71] These later scholars, though not, of course sinless, were all learned, trustworthy and reliable. They cannot be accused of being *ghālī, akhbārī* and *ḥashwī*.

In each of the above three references to *akhbāriyya/akhbārī*, the group is linked with the *ghulāt* and the *ḥashwiyya*. The latter two groups are mentioned together at other points in the *Kitāb al-Naqḍ* without a reference to the *akhbāriyya*.[72] Whilst none of these groups' beliefs are outlined in detail by al-Qazwīnī, they do appear to have shared a deep hatred of Sunni figures such as Abū Bakr and a belief in a recension of the Qur'ān in which ʿAlī and his descendents are explicitly mentioned. These are themes which heresiographers have ascribed to a number of "extremist" groups amongst the early Shīʿa. The terms *ghālī* and *ḥashwī* are always used by opponents. Writers or sects described in these terms do not self-identify as such (a feature which distinguishes them from the Akhbārīs from Astarābādī onwards). *Ghālī* and *ghulāt* are terms which have received extensive treatment.[73] *Ḥashwī*, normally glossed as "literalist" in secondary literature,[74] was a derogative term applied to those who accepted anthropomorphic descriptions of God in the Qur'ān and *ḥadīth*.[75] The *ḥashwiyya* were also accused of accepting traditions which were clearly inauthentic, and hence the *aṣḥāb al-ḥadīth* were sometimes labelled *ḥashwī* by their opponents (particularly the Muʿtazila). In Shīʿī heresiography the term is normally used to describe Sunni *ḥashwī*, though it could conceivably be used to describe Shīʿīs who consider the reports of the superhuman qualities of the Imam to be both unquestioningly authentic and describing real properties of the Imam. Al-Qazwīnī's use of the terms *akhbāriyya* and *akhbārī* in the context of these lists appears, however, to be formulaic and the boundaries he perceives between the groups are not clear. What is clear is that they are all contrasted with the Shīʿah-yi Uṣūliyya (who maintain true Shiʿism).

---

[71] For an analysis of the exegetical method of these scholars, see my "Qur'anic Interpretation". Muḥammad Fattāl al-Nīsabūrī's *tafsīr, al-Tanwīr fī maʿānī al-tafsīr*, appears to have been lost.

[72] See, for example, Qazwīnī, *Naqḍ*, p. 291.

[73] See al-Qāḍī, *al-Kaysāniyya*.

[74] Halkin, "Ḥashwiyya".

[75] Nawbakhtī devotes much space to an exposition of *ḥashwī* doctrines, see *Firaq al-Shīʿa*, p. 6 and passim.

The remaining references in *Kitāb al-Naqḍ*, unfortunately, provide only an incremental increase in our knowledge of al-Qazwīnī's *akhbāriyya*.

4. The *akhbāriyya* are mentioned by al-Qazwīnī as one of the seventy-two sects of the Muslims "who count themselves as Shīʿa".[76] They are mentioned alongside the Zaydiyya, the Faṭaḥiyya and the Kaysāniyya. The reference comes in the course of a reply to the author of *Baʿḍ Faḍāʾiḥ al-Rawāfiḍ* who argues that the geographical spread and numerical superiority of Sunnis ("that they can be found in the West and the East") is evidence of the truth of Sunni belief. Al-Qazwīnī lays out the divisions within Sunnism in order to disrupt this portrayal of Sunni homogeneity, but he also mentions the widespread distribution of groups which consider themselves Shīʿī. There is no indication of what, in this context, the *akhbārīs* are supposed to believe.

5. In "the sixteenth disgrace" of the *rawāfiḍ*, the author of *Baʿḍ Faḍāʾiḥ al-Rawāfiḍ* accuses the Shīʿa of believing that ʿAlī is "better" (*bihtar*) than all Prophets, that God gave ʿAlī understanding of all knowledge given to previous Prophets and that his greatness makes the Prophets redundant. This, of course, renders them unbelievers in the author's opinion. Al-Qazwīnī's reply is that this is not the belief of the Shīʿah-yi Uṣūliyya. The true Shīʿa believe the station of *walī* ("associate", and one of the terms used to describe the position of the Imams) to be lower than that of *nabī* (prophet). Every Prophet is, in this sense, greater (*faḍīlat-i bīshtar*) than ʿAlī. Some of the *akhbārīs* and *ḥashwiyya* (literalists) amongst the early Shīʿa (*az salaf*) may have held the belief that ʿAlī was better than those prophets who did not attain political power (such as Jesus). However, this doctrine is unacceptable (*nā-maqbūl*), pointless (*bī-fāʾidah*), refuted (*mardūd*) and without evidence (*bī-dalīl*).[77]

It should be pointed out that al-Qazwīnī is referring to *akhbārīs* and literalists amongst the early Shīʿa, giving no definite indication here of either their current existence or of their current subscription to this doctrine.

6. In the course of his description of "the thirty-third disgrace" of the *rawāfiḍ*, the author of *Baʿḍ Faḍāʾiḥ al-Rawāfiḍ* accuses the Shīʿa of following a *Sunna* of the Egyptians (*miṣriyān*)[78] in that they make their intention (*niyyat*) to begin their fast two days before Ramaḍān and

[76] Qazwīnī, *Naqḍ*, p. 492.

[77] Qazwīnī, *Naqḍ*, pp. 570–571.

[78] This is most likely a reference to the Fatimids who allegedly used astronomical charts for dating Muslim months, rather than the sighting of the moon. On the modern use of this by the Bohra Ismāʿīlī Shīʿīs see, Amiji, "The Bohras of East Africa", p. 52.

break it two days before the ʿĪd al-Fiṭr. They use methods other than the actual sighting of the moon (such as mathematical tables or the science of stars) to identify the date of ʿĪd. Al-Qazwīnī recognises two elements of the dispute here:

(a) There is "guilt by association" with Egyptian practices. This he rebuffs by saying that elements of the beliefs of the *mujbirān* (such as it not being obligatory to know God until the arrival of a Prophet) are also similar to those of the Egyptians. Simple coincidence between the belief and practice of two groups does not, in itself, indicate heretical imitation.

(b) The second element of the criticism concerns the practice of fasting for two days before the beginning of Ramaḍān. Al-Qazwīnī states that there are pious Shīʿa (*zuhhād va-ʿubbād-i Shīʿah*) who fast for two months prior to Ramaḍān. However, the *niyyat* (intention) here is distinct from the Ramaḍān *niyyat*, and the ensuing fast is wholly superogatory. Furthermore, the jurists of the Shīʿa have clearly declared their views concerning issues such as the sighting of the Ramaḍān moon and the beginning of ʿĪd. They do not use mathematical tables or astronomical calculations.[79] An opponent may argue that the Shīʿa only celebrate ʿĪd on the same day as the Sunnis out of *taqiyya* (dissimulation), but al-Qazwīnī points out that this is a matter where *taqiyya* is not permitted (*taqiyyah natavān kard*).

After this, al-Qazwīnī mentions that there was a group of the *akhbāriyya* who considered themselves Shīʿa, and this was their belief (*jamāʿatī-i akhbāriyyah kih khʷīshtan Shīʿah khʷānand īn maʿnā madhhab-i īshān būdah ast*). However, very few of them have survived and those that have, keep this practice hidden from the Uṣūlīs. Renowned Shīʿī scholars, such as al-Sayyid al-Murtaḍā and al-Ṭūsī, have condemned the *akhbāriyya* for it.

A number of points can be made about this interesting passage. Firstly, it is likely that al-Qazwīnī is not referring to all *akhbārīs* here, and only a section of the *akhbāriyya* are being accused of this practice. Secondly, there is some textual ambiguity. The disreputable practice could be one of a number of things, and the text is not entirely clear. There are at least three possibilities:

(a) The use of mathematical/astrological calculations to determine the beginning of Ramaḍān. This, it is true, has been condemned in Shīʿī

[79] See, for example, Ṭūsī, *Khilāf*, v. 2, p. 169, where the use of mathematics and mathematical tables, stars and the like to determine the first day of Ramaḍān is refuted by reference to "well-attested reports" (*akhbār mutawātira*).

juristic writings, as only the sighting of the moon or two reliable testimonies of the sighting of the moon can impose the obligation to fast during Ramaḍān. If the *akhbāriyya* are supporting the use of such calculations, they would be out of step with the position of most jurists of the time.

(b) Making a *niyyat* (and fasting?) two days prior to the beginning of Ramaḍān. The legitimacy of making an early *niyyat* was debated by Shīʿī jurists, with a significant number (including some, such as Ṭūsī, considered members of the Uṣūliyya by al-Qazwīnī) viewing it a legitimate practice.

(c) Finally, it could be a combination of these, using both the mathematical/astrological tables and beginning one's fast two days before the start of Ramaḍān.

Which of these al-Qazwīnī is condemning is not clear. However the final reference to the *akhbāriyya* in the *Kitāb al-Naqḍ* also concerns fasting and the beginning of Ramaḍān.

7. The "Fifty Seventh disgrace" of the *rawāfiḍ* concerns a view, supposedly linked to al-Murtaḍā, that fasting on the *yawm al-shakk* ("day of doubt") brings a full reward (*faḍīlatī tamām*). As Ramaḍān is signalled by the uncertain criterion of the sighting (or testimony of the sighting) of the moon, the "day of doubt" refers to a day which could be a day of either Ramaḍān or Shaʿbān. This, the author of *Baʿḍ Faḍāʾiḥ al-Rawāfiḍ* maintains, contradicts the Prophet himself. The Prophet reportedly said that whoever fasts on the *yawm al-shakk* has disobeyed the Prophet. For most Sunnis, a *yawm al-shakk*, which is later discovered to be a day of Ramaḍān, requires a compensatory fast (*qaḍāʾ*). The Shīʿa are here accused of arguing that it does not require a compensatory fast since its reward, in relation to the other days of Ramaḍān, is "full" (*tamām*). Al-Qazwīnī argues that fasting on the *yawm al-shakk* is forbidden (*ḥarām*)—by which he means that forming the intention (*niyyat*) to fast for a *yawm al-shakk* (and subsequently fasting on that day) is forbidden. It is permitted to perform a superogatory fast in the month of Shaʿbān. If, on the *yawm al-shakk*, the individual makes an intention to fast a Shaʿbān day, then that fast is valid. If, on the *yawm al-shakk*, one makes an intention to fast a Ramaḍān day, then the fast is also valid. What is not permitted is to fast on a *yawm al-shakk* with the intention to fast on a *yawm al-shakk*. In the context of this discussion, al-Qazwīnī makes reference to the *akhbāriyān* who break their fast on the last day of Ramaḍān (rather than the first day of Shawwāl), and compares them to the Sunnis (*mujbirān*) who, in order to avoid fasting on a *yawm al-shakk*, do not fast on the first day of Ramaḍān. Both are to be cursed as neither fulfils the fast.

The precise reasoning behind the *akhbārī* ruling that one can break the fast on the last day of Ramaḍān is unclear. It presumably is related to the idea that there is a *yawm al-shakk* at the beginning of each month, and if one fasts the *yawm al-shakk* at the end of Shaʿbān, one need not do so at the end of Ramaḍān.[80] This would support the idea that the subject of al-Qazwīnī's earlier condemnation was, in fact, not merely the use of astronomical and mathematical aids to predict the start of Ramaḍān. Rather, it was beginning the fast early, during the last two days of Shaʿbān, which was considered a distinctive, reprehensible *akhbārī* practice.

These scraps of information are hardly sufficient to construct a general understanding of al-Qazwīnī's perception of *akhbārī* doctrine. The *akhbāriyya* appear to be a group with heterodox views (the vilification of the companions and other early figures, superiority of ʿAlī over the Prophets) and hence are often named with the *ghulāt* and *ḥashwiyya*. On the other hand, they are credited with some unusual practices associated with the fast of Ramaḍān, such as the use of astrological and mathematical tools to determine the month's beginning, making one's intention to fast two or so days before the end of Shawwāl (and perhaps fasting for these days also) and breaking the fast one day early (perhaps as compensation for beginning it during Shaʿbān). Such positions do not obviously relate to distinctive doctrines (legal or otherwise) of the later Akhbārī school, and such a relationship can only be shown with some exegetical effort.[81] My conclusion is, then, that whoever the *akhbāriyya* referred to in al-Qazwīnī's text may be, they are not presented with sufficient coherency to confirm their position as precursors or early manifestations of the later Akhbārī school founded by Astarābādī. They seem to share little, except the name.

Another pre-Astarābādī reference to Akhbārīs should be noted here. Fakhr al-dīn al-Rāzī (d. 606/1210) makes mention of the *akhbārī* Imāmīs in his *al-Maḥṣūl*:

---

[80] The later Akhbārīs would not have agreed with this reasoning. As al-Baḥrānī states, "the *yawm al-shakk* is merely a day when there is difference of opinion over whether the *hilāl* of Shaʿbān has been sighted." (Baḥrānī, *al-Ḥadāʾiq*, v. 13, p. 41). One should fast on such occasions, in the interests of caution (*iḥtiyāṭ*), since the day could be one of the days of Ramaḍān.

[81] One could argue, for example, that these earlier Akhbārīs fell into extremist views because they, like the *hashwiyya*, accepted the authenticity of all *ḥadīth*, even those which predicated superhuman qualities to the Imams. This would link with the later Akhbārī doctrine of the unquestioned authenticity of the four books. Unfortunately, al-Qazwīnī makes no mention of the Akhbāriyya arguing for the superiority of ʿAlī or the condemnation of certain *ṣaḥāba* to hell on the basis of reports from the Imams. They appear simply as one element in an almost formulaic list of heretical Shīʿī groups.

> As for the Imāmiyya, the *akhbārīs*—who made up most of the Shīʿa in the earliest times—only rely upon reports which they relate from their Imams in matters of *uṣūl al-dīn*—indeed in *furūʿ* also. As for the *uṣūlīs*, Abū Jaʿfar al-Ṭūsī agrees with us on this point, and there is no one left who denies this other than [al-Sayyid] al-Murtaḍā and a few of his followers.[82]

The passage is found in al-Rāzī's discussion of the probative force of *khabar al-wāḥid*. There are Imāmīs who agree with him (such as al-Shaykh al-Ṭūsī) and argue that *khabar al-wāḥid* has probative force (*ḥujjiyya*) and can be the basis for legitimate action on the part of the believer. There are some who argue against him (namely al-Murtaḍā) who denies that *khabar al-wāḥid* can be used as the basis for action, because it does not bring *ʿilm*.[83] Both are within the Imāmī *uṣūliyya*, and are to be distinguished from the Imāmī *akhbāriyyūn*. The *akhbāriyyūn* argue that only reports transmitted from the Imams can act as proofs. Now a number of ambiguities emerge here about the doctrine of al-Rāzī's "*akhbāriyyūn*". These *akhbārīs* are distinguished from the *uṣūliyyūn* by the fact that they only accept reports from their Imams (or reports about the Prophet transmitted by their Imams—al-Rāzī is not clear here, and it is not clear whether this makes a difference to their position). The *uṣūlīs* appear to accept reports from the Prophet transmitted by others, though they differ over whether those reports need to reach the bar of *tawātur* (which was the Ṭūṣī-Murtaḍā debate). Within the *uṣūlīs* there was the well-known debate over the probative force of *khabar al-wāḥid*, but this appears to refer to reports which are not transmitted by the Imams. How this relates to evidence from within the Shīʿī tradition is discussed below, but there is no harmony between the *akhbāriyyūn*-*uṣūliyyūn* debate described by al-Rāzī here and those found in the later Akhbārī-Uṣūlī conflict. For the later Akhbārīs, reports related by non-Imāmīs—from an Imam, the Prophet or transmitted through an Imam but about the Prophet—are accepted as potential legal proofs. It was the later Uṣūlīs who placed restrictions on the transmission of reports about the Imam or the Prophet (namely, the transmitters must be just Imāmīs).[84] If any marriage is to be made between al-Rāzī's

[82] Rāzī, *al-Maḥṣūl*, v. 4, p. 384.
[83] See Calder, "Doubt and Prerogative", pp. 61–64.
[84] See below, pp. 70–71.

portrayal and the later disputes, then one would link the doctrine of the early *uṣūliyyūn* with later Akhbārī doctrine.[85]

Al-Rāzī's depiction had its influence within the Shīʿī tradition. Al-ʿAllāma, in his work of *uṣūl al-fiqh, Nihāyat al-wuṣūl ilā ʿilm al-uṣūl*, discusses whether isolated reports (*khabar al-wāḥid*) can be relied upon in matters of doctrine and law. He states:

> As for the Imāmiyya, the *akhbāriyyūn* amongst them only rely upon *khabar al-wāḥid* transmitted from the Imams in *uṣūl al-dīn* and its branches. The *uṣūliyyūn* amongst them, such as Abū Jaʿfar al-Ṭūsī and others, agree that the isolated report should be accepted. No one denies this except for al-Murtaḍā and his followers [who deny it] because [*khabar al-wāḥid*] leads to *shubha* for them.[86]

The general wording of this passage is clearly lifted from al-Rāzī's *al-Maḥṣūl*. However, al-ʿAllāma alters al-Rāzī's wording to make the passage refer solely to the debate around the isolated report (*khabar al-wāḥid*). In *al-Maḥṣūl*, the passage related to *akhbārīs* holding a doctrine related to reports from the Imams (about the Prophet or themselves) generally. After al-ʿAllāma's adjustment, the passage imputes a specific doctrine concerning *khabar al-wāḥid* to the *akhbārīs*. The later passage comes in the context of al-ʿAllāma's attempt to establish the acceptability of the isolated report as an indicator of God's law. There had been those (such as al-Sayyid al-Murtaḍā) whose epistemological demands for certainty (*ʿilm*) had led to a rejection of isolated reports as possible legal indicators. Al-ʿAllāma, on the other hand, wishes to argue for an acceptance of isolated traditions

---

[85] Al-Rāzī also makes reference to an Imāmī *akhbārī* doctrine in his theological works—namely the belief that the soul is separate (*mujarrad*) from the body (see Shihada, *Teleological Ethics*, p. 116, n. 36). Such a doctrine was not unique to Akhbārīs in later Imāmī thought, and was not a distinctive Akhbārī doctrine.

[86] The text is also cited in Shaykh Ḥasan, *Maʿālim*, p. 191 and Tūnī, *al-Wāfiya*, pp. 159–160. Stewart has consulted the Princeton manuscript #376 of the *Nihāya* where this passage can be found (see Stewart, *Islamic Legal Orthodoxy*, p. 182, n. 28). The passage does not, however, appear in the new edition of the work (ʿAllāma, *Nihāya*) since the manuscripts used by the editor, al-Bahadurī, are incomplete. Al-Tūnī is talking of isolated reports which have no additional indication (*qarāʾin*) that they have probative force. Shaykh Ḥasan is talking of isolated reports which *by their very nature* do not have additional indications that they have probative force. That is, the conceptions of isolated reports (*akhbār al-āḥād*) are distinct here. Al-Karakī, in his commentary on this passage (Karakī, *Hidāyat*, p. 66), also notes how al-ʿAllāma's portrayal of the Akhbārī-Uṣūlī conflict does not accord with the "real" beliefs of the Akhbārīs and Uṣūlīs.

as legal indicators even though they reach a level of indicatory value which is less than certain. Their probative force (*ḥujjiyya*) is such that whilst they do not establish a ruling with certainty, they do provide evidence of a ruling which is to be assessed along with other legal indicators. As in *al-Maḥṣūl*, both the *akhbārīs* and some *uṣūlīs* differ from al-Murtaḍā, but in al-ʿAllāma's passage they differ in that both accept isolated traditions as indicators of the law. Where they differ amongst themselves is not so clear (presumably the *akhbārīs* consider them to bring *ʿilm*, whilst the *uṣūlīs* consider them to bring *ẓann*, though this may be an anachronistic reading, influenced by the later Akhbārī-Uṣūlī conflict). It would seem also that the use of the construction *lā . . . illā* (translated as "only" above) indicates that, for the *akhbārīs* mentioned here, isolated traditions are the *only* source of theological and legal doctrine.[87] This implies that for the *uṣūlīs* mentioned here, the isolated report is accepted as one (but *not* the only) legal source. There are other sources which can be assessed alongside an isolated report in determining a legal ruling. This interpretation would concur with al-ʿAllāma's assessment elsewhere concerning the probative value of isolated traditions, and the need to evaluate them alongside other possible indicators of the law.[88] Calder cites this passage as it is quoted in Shaykh Ḥasan's *Maʿālim al-dīn* and adds the note:

> It is my belief that the terms used in this passage do not have a "party" or "sectarian" significance but a much vaguer sense of "traditionist" and *jurisprudens*. I think the Akhbārī movement is best understood as

[87] That this cannot be, strictly speaking, their position is clear in that whilst these Akhbārīs may not accept the Qur'ān as an indicator in the absence of its interpretation in the *akhbār* (a common later Akhbārī position), they surely would accept a *khabar mutawātir* as having probative force, even if its probative force is identical with that of an isolated tradition. The presence of the *lā...illā* phrase is a hangover from the al-Rāzī passage which al-ʿAllāma has copied and adjusted to suit his own purposes.

[88] If this is, indeed, al-ʿAllāma's intention by this statement, his association of this position with al-Shaykh al-Ṭūsī is accurate in so far as isolated traditions are indicators alongside others (Qur'anic verses, *al-khabar al-mutawātir* and, under very restrictive conditions, *ijmāʿ*). As is clear from al-Ṭūsī's major work of *uṣūl al-fiqh, ʿUddat al-uṣūl*, he reaches this position on the probative force of isolated traditions on the basis of an identical epistemology to that of al-Murtaḍā. Both al-Ṭūsī and al-Murtaḍā consider certainty as crucial to *uṣūl al-fiqh*; they differ over whether *khabar al-wāḥid* bring certainty or not. The description of Shaykh Ṭūsī as an *uṣūlī* is a mere appropriation of the past, rather than a considered depiction of Ṭūsī's position.

> beginning where Shīʿī historians see its beginnings, with the attack of Astarābādī on ʿAllāma's innovation.[89]

Now this may or may not be an accurate assessment of the implications of this passage for our depiction of the Akhbārī-Uṣūlī conflict. What is clear is that al-ʿAllāma's statement is, to an extent, hyperbole (the Akhbārīs do not *only* accept *akhbār al-āḥād*—what of the *khabar mutawātir*?), and that his maintenance of al-Ṭūsī as an *uṣūlī* is polemic. However, that isolated traditions bring knowledge of the law was, indeed, a view which formed a major element of Akhbārī doctrine after Astarābādī. Al-ʿAllāma's depiction of Akhbārīs does accord with an element of post-Astarābādī Akhbarism. However, al-ʿAllāma is also arguing, with the Akhbārīs, that *khabar al-wāḥid* can act as a source of law.

Furthermore, this is the only occasion—to my knowledge—on which al-ʿAllāma uses the name *akhbārīs* to describe those who hold this position. Elsewhere in the *Nihāya*, those who hold positions which chime with *some* later Akhbārī doctrines are either given no name ("a group believe...") or their position is characterised as that of the Sunnis. Outside of the *Nihāya* (such as in al-ʿAllāma's *Mabādiʾ al-uṣūl*), those who consider isolated traditions to have probative force—be it certain or presumptive—are lumped together and their position *together* is described as an *ijmāʿ* which Murtaḍā has failed to observe. My point here is that al-ʿAllāma uses the term *akhbāriyyūn* only once (here in the *Nihāya*), in a passage he has lifted (and altered) from al-Rāzī, and in relation to a specific doctrine concerning isolated reports. He does not use it elsewhere in his works of *uṣūl al-fiqh* (nor, as far as I can tell, in his works of *furūʿ* or *ṭabaqāt*).

It may be possible to explain the lack of Akhbārī references by al-ʿAllama outside of the *Nihāya* passage (plagiarised from al-Rāzī). It may be the case that al-ʿAllāma is aware of the revolution in Shīʿī legal thinking he is attempting in both the *Mabādiʾ* and the *Nihāya*, and this explains his unwillingness to identify, regularly and overtly, the more conservative forces in Shīʿī thought and call them Akhbārī. It may have been that there was a definable group of Shīʿī jurists called Akhbārīs, and this was their only *uṣūl* doctrine. These possible explanations, however, seem to me forced. It is more plausible

[89] Calder, "Doubt and Prerogative", p. 68, n. 31.

to consider the term *akhbāriyyūn* here as a remnant of al-ʿAllāma's plagiarism from al-Rāzī. He edited al-Rāzī's passage, missing out (for obvious reasons) the phrase which states that most early Shīʿīs were *akhbārīs*.[90] The *akhbārī* position on this issue is not repeated in other areas of al-ʿAllāma's *uṣūl* or *furūʿ* and they cannot, therefore, be said to form a definable group of scholars with a common approach.

Conclusions

My own position, outlined in subsequent chapters, is that the existence of a broadly "traditionalist" trend in Twelver Shiʿism before Astarābādī is indisputable. Some of the elements of this trend were taken up by later Akhbārīs (particularly those based in Safavid Iran), and used as historical precedent for their own (innovative and distinctively Safavid) doctrines. Just as Uṣūlīs before Astarābādī jumped on every scrap of evidence to justify their own position concerning *ijtihād*, Akhbārīs (after Astarābādī) wished to avoid the accusation of *bidʿa* (or innovation). However, neither the occasional use of the term *akhbārī* in pre-Astarābādī texts, nor the fact that al-ʿAllāma has introduced a novel element (*ijtihād* and the epistemology which accompanies its operation) into Shīʿī *uṣūl al-fiqh*, can divert us from the fact that Astarābādī's critique of the *mujtahid* position was unprecedented in its sophistication and nuance. Whilst the term *akhbārī* may have been employed very occasionally by writers before Astarābādī, its usage was undefined and lacked consistency. It clearly did not always refer to doctrines which the later Akhbārīs considered important or defining. After Astarābādī, the term was used as a proper name, for a particular (legal) doctrinal position and its plural (Akhbāriyyūn/Akhbāriyya) came to describe an identifiable group of scholars who put forward definable doctrines. The principal elements of this position were the rejection of *ijtihād* and the assertion of a binary epistemology between certain knowledge and ignorance.

---

[90] Al-ʿAllāma does not even condemn them. His real criticism is reserved for al-Murtaḍā who rejects the probative force of isolated traditions entirely.

CHAPTER TWO

MUḤAMMAD AMĪN AL-ASTARĀBĀDĪ AND THE FORMATION OF THE AKHBĀRĪ SCHOOL

Muḥammad Amīn al-Astarābādī is sometimes described as either the "founder" (*mu'assis*) or the "propagator" (*murawwij*) of the Akhbārī school. Which of these epithets one selects depends, of course, on whether one traces Akhbarism to the earliest period of post-ghayba Twelver Shiʿism or one considers it a later phenonemon.[1] Whichever is the most appropriate description, it is clear that the Safavid Akhbārī school emerged as a challenge to the dominant legal methodology of the day, and this was primarily due to the popularity of Astarābādī's ideas, in particular, those found in his *al-Fawā'id al-Madaniyya*. This work, composed in the later part of his life, represents Astarābādī's attack on the jurisprudence common amongst his contemporary Shīʿī jurists (whom he identifies as Uṣūlīs or *mujtahids*). However, within the *Fawā'id*,[2] there are also extensive discussions of the validity or otherwise of philosophy, theology and mysticism, with Astarābādī's verdict on the acceptability of these scholastic disciplines. Astarābādī's intellectual preoccupations, then, were not limited to *uṣūl al-fiqh*. Whilst the *Fawā'id* is undoubtedly his best-known work, it by no means exhausts his intellectual output. A number of other works, including both concise *risālas* and extended discussions of theological, philosophical and legal topics, have survived, and are useful, not only for an account of the development of Astarābādī's thought, but also for a reconstruction of his biography. His legal theory, his exegetical method and his theological ideas are the subject of subsequent chapters.[3] This chapter is primarily concerned with establishing

[1] The doctrinal (rather than strictly historical) nature of the debate over whether or not the Akhbārī school predates Astarābādī was discussed in chapter 1 above (pp. 1–30).

[2] Since this is Astarābādī's major work, I refer to it hereon by either its full title or simply as the *Fawā'id*. When referring to other *fawā'id* works (such as *al-Fawā'id al-Makkiyya* or *Fawā'id daqā'iq al-ʿulūm*) I use their full titles.

[3] Chapter 3 below (pp. 61–101) charts his legal methodology and Chapter 4 (pp. 102–139) describes his theological and philosophical thought.

Astarābādī's life history, intellectual output and his portrayal in subsequent Imāmī tradition.

Whilst Astarābādī's date and birthplace are not known, his family had its roots in Astarābād, northern Iran.[4] He travelled first to Najaf to study under one of the leading scholars of the day, Muḥammad b. ʿAlī b. al-Ḥusayn al-ʿĀmilī "Ṣāḥib al-Madārik" (d. 1009/1600). Astarābādī states in the *Fawāʾid* that he was "in the bloom of youth" (*ʿunfuwan shabābī*) when he gained his *ijāza* (a licence to teach or transmit *ḥadīth*[5]) from Muḥammad Ṣāḥib al-Madārik, (whom he describes as "my first teacher in the disciplines of *ḥadīth* and *rijāl*") in 1007.[6] Most likely, this occurred in Astarābādī's twenties (any younger seems presumptuous for an *ijāza*), making his birth date sometime in the 980s.[7] In an otherwise condemnatory notice, Yūsuf al-Baḥrānī praises Astarābādī's commentary (*ḥāshiya*) on his first teacher's famous work of *fiqh, Madārik al-aḥkām*. This *ḥāshiya* was probably written whilst studying under Ṣāḥib al-Madārik's tutelage and contributed towards the award of an *ijāza*.[8] Astarābādī is also said to have gained an *ijāza* from al-Ḥasan b. al-Shahīd al-Thānī (Ḥasan b. Shahīd II) Ṣāḥib al-Maʿālim,[9] who was teaching in Najaf at the time. If true, this would have come after his first *ijāza* in 1007 and before Ṣāḥib al-Maʿālim returned to Lebanon (where he died in 1011/1602). In 1010 Astarābādī relocated to Shiraz where he studied under Muḥammad Shāh Taqī al-Dīn al-Nassāba (d. 1019/1610–11).[10] He remained in Shiraz for at least four years, during which time he probably wrote his first independent academic work *al-Mabāḥith*

[4] The biographical sources do not name his birthplace. Muḥsin Fayḍ describes him as "one of our colleagues from Astarābād", though this may be a reference to his *nisba* (Kāshānī, *al-Ḥaqq al-Mubīn*, p. 12).

[5] On which, see below, pp. 141–148 and the references cited there.

[6] The *ijāza*, found within his commentary on the *ḥadīth* collection *Tahdhīb al-Aḥkām* (Astarābādī, *Sharḥ al-Tahdhīb*, MS#4064/2, fs.109–112, on which see *Marʿashī Fihrist*, v. 10, p. 174), was issued on 17 Jumādā al-Thānī 1007 (see Astarābādī, *Sharḥ al-Tahdhīb*, f.112; the date corresponds to 14 January 1599). On the *ijāza*, see Fāḍilī, "Ijāzāt", pp. 519–522.

[7] Astarābādī, *al-Fawāʾid*, p. 17.

[8] "I have seen a *ḥāshiya* on the commentary on the *Madārik* written in [Astarābādī's] own hand scribbled on a section of the book concerning purity laws. It demonstrates his expertise, his precision and the good orderliness of his mind." Baḥrānī, *Luʾluʾa*, p. 119.

[9] Khwānsārī, *Rawḍāt*, 1, p. 129.

[10] On this relatively unknown scholar see Madanī, *Sulāfat*, p. 498; Ḥurr, *Amal*, v. 2, p. 309; Afandī, *Riyāḍ*, v. 5, p. 194; Kashmīrī, *Nujūm*, pp. 18–19 and his mention in the Persian chronicle of Iskandar Bayg, *Tārīkh-i ʿĀlam-Ārā*, v. 1, p. 235.

al-Thalātha.[11] Some time in 1014, he moved to Mecca, arriving at the beginning of 1015. There he met and studied with his "last teacher in the science of *fiqh, ḥadīth* and *rijāl*",[12] Mīrzā Muḥammad b. ʿAlī "Ṣāḥib al-Rijāl" (d. 1028/1619).[13] Some sources record that Astarābādī married one of Ṣāḥib al-Rijāl's daughters,[14] and it may have been his daughter from this marriage who in turn married Muḥammad Muʾmin b. Dawst (or Dūst) Muḥammad al-Ḥusaynī (d. 1087/1676–77), described in various sources as Astarābādī's son-in-law.[15] It was Ṣāḥib al-Rijāl who, according to Astarābādī, told him to revive the school of the Akhbārīs. Astarābādī's account of this life-changing encounter deserves an extended citation. He states that the Imāmī *fuqahā'* had strayed from the path of the Imams in their legal methodology:

> ...until the time of the most learned of the modern scholars in the sciences of *ḥadīth* and *rijāl*, and the most pious of them all, the teacher of all in all subjects Mīrzā Muḥammad al-Astarābādī [i.e. Ṣāḥib al-Rijāl]—may God bless his grave—came. After he had taught all the disciplines of the *ḥadīths* to this lowly person [that is, Muḥammad Amīn al-Astarābādī], he directed him to "Revive the way of the Akhbārīs (*iḥyā-yi ṭarīqah-yi Akhbāriyyīn bikun*).[16] Dispel the doubts of those who oppose this way.

[11] The title was given to the work by the cataloguists at Āstān-i Quds Library (see *Mashhad Fihrist*, v. 13, p. 490). The colophon of *al-Mabāḥith* records that it was written in Shiraz at the beginning of Jumādā al-awwal, 1014 (corresponding to September 1605). (Astarābādī, *Mabāḥith*, f. 45a.1–4).

[12] Astarābādī, *al-Fawā'id*, pp. 17–18.

[13] So named because of his work of *rijāl*, *Manhaj al-Maqāl*, and also known as Mīrzā Muḥammad Astarābādī. In order to avoid confusion, I refer to him as Ṣāḥib al-Rijāl.

[14] He is introduced as Ṣāḥib al-Rijāl's son-in-law in Madanī, *Sulāfat*, p. 499.

[15] See Madanī, *Sulāfat*, p. 449 and Afandī, *Riyāḍ*, v. 5, p. 154. On this interesting figure, see generally Ṭihrānī, *Ṭabaqāt*, v. 5, p. 592 and below p. 155. Muḥammad Muʾmin's *nisba* is "al-Astarābādī"; the intermarriage amongst the Astarābādī community in Mecca described in these marriage relationships seems plausible.

[16] Newman indicates that the reference to Ṣāḥib al-Rijāl's command to revive "the way of the Akhbārīs" in the *Dānishnāmah-yi Shāhī* may have been instituted in "an attempt to bolster the legitimacy of *al-Fawā'id al-Madaniyya's* criticisms" of other scholars (Newman, "Akhbārī-Uṣūlī dispute, pt. 2", p. 253, n. 11). However, this can be set to one side by the fact that Ṣāḥib al-Rijāl's *ijāza* to Astarābādī has survived, included in Astarābādī's commentary upon the *Tahdhīb al-Aḥkām* (see the edition of these *ijāzāt* in Fāḍilī, "Ijāzāt"). The relationship is also confirmed by the earliest biographical entry on Astarābādī, namely al-Madanī's *Sulāfat al-ʿaṣr*, on which see below (p. 41 and p. 43). Furthermore, Astarābādī refers to Ṣāḥib al-Rijāl as his teacher, not only in the *Dānishnāmah*, but in other writings also (such as *al-Fawā'id al-Madaniyya*, p. 17 and p. 185). Finally, whether or not Ṣāḥib al-Rijāl was an Akhbārī seems irrelevant to whether or not Astarābādī was the founder of Akhbarism. Indeed the fact that he

> These ideas may have been lost to [their] minds, but God has decreed that these ideas flow from your pen!" After I had acquired all the disciplines of knowledge from him—the most learned of scholars—I spent a number of years (*chandīn sāl*) in Madīna meditating, petitioning at the threshold of the Almighty, gaining access to the Muḥammadan spirits of the companions.[17] I returned once again to the *ḥadīths* and the books of the Sunnis and of the Shīʿa with complete contemplation and consideration. Then, by the grace of God and the blessings of the Prophet and the Sinless Imams who had indicated that it was necessary for me to obey [Ṣāḥib al-Rijāl's order], I did in the end obey and decided to write *al-Fawāʾid al-Madaniyya*. [This work] was blessed by him [Ṣāḥib al-Rijāl] reading it, and then approving of its composition and praising the author.[18]

Astarābādī, then, did not immediately take up the command from Ṣāḥib al-Rijāl to revive the Akhbārī school. Only after some years of meditation in Madīna did he agree. We know he was in Mecca both in 1017 and in 1018 from the colophons of two works on *kalām*.[19] In neither of these works does he refer to Akhbarism or, indeed, his studying with Ṣāḥib al-Rijāl. It seems most likely, then, that after his arrival in Mecca, he studied with Ṣāḥib al-Rijāl until at least 1018 before retiring to Madīna for a number of years to consider the command to revive Akhbarism. After these years of meditation, he began writing *al-Fawāʾid al-Madaniyya* in compliance with Ṣāḥib al-Rijāl's command and returned to Mecca. Ṣāḥib al-Rijāl seems to have read at least sections of the *Fawāʾid* and praised Astarābādī for its contents. Ṣāḥib al-Rijāl died in 1028/1619, meaning these events (from his retirement to his return and writing of the *Fawāʾid*) occurred in a ten year period. Astarābādī refers in the *Fawāʾid* to Ṣāḥib al-Rijāl as having died in Mecca and having been buried there, and Āghā Buzūrg al-Ṭihrānī records *al-Fawāʾid al-Madaniyya* being completed in 1031 in Mecca[20]—some three years after Ṣāḥib al-Rijāl's death. It seems clear, then, that *al-Fawāʾid al-Madaniyya* was composed over

---

was not supports the proposition that the Akhbārī school was not a current option for a jurist in the period just before Astarābādī.

[17] The citation of this passage in Khʷānsārī, *Rawḍāt*, 1, p. 130 has not *tawassul bih arwāḥ-i Muḥammadiyyah-yi aṣḥāb* but *tawassul bih arwāḥ-i ahl-i ʿiṣmat* meaning "gaining access to the souls of the Sinless Ones".

[18] Astarābādī, *Dānishnāmah*, f. 3a.12–3b.3. The pages are numbered in a later hand as 5 and 6.

[19] Astarābādī, *al-Fawāʾid al-Iʿtiqādiyya* f. 33b.7 (*Marʿashī Fihrist*, v. 2, p. 97 MS#488/1) and Astarābādī, *al-Fawāʾid al-Makkiyya*, f. 26a.5–6. I am assuming he was resident there throughout this period.

[20] Ṭihrānī, *al-Dharīʿa*, v. 16, p. 358, #1663. Al-Ṭihrānī presumably based this on a manuscript copy he had seen.

a number of years, during which time Astarābādī showed portions of it to Ṣāḥib al-Rijāl for the latter's approval. Astarābādī says, for example:

> I showed [Ṣāḥib al-Rijāl] what I am going to present: that is, that the way of the ancients is to be preferred and the way of the moderns rejected. He approved of it (*istiḥsanahu*) and praised me for it.[21]

The *Dānishnāmah-yi Shāhī*, Astarābādī's other major work, was completed sometime after the *Fawā'id* (the *Dānishnāmah* includes references to the *Fawā'id* as completed on a number of occasions). It was sent as a gift to the Deccan ruler Muḥammad Quṭb Shāh who ruled 1020/1611–2 to 1035/1626. The gift must have been given towards the end of Muḥammad Quṭb Shāh's reign.

We also know that Astarābādī visited Ṭā'if at some point, though we do not know when. He refers to his books being in Ṭā'if on two occasions, using this as an excuse for not giving fuller answers to questions.[22] Whilst all the sources agree that Astarābādī died and was buried in Mecca, there is a difference of opinion concerning his death date. The earliest biographer (al-Madanī in *Sulāfat al-ʿaṣr*) records 1036,[23] though Yūsuf al-Baḥrānī, and most biographers subsequent to him, record 1033, describing the later date as "clearly wrong".[24] If, as suggested above, Astarābādī was born in the 980s, he would have been 50 years old at the time of his death. From what we know of his personal life, he seems to have been part of an expatriate Astarābādī community in Mecca, marrying the daughter of an Astarābādī and ensuring that his own daughter married an Astarābādī.

Astarābādī, then, had scholarly links with the major Shīʿī intellectual figures of his day: Muḥammad Ṣāḥib al-Madārik, (possibly) Hasan Ṣāḥib al-Maʿālim,[25] Muḥammad Ṣāḥib al-Rijāl, as well as less well-known scholars such as Muḥammad al-Nassāba. However, only Ṣāḥib al-Rijāl seems to have approved of (and indeed inspired)

---

[21] Astarābādī, *al-Fawā'id*, p. 18.

[22] See Astarābādī, *Jawāb*, f.5b.8 (see also Gleave "Questions and Answers") and the *fatwā* concerning wine (in the appendix to this book, pp. 315–319 below).

[23] Madanī, *Sulāfat*, p. 499.

[24] Baḥrānī, *Lu'lu'a*, p. 119. Stewart (*Islamic Legal Orthodoxy*, p. 181, n. 22) notices that al-Afandī (*Riyāḍ*, v. 5, p. 36) states that the *risāla* on the purity of wine was completed in 1034. If correct, the later date is, of course, the more likely. The extant copy of the *risāla* (see below, n. 32) is undated.

[25] Al-Fāḍilī considers it unlikely that Astarābādī received an *ijāza* from Ḥasan b. Shahīd II (Fāḍilī, "Ijāzāt", p. 233), though he does think that he studied with him.

his development of an Akhbārī position, his other teachers having died before his response to Ṣāḥib al-Rijāl's call to revive the school. Astarābādī's writings stretch across a thirty-year period, and cover a number of Islamic sciences. An approximate chronology of his extant writings can be constructed as follows:

1. *Al-Mabāḥith al-Thalātha* concerning three issues of philosophical theology: this work does not appear in any of the early biographical lists of Astarābādī's works, but has survived in a single manuscript held in Mashhad, dated 1014 and composed in Shiraz.[26]
2. *Al-Fawā'id al-I'tiqādiyya*, also on philosophy and theology and covering some of the same issues as *al-Mabāḥith al-Thalātha*. It also is not listed in biographical sources, but survives in a single manuscript dated 1017, written in Mecca.[27]
3. *Al-Fawā'id al-Makkiyya*, so named by the cataloguists at the Āstān-i Quds Library,[28] and also on philosophy and theology: this has survived in a single manuscript, dated 1018, written in Mecca.[29]
4. *Al-Fawā'id al-Madaniyya*: his first and major Akhbārī work, completed in Rabī' al-awwal in 1031 (January, 1622), although as discussed above, some sections of it existed during the lifetime of Ṣāḥib al-Rijāl (d. 1028/1619). There are numerous manuscripts of this work, a lithograph edition and a recent, typeset version of the lithograph.

In addition to the above, there are works which are of unsure date, but which were written after the completion of *al-Fawā'id al-Madaniyya* (that is, during the last four years of Astarābādī's life):

5. *Dānishnāmah-yi Shāhī*, Astarābādī's only major Persian composition and concerned with theological and philosophical issues. It was written for Muḥammad Quṭb Shāh of the Deccan and can be dated after the *Fawā'id* since it includes the account cited above concerning the composition of the *Fawā'id*. A number of manuscripts of this work have survived.[30]

---

[26] See *Mashhad Fihrist*, v. 13, p. 490, MS#132.

[27] See *Mar'ashī Fihrist*, v. 2, p. 97, MS#488/1.

[28] Astarābādī mentions that he "completed these *fawā'id* in Mecca", and it is from this that the cataloguists named the work *al-Fawā'id al-Makkiyya*. Astarābādī, *al-Fawā'id al-Makkiyya*, f.26a. See below n. 38 on the distinction between this and Astarābādī's commentary on *al-Istibṣār*.

[29] *Mashhad Fihrist*, v. 13, p. 437, MS#213/4.

[30] The manuscript used here is *Majlis*, MS#3071/1, on which see *Majlis Fihrist*, v. 10.2, p. 621.

6. *Al-Risāla fī ṭahārat al-khamr wa-najāsatihā*: a treatise on the issue of whether or not wine is to be considered an impure substance, surviving in an undated manuscript in Tehran. It was clearly written after the *Fawā'id*, as it mentions the latter as being complete. It also mentions *al-Fawā'id al-Makkiyya*, though it is not clear which book is being referred to by this title.[31] A manuscript of this work is to be found in the University of Tehran Central Library.[32]

7. *Al-Risāla fī'l-Badā'*, on God's ability or otherwise to alter his will. This can also be dated after the completion of the *Fawā'id*, as it is recorded as having been written on the *Fawā'id*'s margins in Astarābādī's own hand.[33]

8. *Jawāb masā'il Shaykhinā Ḥasan al-Ẓahīrī al-ʿĀmilī*, a series of answers to questions by his pupil Ḥasan al-Ẓahīrī.[34]

There are also works by Astarābādī which have survived, but cannot be dated. They were probably written during his later life, and include commentaries on canonical books of *ḥadīth*:

9. A commentary (*sharḥ*) on *al-Kāfī* (of Kulaynī; the commentary seems to be restricted to the first section of the work, *Uṣūl al-Kāfī*).[35]

10. A commentary (*sharḥ*) on *al-Tahdhīb* (of al-Ṭūsī).[36]

11. A commentary (*ḥāshiya*) on *Man lā Yaḥḍuruhu al-Faqīh* (of Ibn Bābūya).[37]

---

[31] See above, n. 28 and below, n. 38.

[32] See *University Fihrist*, v. 7.3, p. 2667, MS#1257/12. There is also a copy of Astarābādī's answer to questions from Shāh ʿAbbās al-Ṣafavī in the Mālik Library (*Malik Fihrist*, v. 9, p. 203 MS#1563) which predates this *risāla* and is a separate work. An edition of the *fatwā* is found in Appendix 3 below, pp. 315–319.

[33] A manuscript of this work is to be found in Mashhad (*Mashhad Fihrist*, v. 13, p. 84, MS#6543), though this is a collection of both Astarābādī's own comments on *badā'* and those of other Safavid writers.

[34] An edition of this work, with the questions which prompted it, is included in Gleave, "Questions and Answers" (Another edition can be found in *al-Fawā'id al-Madaniyya*, pp. 568–575). The questions relate primarily to *fiqh*.

[35] This, or part of it, has been published, Astarābādī, *al-Ḥāshiya ʿalā Uṣūl al-Kāfī*. See also *Marʿashī Fihrist*, v. 17, p. 232, MS#6665/3 and v. 12, p. 163, MS#4594/1. The work was clearly written after *Dānishnāmah-yi Shāhī* as the latter work records, "we have already clarified this matter [cf. the question of God's attributes] in our comment (*ḥawāshī*) on the *Uṣūl-i Kāfī*" (Astarābādī, *Dānishnāmah*, f.31b.16, p. 62.16). The doctrine then proposed in the ensuing discussion in *Dānishnāmah-yi Shāhī* is identical to that of Astarābādī's commentary on *Uṣūl al-Kāfī* (Astarābādī, *al-Ḥāshiya ʿalā Uṣūl al-Kāfī*, p. 309). For a discussion of the doctrines expressed in these works, see below, p. 115 n. 42.

[36] See *Marʿashī Fihrist*, v. 10, p. 174, MS#3789 and v. 112, p. 171, MS#4704.

[37] This has been edited and published by ʿAlī Fāḍilī, see Astarābādī, *al-Ḥāshiya*

12. A commentary (*sharḥ*, sometimes entitled *al-Fawāʾid al-Makkiyya*) on *al-Istibṣār* (of al-Ṭūsī).[38]

Numbers 9, 11 and 12 are mentioned in *al-Fawāʾid al-Madaniyya*.[39] Fāḍilī has argued that Astarābādī wrote two commentaries on each of *al-Kāfī*, *al-Tahdhīb* and *al-Istibṣār* (one being a *sharḥ* and the other a *ḥāshiya*), not all of which have survived.[40]

Also extant, but of unknown date, are:

13. A commentary (*ḥāshiya*) on *Maʿārij al-Aḥkām* of al-Muḥaqqiq al-Ḥillī.[41]
14. A collection of poetry, attributed to Astarābādī.[42]
15. A work given the title *Taḥqīq al-Amīn* by the cataloguists of Āstān-i Quds Library. It is a commentary on a section of Dawwānī's work on logic.[43]

There are also a number of works which have not survived (or have not yet come to light). They are, however, mentioned in the biographical sources:

16. *Fawāʾid Daqāʾiq al-ʿUlūm wa Ḥaqāʾiqihā*, which is reported to have been a study of the Arabic Language and is mentioned in *al-Fawāʾid al-Madaniyya*.[44]

ʿalā Man lā yaḥḍuruhu al-Faqīh. The manuscript can be found in the library of the Markaz Iḥyāʾ al-turāth al-Islamī in Qum, MS#2750.

[38] Ṭihrānī, *al-Dharīʿa* v. 13, p. 83, #264. This work should be distinguished from *al-Fawāʾid al-Makkiyya* (concerning theological issues) mentioned above (no.4), which is the cataloguists' title only. Two copies are extant (*Marʿashī Fihrist*, v. 12, pp. 170–1, MS4604/1). Fāḍilī also claims it is a different work from Astarābādī's *al-Ḥāshiya ʿalā al-Istibṣār*. A copy of this *ḥāshiya* is found in the Markaz Iḥyāʾ al-turāth al-Islāmī in Qum in the same *majmūʿ* (MS#2750) as the commentary on *Man lā Yaḥḍuruhu al-Faqīh* (see above, n. 37). See Astarābādī, *al-Ḥāshiya ʿalā Uṣūl al-Kāfī*, pp. 241–243 (editor's introduction).

[39] Astarābādī, *al-Fawāʾid*, p. 4. Al-Ḥurr al-ʿĀmilī records that nos 10 and 12 are unfinished (Ḥurr, *Amal*, v. 2, p. 246).

[40] Astarābādī, *al-Ḥāshiya ʿalā Uṣūl al-Kafi*, pp. 241–253. Until the respective *sharḥs* and *ḥāshiyas* are subjected to detailed comparison, it is diącult to confirm whether these are separate works, extracts from a single work or notes made by pupils during teaching sessions.

[41] See *Malik Fihrist*, v. 5, p. 222, MS#111/1013.

[42] *Mellī Fihrist*, v. 9, p. 230, MS#22521. Fāḍilī expresses reservations about this attribution, see Astarābādī, *al-Ḥāshiya ʿalā Uṣūl al-Kāfī*, p. 253 (editor's introduction).

[43] *Mashhad Fihrist*, v. 13, p. 118, MS#979.

[44] Astarābādī, *Fawāʾid*, p. 4 (recorded in the lithograph as *daqāʾiq al-funūn* rather than *daqāʾiq al-ulūm*). See also Ṭihrānī, *al-Dharīʿa*, v. 16, p. 336, #1555. The suggestion that the manuscript MS#8816 in the University of Tehran Library (*University*

17. *Al-Kitāb fi'l-Radd mā aḥdathahu al-Fāḍilān*, a refutation of the commentaries of Ṣadr al-Dīn al-Dashtakī (d. 901/1497) and Jalāl al-Dīn al-Dawwānī (d. 908/1502–3) on the *Tajrīd al-I'tiqād* of Naṣīr al-Dīn al-Ṭūsī (d. 672/1274).[45] Whilst this work has not survived as a separate volume, Astarābādī's criticisms of both al-Dashtakī's and al-Dawwānī's interpretations of the *Tajrīd* are found in his various extant theological works. It is al-Dawwānī who seems to have been the main target of Astarābādī's criticism in these writings.[46]
18. A commentary (*ḥāshiya*) on the *Tamhīd al-Qawā'id* of Shahīd II (d. 965/1558) which is mentioned in *al-Ḥāshiya 'alā Uṣūl al-Kāfī*.[47]
19. *Al-Ḥāshiya 'alā al-Madārik*, referred to above. This is perhaps Astarābādī's earliest work, though possibly not a separate volume as such, consisting of marginal comments upon a manuscript of the *Madārik al-aḥkām*. It does not appear to have survived, though Yūsuf al-Baḥrānī (d. 1186/1772) claims to have seen a copy in Astarābādī's hand.[48]

In this list, there are works which span a spectrum of Islamic disciplines including philosophy, theology, legal theory, treatises on discrete issues of law and commentaries on *ḥadīth* and *fiqh* works. The major lacunae for a scholar of Astarābādī's wide interests are a compendium of legal rulings (*fiqh*) and a commentary upon the Qur'ān (*tafsīr*).[49] It seems he intentionally did not write a work of *fiqh*. In his answers to Ḥasan al-Ẓahīrī he is asked to write a work of *fiqh*, and replies that he follows the way of the ancients. As they

Fihrist, v. 17, pp. 226–227) could be this work can be rejected since, on inspection, this work is actually a (damaged) copy of *Dānishnāmah-yi Shāhī*.

[45] In the introduction to the latest lithograph edition of *al-Fawā'id al-Madaniyya*, Abū Aḥmad b. Khalaf Āl Aṣfūr al-Baḥrānī mentions that a refutation of Mullā Ṣadrā is also amongst Astarābādī's works, and that Astarābādī refers to it in the *Fawā'id*. I have found no reference to such a work in *al-Fawā'id*, though there is a reference to Astarābādī's refutation of Ṣadr al-Dīn al-Dashtakī al-Shīrāzī, referred to as al-Fāḍil al-Shīrāzī in *al-Fawā'id al-Madaniyya* (Astarābādī, *Fawā'id*, p. 272). Dashtakī is sometimes confused with Mullā Ṣadrā. This is most likely identical to no. 17 listed here.

[46] An analysis of Astarābādī's theological writing is presented below, pp. 102–139.

[47] Astarābādī, *al-Ḥāshiya 'alā Uṣūl al-Kāfī*, p. 279. Al-Afandī records that Astarābādī's *al-Ḥāshiya* acted as the starting point for the *Fawā'id* (al-Afandī, *Riyāḍ*, v. 5, p. 246).

[48] Baḥrānī, *Lu'lu'a*, p. 119. It only covers part of the book on purity, and is the only one of Astarābādī's works of which al-Baḥrānī seems to have approved (*tashhadu bi-faḍlihi wa-diqqatihi wa-husn taqrīrihi*). See above, n. 8.

[49] Commentaries on the Qur'ān were written by subsequent Akhbārīs, and an introduction to them is given by Lawson ("Akhbārī tafsir").

did not write works of *fiqh*, but merely collected or commentated on *ḥadīth*, he intends to do the same.[50]

Astarābādī's ideas, particularly his criticism of *ijtihād* and the hermeneutic methodology of the *mujtahids*, spread primarily through the dissemination of the text of *al-Fawāʾid al-madaniyya*. His pupils, a number of whom are recorded in the biographical dictionaries of the Safavid period, took on this task, and the growth and development of the Akhbārī school was in no small measure the result of their efforts.[51]

Astarābādī and the Shīʿī Biographical Tradition

An assessment of Astarābādī's position within the Shīʿī tradition can be gleaned from an analysis of the entries devoted to him found in biographical compendia (*ṭabaqāt* or *tarājim*)[52] of the Safavid period and later. Newman has argued that the lack of mention of a connection between Astarābādī and Akhbarism in the *ṭabaqāt* works closest to his death date indicates that he was not associated with a revival or founding of the Akhbārī school.[53] If true, this would support the view that Astarābādī, at least initially, was not considered the founder of Akhbarism. The connection was only made when it became an element of anti-Akhbārī polemic later in the Safavid period. It could be argued, then, that since Astarābādī does not appear as the founder of the Akhbārī school until sometime after his death, he should not be considered its founder. Opponents of Akhbarism wished to present Akhbarism as an innovation (*bidʿa*), contrasting it with their own long tradition of scholarship. In Astarābādī, it might be argued, the Uṣūlī/*mujtahid* scholars found an appropriately late "founder" of the heretical Akhbārī school, enabling them to label the Akhbārīs as innovators (*mubtadiʿūn*).

---

50 Astarābādī, *Jawāb*, f.1b.5–9.

51 The spread of Akhbarism, and an account of the activities of Astarābādī's pupils is given in Chapter 5, below, pp. 140–176.

52 Works of *ṭabaqāt* are strictly speaking arranged according to date, whilst works of *tarājim* are arranged alphabetically—for a general overview see Hafsi, "Recherches sur le genre '*Tabaqat*' " parts 1, 2 and 3. However, biographical compendia have comè to be known as *ṭabaqāt* and hence I use the term in what follows to refer to biographical compendia generally, irrespective of arrangement.

53 Newman, "Akhbārī-Uṣūlī Dispute, Part 2", pp. 250–253, p. 260, n. 32.

Newman raises a number of important points which deserve attention if my analysis presented in the previous chapter, is to be defended.[54] My own view, as outlined earlier, is that Astarābādī should be considered the founder of the Akhbārī movement. Whilst "school" (*madhhab*) is, perhaps, too strong a term to describe the early Akhbārīs, there was something original in Astarābādī's critique of the dominant method of jurisprudence at the time, and his own use of the term Akhbārī to designate this new approach brought a new level of specificity to the use of the term. This specificity comprised of a set of epistemological presumptions which bolstered distinctive positions regarding the role of the Qur'ān as a source of law, the criticism of *ḥadīth* and the hermeneutic methods to be applied to revelatory texts. If this is the case, the lack of any hint of Astarābādī's "originality" in *ṭabaqāt* works requires explanation.

It is indeed true that early biographical entries on Astarābādī do not identify him as an Akhbārī. The earliest source, the *Sulāfat al-ʿAsr* of Sayyid ʿAlī al-Madanī (who died in 1120/1709, though the work was completed sometime earlier in 1082)[55] includes Astarābādī in its list of "the outstanding Persians [of this period]...amongst the best of the most excellent of them, amongst the greatest of the outstanding ones amongst them".[56] Another early source is the well-known *ṭabaqāt* work, *Amal al-Āmil* of al-Ḥurr al-ʿĀmilī (d. 1104/1693, himself an Akhbārī).[57] Al-Ḥurr also makes no mention of Astarābādī's Akhbarism, though he does describe him as a "scholar, expert, theologian, jurist and *ḥadīth* expert (*muḥaddith*)".[58] The *Riyāḍ al-ʿulamāʾ* of al-Afandī (d. around 1130/1718) omits any reference to Astarābādī's Akhbarism. He is content to cite al-Ḥurr's entry in *Amal al-Āmil*, adding his own corrective remarks.[59] Indeed, the first work of *ṭabaqāt* I have found which mentions Astarābādī's Akhbārī activities is Yūsuf al-Baḥrānī's *Lu'lu'at al-Baḥrayn* where al-Baḥrānī,

[54] Newman extends his analysis to other Safavid scholars labelled Akhbārīs in his "Anti-Akhbārī Sentiments", pp. 155–162 and links this to al- Khʷānsārī's own family history (see, pp. 166–169), in which anti-Akhbārī scholars loom large.
[55] Ṭihrānī, *al-Dharīʿa*, v. 12, p. 212, #1401.
[56] Madanī, *Sulāfat*, p. 398 (the Astarābādī passage is on p. 499).
[57] See below, p. 158. The work was completed in 1096 (Ṭihrānī, *al-Dharīʿa*, v. 2, p. 350, #1400).
[58] Ḥurr, *Amal*, v. 2, p. 246, #725.
[59] Afandī, *Riyāḍ*, v. 5, pp. 35–37.

himself an Akhbārī of sorts,[60] writes that Astarābādī was "the first to open the door of slander against the *mujtahids* and divide the saved sect [that is, the Imāmī Shīʿa] into Akhbārī and *mujtahid*."[61] After al-Baḥrānī made this association, it became a regular element in *ṭabaqāt* works, and one finds it repeated in the major subsequent Imāmī works of *ṭabaqāt*, including:

1. The *Qiṣaṣ al-ʿulamāʾ* of Muḥammad b. Sulaymān al-Tanukābunī (d. 1302/1884–5). Al-Tanukābunī reproduces al-Baḥrānī's assessment, and adds that insulting the *mujtahids* "arose from [Astarābādī's] stupidity" (*nāshiʾ az ghabāvat-i ū*).[62]
2. The *Nujūm al-samāʾ* of Muḥammad ʿAlī al-Kashmīrī (d. 1309/1891–2) records that Astarābādī was "the leader of the Akhbārī sect" (*raʾīs-i firqah-yi Akhbāriyyīn*).[63]
3. The *Rawḍāt al-Jannāt* of Muḥammad Bāqir al-Khʷānsārī (d. 1313/1895) contains a lengthy entry on Astarābādī, in which the former attributes to the latter all manner of sins, including being an agent of Sunnism.[64]

The identification continues into the Twentieth Century,[65] and the western language scholars who relied upon these Qajar (and later) *ṭabaqāt* works reproduce the association in their descriptions of the Akhbārī-Uṣūlī conflict.[66]

---

[60] Whilst he criticises Astarābādī in his *Luʾluʾa*, I have argued that al-Baḥrānī's legal methodology is thoroughly Akhbārī (see Gleave, *Inevitable Doubt*, pp. 251–253 and passim). Indeed his *uṣūl al-fiqh*, as found in *al-Ḥadāʾiq al-Nāḍira*, whilst maintaining the pretence of moderation, is perhaps the most sophisticated surviving exposition of Akhbārī methodology.

[61] Baḥrānī, *Luʾluʾa*, p. 117.

[62] Tanukābunī, *Qiṣaṣ*, pp. 321–322 # *ʿayn-dāʾ*. In al-Tanukābunī's more orthodox work of *ṭabaqāt*, *Tadhkirat al-ʿulamāʾ*, there is no mention of Astarābādī, and though other Akhbārīs are identified, there is little critical comment.

[63] Kashmīrī, *Nujūm*, pp. 41–42. Most of the entry is a translation into Persian of al-Baḥrānī's entry on Astarābādī.

[64] Khʷānsārī, *Rawḍāt*, v. 1, pp. 129–148, #33. Al-Khʷānsārī positions his entry out of place in the overall scheme of his work (under *alif* not *mīm*) because he considers it inappropriate to sully the section on scholars named Muḥammad with someone like Astarābādī (pp. 147–148). The entry as a whole provided Khʷānsārī with an excuse to refute and vilify Akhbarism generally, and the section relating to Samāhījī's "forty points" is analysed in Newman, "The Akhbārī-Uṣūlī Dispute, Pt.2", pp. 253–261.

[65] See, for example, Amīn, *Aʿyān al-Shīʿa*, v. 3, p. 222 and Muddarris, *Rayḥānat al-adab*, v. 1, p. 65 (who talks of Astarābādī "constructing"—*tashyīd*—the bases of the Akhbārī school), Qummī, *Favāyid*, p. 398.

[66] For a list of these western language scholars, see Newman, "Akhbārī-Uṣūlī Dispute, Pt. 2", p. 251n. 3.

It is, then, strange (perhaps suspicious) that Astarābādī is not mentioned as the founder of Akhbarism in *ṭabaqāt* works until around 150 years after his death, and Newman's implication that the emergence of this motif in the Shī'ī *ṭabaqāt* entries is in some sense polemically motivated is surely correct. An explanation of this phenomenon need not, however, lead one to conclude that the portrayal of Astarābādī as the founder (or, indeed, reviver) of Akhbarism was a mere invention of later Shī'ī writers.

Firstly, it is not merely that there is no connection between Astarābādī and Akhbarism in *ṭabaqāt* works prior to the *Lu'lu'a* of al-Baḥrānī. There are very few explicit references to the Akhbārī-Uṣūlī dispute in any Safavid era biographical compendium.[67] For this *argumentum e silentio* to be convincing, the general lack of mention of the Akhbārī-Uṣūlī dispute in Safavid *ṭabaqāt* literature could be evidence that the dispute itself was an invention of later scholarship.[68] Al-Madanī's position (if he had one) on the dispute is not known. His *Sulāfat al-'Aṣr* is a collection of biographical entries on learned men of various sectarian persuasions and his lack of reference to the Akhbārī-Uṣūlī dispute is not particularly surprising. It is perhaps more surprising that al-Ḥurr al-'Āmilī who explicitly identifies himself as an Akhbārī,[69] makes no mention of the dispute in relation to Astarābādī (or anyone else) in his *Amal al-Āmil*. This work is a biographical dictionary in two volumes, each volume arranged alphabetically by scholar's name. The first volume concentrates on scholars from the Jabal Āmil region of Lebanon, whence al-Ḥurr himself came; the second has a more general remit, and it is in this section that Astarābādī's entry is located. If one was to argue that there is reference to the dispute here, it would have to be through coded descriptions of scholars being *muḥaddiths* (or from the *ahl al-ḥadīth*) or *mujtahids* (the latter used as a title signifying scholarly distinction as well as proponents of

[67] The only reference I have found to date is the description of Mullā Khalīl al-Qazwīnī (d. 1089/1678) as an Akhbārī in Afandī, *Riyāḍ*, v. 2, p. 261. Al-Afandī's entries on all other Safavid Akhbārīs make no explicit mention of Akhbarism.

[68] Naturally, this view has not been put forward in these terms, though there are those who consider the significance of the dispute to have been rather exaggerated by later tradition, and this has led to an unnecessary emphasis on the dispute in Western accounts of late classical Shī'ī jurisprudence. See for example, Cole, "Ideology, Ethics and Philosophical Discourse", pp. 32–34.

[69] See, for example, Ḥurr, *Wasā'il*, v. 20, p. 106 and Ḥurr, *al-Fawā'id al-Ṭūsiyya* generally. The latter is the closest we have to an Akhbārī work of *uṣūl al-fiqh* by al-Ḥurr.

the theory of *ijtihād*). Perhaps part of the reason for there being no mention of Astarābādī "founding" Akhbarism lies in al-Ḥurr's own Akhbārī position. From an Akhbārī perspective, Astarābādī was not the "founder" of the Akhbārī movement. Instead, he re-established the true Shiʿism of the Imams and revived the method of the first Shīʿī scholars. It is this original Shiʿism which is the "way of the Imams", or "the method of the early scholars (*al-mutaqaddimīn*)". This could explain why an Akhbārī scholar, such as al-Ḥurr al-ʿĀmilī, did not describe Astarābādī as the founder of Akhbarism in his *Amal al-Āmil*. Doing so would contradict one of the main Akhbārī criticisms of the *mujtahids*. This does not, admittedly, explain why there is no reference to any association with Akhbarism, though it might explain why Astarābādī is not identified as the founder.

The same point cannot be made, however, about al-Afandī's *Riyāḍ al-ʿulamā'* (completed in 1130).[70] This work includes entries on scholars who self-identify as Akhbārīs (such as al-Ḥurr), but is almost silent on the Akhbārī-Uṣūlī dispute.[71] Al-Afandī himself was a pupil of Muḥammad Bāqir al-Majlisī (d. 1110/1699, hereon al-Majlisī II, a scholar whom both Akhbārīs and Uṣūlīs claim as one of their own)[72] and though al-Afandī does not explicitly claim to be an Uṣūlī, his association with Uṣūlī scholars such as Muḥammad Bāqir al-Sabzawārī (d. 1090/1679) and philosophers (with Uṣūlī tendencies) such as Ḥusayn al-Khʷansārī and Muḥammad b. Ḥasan al-Shīrwānī (d. 1098/1686) hint at his position.[73] Such relationships increase the likelihood of him being sympathetic to the Uṣūlī position, as does a passage in the *Riyāḍ* where he criticises (albeit mildly) Akhbarism.[74] If he was an Uṣūlī, it is surprising that he, like al-Ḥurr al-ʿĀmilī the generation before, does not give the Akhbārī-Uṣūlī dispute a high profile in his work. Scholars are identified as *mujtahids* (and

[70] Ṭihrānī, *al-Dharīʿa*, v. 11, p. 331, #1981.

[71] See above, n. 67.

[72] On Majlisī's position within the dispute, see below, pp. 241–244 and pp. 264–266.

[73] He names these, with Majlisī II, as his teachers in the *Riyāḍ* through honorific titles (see Ṭihrānī, *al-Dharīʿa*, v. 11, p. 331 #1981 for the interpretation of these honorifics).

[74] See above, n. 67 and the reference there, where he states that Khalīl was "one of the Akhbārīs who strongly denied [the legitimacy of] *ijtihād*. He went too far and was extreme in this." This, of course, is mild criticism compared to that expressed by al-Khʷānsārī a century later (on which see Newman, "Anti-Akhbārī Sentiments").

very rarely as *uṣūlīs*),[75] though this appears to be a measure of their scholarly prowess and not their position in any intra-sectarian dispute. The reason why al-Afandī declined to use the *Riyāḍ al-ʿUlamāʾ* as an opportunity to highlight the Akhbārī-Uṣūlī dispute is, initially, puzzling. One possible explanation is that since the dispute was so heated at the time of writing, he wished to avoid controversy by down playing it.

That Safavid *ṭabaqāt* works do not describe Astarābādī as the founder, reviver or even having any particular connection with Akhbarism stands in marked contrast to mention of him in other literary contexts. Astarābādī himself clearly considered himself to be the reviver of the "Akhbārī way", as he laid claim to this role in the passage from the *Dānishnāmah-yi Shahi*.[76] Whilst Ḥusayn b. Shihāb al-Dīn al-Karakī al-ʿĀmilī (d. 1076),[77] himself an Akhbārī, makes only an oblique reference to Astarābādī,[78] a source which can be dated to within fifty years of Astarābādī's life, the *Dabistān-i Madhāhib*,[79] contains a lengthy passage concerning Astarābādī and the Akhbārī school, which begins:

> The path of the Akhbārīs: Mullā Muḥammad Amīn al-Astarābādī is the propagator (*murawwij*) of this group in this era. They say that after acquiring both rational and transmitted sciences in Holy Mecca, and after comparing [them] with the *ḥadīth*, he reached this position [that is, that of the Akhbārī school] and he wrote the *Favāyid-i Madanī* [sic].[80]

Similarly, Muḥammad Taqī al-Majlisī (d. 1070/1659–60, hereon Majlisī I) refers to Astarābādī in his Persian commentary on Ibn

---

[75] Describing a person as an Uṣūlī does not preclude his categorisation as an Akhbārī in al-Afandī's *Riyāḍ al-ʿulamāʾ*. Khalīl al-Qazwīnī has both distinctions (Afandī, *Riyāḍ*, v. 2, p. 261), which demonstrates that *uṣūlī* here probably refers to one who writes about issues of *uṣūl al-fiqh* rather than one who denies the legitimacy of *ijtihād*.

[76] See above, pp. 33–34.

[77] On whom, see below, p. 166.

[78] Karakī, *Hidāyat*, p. 172. The description of Astarābādī's position is accurate, though the citation is not from *al-Fawāʾid al-Madaniyya*.

[79] The authorship of this text is debated, though it can be dated to circa 1064. See, generally, Athar Ali "Dabistan-e Mazahib".

[80] Fānī (attributed), *Dabistān*, p. 221. The occasional similarity in wording between this passage and Astarābādī's *Dānishnāmah* cited above does not preclude the Indian author of the *Dabistān* having access to the *Dānishnāmah*—the latter had, after all, been sent to India as a gift for the Deccan Muḥammad Quṭbshāh. The *Dānishnāmah* is referred to in the *Dabistān* as being titled *Dānishnāmah-yi Quṭbshāhī* and the author shows a knowledge of its contents (Fānī (attributed), *Dabistān*, p. 7).

Bābūya's *Man lā Yaḥḍuruhu al-Faqīh*.[81] After describing the disappearance of some books of *akhbār* and the ensuing reduction in knowledge of *ḥadīth* amongst the Imāmī *'ulamā'*, Majisī I mentions the differences of opinion over the law which emerged (implying some causal relationship between the loss of books and the emergence of *ikhtilāf*). Then he states:

> Around thirty years ago, the excellent and learned, our *mawlā*, Muḥammad Amīn al-Astarābādī (may God's mercy be upon him) busied himself with collating and studying the *akhbār* of the sinless Imams. He studied the refutation of mere opinion (*ārā'*) and analogies. He knew the way of the companions of the holy, sinless Imams and wrote the *Favāyid-i Madaniyya*, and sent it to this land. Most of the people of Najaf and the Holy 'Atabāt approved of his method and returned to the *akhbār*. In truth, most of what our Mawlā Muḥammad Amīn said was the truth.[82]

Astarābādī is portrayed here as calling on the *'ulamā'* to return to the true path of the Imams, and as being at least partially successful in this effort. That is, he heads and inspires a movement for the re-establishment of the "way of the companions of the holy, sinless Imams" through his book. Muḥsin Fayḍ al-Kāshānī (d. 1091/1680) refers to having studied with Astarābādī, and approving of his method of rule derivation. Fayḍ then states: "He opened the door of this [way] to us and led us to the right path",[83] implying that Astarābādī's ideas were distinct from those of other scholars of the time. Majlisī II mentions Astarābādī as "the head of the *ḥadīth* experts" (*ra'īs al-muḥaddithīn*)[84] and says he makes use of *al-Fawā'id al-Madaniyya* and *al-Fawā'id al-Makkiyya*[85] in the compilation of his *Biḥār*

[81] Entitled *Lawāmi' ṣāḥibqirānī* and written for Shāh 'Abbās II, it is also known as *al-Lawāmi' al-Qudsiyya* and was completed in 1066. See Ṭihrānī, *al-Dharī'a*, v. 18, p. 369, #.500.

[82] Majlisī I, *Lawāmi'*, v. 1, p. 47. Majislī I also makes reference to Astarābādī in his Arabic commentary on *Man lā Yaḥḍuruhu al-Faqīh* (Majlisī I, *Rawḍat*, v. 1, p. 20), saying that the rational indicators used by many contemporary jurists are "weak innovations", and the correct opinion with respect to most of them is that of "al-Fāḍil al-Astarabādī".

[83] Fayḍ, *al-Ḥaqq al-Mubīn*, p. 12. Fayḍ goes on to criticise elements of Astarābādī's attack on the *fuqahā'*.

[84] Majlisī I, *Biḥār*, v. 1, p. 20. Al-Ḥurr al-'Āmilī disapproves of calling Astarābādī "*ra'īs al-muḥaddithīn*" since this is a title, he says, reserved for the Prophet and Imams (al-Ḥurr, *al-Fawā'id al-Ṭūsiyya*, p. 446).

[85] He most likely means the commentary on *al-Istibṣār* and not the work of theology (see above, p. 38 n. 38).

al-Anwār. Finally, al-Ḥurr al-ʿĀmilī, in his *al-Fawāʾid al-Ṭūsiyya*, defends Astarabādī's position from attack by Uṣūlī scholars, arguing that his position corrects the errors made by al-ʿAllāma.[86]

One could add to these references, but it seems clear that within a short time of his death (around 30 years), Astarābādī was recognised by both his followers, his opponents and those outside of Imāmī Shiʿism (such as the author of the *Dabistān*) as either the founder or the reviver of a method of jurisprudence which demanded devotion to the texts of the Imams' sayings, and that this new (or revived) methodology was given the name "the Akhbārī way" (or variant locutions thereof). As we have intimated before, which of these terms (that is, founder/propagator) an author uses is much influenced by his view of both Shīʿī juristic history and the authenticity or heresy of the Akhbārī message.[87] The question raised by Newman's argument concerns whether or not the lack of association between Astarābādī and Akhbarism in Safavid biographical dictionaries indicates that he was not, at first, thought of as the founder/propagator of Akhbarism. Here I merely make note of the fact that near contemporary citations from other genres of Imāmī literature (most earlier than the first *ṭabaqāt* works in which Astarābādī is mentioned), and even references from outside of the Shīʿī tradition (cf *Dabistān-i Madhāhib*), indicate that he was thought of as either reviving the way of the ancients (*ṭariqat al-qudamāʾ*/*ṭarīqat al-akhbāriyyīn*) or inventing a new and heretical innovation in Shīʿī jurisprudence. This new movement—the Akhbāriyya—maligned the Uṣūlī scholars and accused them of deviating from the true path. In both assessments, Astarābādī is considered as representing a challenge to the dominant jurisprudence of the day and contributing something novel to the tradition. The lack of an association of Astarābādī and Akhbārī ideas in *ṭabaqāt* literature is not, then, evidence that the Astarābādī-Akhbārī connection is a later invention when other near contemporary sources are taken into account.

There remains, however, the rather puzzling absence of any mention of an Astarābādī-Akhbarism connection in the biographical compendia written closest to Astarābādī's death (that is, *Sulāfat al-ʿAṣr*, *Amal al-Āmil* and *Riyāḍ al-ʿUlamāʾ*). As already mentioned, the *Luʾluʾat*

---

[86] See al-Ḥurr, *al-Fawāʾid al-Ṭūsiyya*, pp. 417–458.

[87] See above, pp. 31–32.

al-Baḥrayn of Yūsuf al-Baḥrānī (written around 150 years after Astarābādī's death) is the first *ṭabaqāt* work to make mention of this connection, and al-Baḥrānī uses the opportunity to criticise Astarābādī for bringing disharmony into the community. Once established as the principal "Astarābādī" motif, the association is repeated in most subsequent works of *ṭabaqāt*. It was these later *ṭabaqāt* works which Western commentators relied on in their assessments of the Akhbārī-Uṣūlī dispute, and in this, it has been argued, they were mistaken. How is the *ṭabaqāt* authors' delay in describing Astarābādī as the founder/propagator of Akhbarism to be explained? My argument in what follows is that the time lapse is best explained by generic considerations linked to the development of late classical Shīʿī *ṭabaqāt* literature. In order to demonstrate this, an examination of the role and nature of *ṭabaqāt* literature generally, and late Imāmī *ṭabaqāt* literature in particular, is necessary.

Ṭabaqāt literature is used extensively by commentators in the analysis of Muslim religious developments. Biographical information is usually extracted from *ṭabaqāt* works in the course of an analysis of a particular intellectual movement or geographical area. Little attention is paid to the structure and underlying aims of the genre. The information can be used in quantitative analysis in order to construct a picture of the social make up of society generally,[88] and the scholarly class in particular.[89] *Ṭabaqāt* works are also used to build up narrative accounts of a scholar's life and influence, and to locate him (or occasionally, her) in a particular historical setting. Reflection upon *ṭabaqāt* as a genre of literature is less common, though this approach has produced a number of interesting studies in recent years.[90] In what follows, I argue that an understanding of the development of the scholarly compendia of late classical Imāmī Shiʿism enables us to explain the apparent contradiction noted above: namely, the absence of an association between Astarābādī and Akhbarism in *ṭabaqāt* works until 150 years after Astarābādī's death, and the vehement assertion of this association (as founder, reviver or innovator) in other genres of literature (*uṣūl al-fiqh*, commentary

[88] See Bulliet, "A Quantitative Approach" and his monograph *Nishapur*.

[89] See, for example, Nawas and Bernards, "Geographic Distribution" and "Netherlands Ulama Project".

[90] See Mojaddedi, *Biographical Tradition*; al-Qadi, "Biographical Dictionaries"; Stewart, "Captial, Accumulation".

on *ḥadīth* collections, *firaq* literature and *masā'il*)[91] composed nearer to Astarābādī's time.

A reading of late classical Imāmī biographical compendia as literature reveals them to be highly formalised works. Entries on scholars are normally arranged alphabetically by scholars' surnames,[92] and each entry follows a reasonably consistent model of presentation. The model comprises of a number of elements:

1. Epithets of praise for the scholar.
2. A list of works by the scholar.
3. A list of teachers (and sometimes teachers of teachers), usually comprising of the formula "[pupil] relates from [teacher]" though occasionally accompanied by references to *ijāzas* received.
4. A list of pupils (with similar variations to the list of teachers).
5. Minimal details of the scholar's travels .
6. Dates and places of birth and death of the scholar.[93]

The order of these elements in the biography varies between authors and between entries within a single author's *oeuvre*. If the author is unable to discover a particular piece of information, an element is omitted. Furthermore, the inclusion of these elements is often achieved by the citation of the entry on a scholar from a previous work of *ṭabaqāt*.[94] Amongst these works are pieces of information which prove to be of great utility to the historian. However, there are limitations upon the use of such material when compared to the idea of "biography" within Western literary tradition. Information

[91] By *masā'il* literature I refer to the genre of collections of discussions on specific topics. At times these are presented as questions and answers (and *fatāwā* may be included in this category). At other times this appears as a work in a collection of extended discussions of particular issues, which may have originally been separate treatises (*rasā'il*), collected together without excessive care for order or overall structure (such as al-Ḥurr's *al-Fawā'id al-Ṭūsiyya*).

[92] The exception to this generalisation is Madanī's *Sulāfat al-ʿaṣr* which is arranged geographically in five sections (scholars of: i. the Holy Cities of the Hijāz, ii. Syria and Egypt and their environs, iii. Yemen, iv. Iran, Baḥrayn and ʿIrāq, v. the Maghrib). Astarābādī's entry is in the first section.

[93] On occasions this list is supplemented with information such as his family connections, ethnic identity, employment by the state or other authorities and citations from the subject's works (especially poetry).

[94] Hence al-Afandī quotes all of al-Ḥurr's entry on Astarābādī, giving additional pieces of information subsequently. Compare Ḥurr, *Amal*, v. 2, p. 247 and Afandī, *Riyāḍ*, v. 5, pp. 35–37. The *Amal al-Amil* is clearly one of the major sources for Afandī.

concerning the scholar's personality and non-academic relationships (for example, friendships, family, business) is minimal. Judgements concerning a scholar's contribution are introduced by a list of panegyric formulae at the outset of the entry, and are almost universally positive. Occasionally there will be further assessments of the scholar's merit (for example, praise of a particular book, treatise or poetry), but these are, on the whole, afterthoughts, disrupting the (rather repetitive) reproduction of the above model. In short, they contrast with Woolf's assessment of biography, based on the premise that "we... can no longer maintain that life consists in actions only, or in works. It consists in personality".[95] They do bear some similarity to biographical compendia developed in the Nineteenth Century, and their use (primarily as reference works) may have been similar.[96]

Works such as these may not be designed to be read in the manner of modern biographies (i.e. as complete entities); they do, however, reveal the value system of the *ʿulamāʾ*. A scholar's position is determined by his literary achievements and his place within the scholarly network. His most notable characteristics are his output, his teachers and his pupils. These represent his contribution to the maintenance of the scholarly tradition.[97] Issues of "self" (personal characteristics, non-academic relationships, even innovatory ideas) are mentioned only when they impinge upon the description of the scholar's role as an *ʿālim*. The scholar is, in a sense, de-personalised by the imposition of the model of entry presentation, and this is almost an intentional outcome of the process of composing a biographical compendium. The regularity and uniformity of presentation reflects the overall kerygma of the work: the scholarly class (in this case, the Imāmī *ʿulamāʾ*) are unified in purpose and participate in the common task of promoting learning, thereby sharing the privileges (respect and class honour) which accrue to a person as a result of a scholarly education. The organisational scheme of the work (be it alphabetical, chronological, by teacher-pupil relationship or geographical area) is declared and

[95] Woolf, "The New Biography", v. 4, p. 230.

[96] The biographical dictionaries published since the Eighteenth Century CE, both religious (e.g. *Butler's Lives of Saints*) and profane (*American Dictionary of Biography* and later *Who's Who*) can be viewed as the literary reflection of Carlyle's conception of history being "the essence of innumerable biographies" (Carlyle, "On History").

[97] Jaques, *Authority, Conflict*, pp. 12–17.

adhered to, further projecting the regimented nature and homogeneous excellence of the scholarly class.

These features of works such as *Sulāfat al-ʿAṣr, Amal al-Āmil* and *Riyāḍ al-ʿUlamāʾ* highlight another important desideratum of *ṭabaqāt* literature generally. The ordered presentation of a scholarly elite must not be disrupted by the intrusive presence of intra-sectarian dispute. Unlike the theories of *ijtihād*, which enabled the scholarly elite to present their prerogative through the institutionalisation of difference of opinion,[98] late Imāmī *ṭabaqāt* literature aims to establish prerogative through reducing the perception of dispute between scholars and scholarly groups. Of course, this cannot be achieved completely.[99] Amongst the scholar's writings there may be a work which highlights dispute with another scholar, such as explicitly titled "refutation" works (*radd*).[100] Occasionally, then, a *ṭabaqāt* work such as those under examination here unavoidably refers to disputes within the community. This is not usually by design, but instead is forced upon the compiler by the requirements of the model.

Having briefly recognised these features, it is now possible to delineate the function of the various elements of a typical biographical entry. Praise and panegyric (item 1 above) serve to incorporate the scholar in question into the scholarly class and highlight the scholarly qualities and the learning-related tasks he performed, all of which justify this incorporation. The scholarly output (item 2: lists of works) is a recognition of the importance of a scholar's literary

[98] See Calder "Doubt and Prerogative".

[99] Occasional criticism of a scholar (for example, for having corrupt or deviant beliefs or more likely of being suspected of such) is not unknown. In Safavid *ṭabaqāt* works, these criticisms are restricted to occasional comments by al-Afandī in his *Riyāḍ*. Al-Afandī's criticism of Ṣūfi-leaning members of the *ʿulamāʾ* is one example. In Afandī, *Riyāḍ*, v. 2, p. 283, Rajab al-Bursī is condemned, as is his pupil, Qawwām al-Dīn al-Iṣfahānī (Afandī, *Riyāḍ*, v. 2, pp. 315–6). It is noteworthy though that despite the criticism of these scholars (as being ignorant of religious sciences and of corrupt beliefs), their entries begin with the standard assessments (*fāḍil, māhir*) together with general descriptions (*hakīm, ṣūfī*). Instances of criticism are, however, rare; mostly opprobrium is expressed by exclusion from the compendium completely. Turner has made some comments regarding al-Afandī's approach in his *Islam without Allah* (pp. 104–116), though his rather rigid externalist/internalist division detracts from the clarity of his analysis. The interesting suggestion concerning coding in these panegyrics (such as *fāḍil* being applied to scholars having reached a particular educational level etc) would need additional evidence from the *Riyāḍ* and other works to be entirely convincing.

[100] For example, Astarābādī's refutation of al-Dawwānī and al-Dashtakī referred to above (p. 39 n. 45).

contribution, as it is these works which are studied, copied and memorised in educational establishments across the Imāmī community. Teachers and pupils (items 3 and 4) are a measure of a scholar's connections, whose views he might be said to represent accurately, further embedding him within the network. The extraneous facts of a scholar's life (travels, birth and death dates etcetera—items 5 and 6) are confirmations of the validity of the above information since they confirm the feasibility of teacher-pupil relationships by location and time period.

The question of how the genre came to take this form is, to an extent, tangential to my purpose here. However, it is worth noting that suggested origins include pre-Islamic conceptions of the individual,[101] general Arab interest in genealogy[102] and in the needs of *isnād* criticism.[103] The last of these certainly explains the function of at least some early *ṭabaqāt* works. Transmitters of information regarding the Prophet, his companions and other early luminaries were tested for reliability by an analysis of the chains of transmission (*asānīd*) which prefaced reports. The *isnād* was scrutinised for the plausibility of the linkages within it (in terms of coincident times and places between transmitters). An assessment of the moral probity of the transmitters was also carried out. In order to complete this procedure, a body of biographical information relating to the transmitters (*rijāl*) was necessary, and hence such reference works became essential tools for validating *ḥadīth*s. When this is combined with the conceptions of knowledge transmission within the scholarly class, and the tendency to valorise the *ʿulamāʾ* as the guardians of Islam against potentially iniquitous rule, the *ṭabaqāt* genre (and the related genre of *manāqib*)[104] can be seen as natural outgrowths of existing biographical compendia. Of course *rijāl* works (which primarily referred to *ḥadīth* transmitters) contained not only praiseworthy transmitters, but also those not to be trusted.[105] However, as the genre became detached

[101] See Khalidi, *Arabic Historical Thought*, p. 205.

[102] See Heffening, "Ṭabaḳāt" and Makdisi, "*Tabaqāt*-biography", p. 372.

[103] Loth, "Die Ursprung"; Hafsi, "Recherches sur le genre *ṭabaqāt* pt. 1", pp. 227–229.

[104] *Manāqib* literature was devoted to relating the exceptional achievements of well-known figures of the past, and is perhaps best viewed as being hagiography rather than biography.

[105] The genre of biographical compendia devoted exclusively to "weak" transmit-

from its original function of *ḥadīth* validation, the need to include figures who did not play a part in the transmission of knowledge diminished.[106] Whether the genre's origins lie in the *ḥadīth* sciences or somewhere else, the genre became streamlined and rationalised. Material considered extraneous (accounts of the personal experiences of the scholars, assessments of their personalities, relationships and ideas) was excised. The highly formalised character of late classical Imāmī *ṭabaqāt* works stands at the end point of this development. Such works represent the most refined, and hence excessively formal, expression of the genre. The power of generic constraints upon writers in the late classical period in most genres of Muslim literature produced works with a high degree of regularity and consistency of structure. Authors of *ṭabaqāt* works (and indeed those writing *fiqh, tafsīr* or *kalām*) were aware of the demands of their literary tradition. They rarely deviated from the model.[107] It is, then, unsurprising that one finds extensive citations from earlier *ṭabaqāt* works in later compendia, as the author pays homage to the tradition in which he is writing. This is not to say, of course, that such works lack originality or are mere reproductions of a tried and tested formula. There is, undoubtedly, room for innovation and development in the genre (the most radical of which, beginning in the Twelfth Century AH, is discussed below), and writers did criticise and correct earlier authors.[108] However, the general features of Imāmī *ṭabaqāt* literature

ters (*ḍuʿafāʾ*) is a specialised sub-genre of *rijāl* works. See, for example, Ibn ʿĀdī's *al-Kāmil fīʾl-ḍuʿafāʾ* and Dhahabī's *Kitāb al-ḍuʿafāʾ*.

[106] Interestingly, as late Imāmī *ṭabaqāt* writers developed a greater interest in the personalities of scholars, non-Imāmī scholars (and disreputable Imāmī scholars) were, once again, included in the compendia. See, for example, al-Khʷānsārī's entries on various Sunni scholars including al-Shafiʿī (Khʷānsārī, *Rawḍāt*, v. 7, p. 245), Mālik b. Anas (*Rawḍāt*, v. 7, p. 211) and Daʾūd al-Ẓāhirī (*Rawḍāt*, v. 3, p. 289). That the inclusion of such figures signals a change in function for *ṭabaqāt* literature is argued below.

[107] In the religious sciences, this entrenchment of literary forms has become associated with the demise of (Sunni) philosophical writings after al-Ghazālī, the lack of theological adventure in the post-classical period, the closing of the gate and *ijtihād* and the derivative nature of later Quranic exegesis. That this contributed to the emergence of reform and modernist movements, which were frustrated with established presentation of Islam, is clear. On this generally, see Kerr, *Islamic Reform*.

[108] Al-Afandī's corrections of the errors of al-Ḥurr's *Amal* are found throughout the *Riyāḍ*. He also authored an independent work in which the corrections were catalogued (Afandī, *Taʿlīqat Amal al-Amil*).

were much as described above, and remained stable throughout the Safavid period.

Given these literary factors, it becomes less remarkable that Astarābādī's Akhbarism is not the subject of comment in Safavid *ṭabaqāt* works. The aim of such works was to present a uniform and unified picture of the scholarly class involved in learning and the transmission of knowledge. Discussing internal disputes had its proper place in other types of literature, but *ṭabaqāt* were designed to record the achievements of the *ʿulamāʾ* as a class through history. The contributions of this class to the maintenance of the valid interpretation of Islam were recorded, and their record reflected the prerogative they considered their due (a prerogative which was, perhaps, more distinctly marked in Imāmī Shiʿism than in other Muslim traditions).[109] Associating Astarābādī with Akhbarism, and making this the focus of attention within an entry concerning his achievements would have violated the boundaries of *ṭabaqāt* writing.[110] Of course, one could have simply omitted reference to him altogether, and use censorship to express a doctrinal orthodoxy. However, as has already been pointed out, Astarābādī was well connected both in terms of teachers and pupils, and his ideas gave rise to a number of Akhbārī advocates in late Safavid Iran, including some writers of *ṭabaqāt* works. Ignoring him, with the excuse of him being of lowly academic stature or outside of the Imāmī tradition, does not seem to have been a viable option. The solution was to include him within the tradition, but reduce the disruptive effect his contribution had clearly made.

Astarābādī was not alone in this treatment. An examination of figures from other controversial Imāmī traditions confirms the occurrence of this process. For example, al-Ḥurr al-ʿĀmilī who, as a good Akhbārī, criticises al-ʿAllāma al-Ḥillī as erring from the true path,[111] but describes him in *Amal al-Āmil* in nothing but glowing terms with the usual elements (panegyric epithets, teachers and pupil, lists of works and basic death details).[112] Similarly, al-Afandī, who prob-

[109] On this, see Chapter 9 below pp. 268–296.

[110] Note Cooperson's assertions (in his *Classical Arabic Biography*, pp. 1–8) that Muslim biography records the transmission of knowledge, not doctrine.

[111] See Ḥurr, *al-Fawāʾid al-Ṭūsiyya*, p. 423. There are also criticisms of al-ʿAllāma's legal views found in al-Ḥurr al-ʿĀmilī's *Wasāʾil* (e.g. v. 7, p. 488 and v. 10, p. 24).

[112] Ḥurr, *Amal al-Āmil*, v. 2, pp. 81–85, #224.

ably leaned towards an Uṣūlī perspective, and even devoted a whole volume to correcting al-Ḥurr's perceived mistakes,[113] makes nothing but positive comments in his entry on al-Ḥurr. This analysis could be expanded, and full evidence for my characterisation of these works would require an analysis beyond the space available here. However, it is against this background that the lack of reference to Astarābādī and Akhbarism in Safavid *ṭabaqāt* works is best understood.

Ṭabaqāt works which continued this tradition of scholarship are to be found in the Nineteenth Century C.E.[114] If there is an intrusion of doctrinal criticism before al-Baḥrānī, it appears in very occasional comments in al-Afandī's work.[115] However, the appearance in *ṭabaqāt* works of *ad hominem* comments on the scholarly worth of previous Imāmī scholars makes a full-blooded appearance in al-Baḥrānī's *Lu'lu'at al-Baḥrayn*. This work, which is innovative in a number of ways, displays a marked use of the *ṭabaqāt* genre for polemic purposes. The work itself is a *ṭabaqāt* work in the form of an *ijāza* to al-Baḥrānī's two sons. As with most *ijāzas*, extended *isnāds*, comprising of teacher-pupil (*mujīz—mujāz lahu*) relationships, are traced back to early figures, thereby demonstrating the continuity of the learning being passed on to the recipient of the *ijāza*. I have analysed the network of scholars created by al-Baḥrānī's transmission chains elsewhere.[116] In the *Lu'lu'a*, al-Baḥrānī uses the transmission chains as an opportunity to interject biographical information upon the first mention of a scholar's name. The result is a compendium arranged in approximate reverse chronological order, from al-Baḥrānī's own teachers to the companions of the Imams. The reverse chronological

[113] Afandī, *Taʿlīqāt Amal al-ʿĀmil*, where he regularly corrects al-Ḥurr's errors in death dates and chronological order.

[114] Examples include al-ʿĀmilī's *Muntahā al-Maqāl*, Al-Tanukābunī's *Tadhkirat al-ʿulamā'* and al-Qummī's *al-Fawā'id al-Raḍawiyya*.

[115] There also appears to be considerations of chronological distance in the insertion of critical comments. For some writers (such as al-Madanī and al-Ḥurr), scholars of the more distant past whose strange opinions have been incorporated into the difference of opinion (*ikhitlāf*) of the *ʿulamā'* can be mentioned as contentious without threatening to disrupt scholarly authority (see, for example, al-Ḥurr al-ʿĀmilī's comments on Ibn Junayd; *Amal*, v. 2, pp. 236–238). Later writers (such as al-Afandī) seem willing to express mild reservations about more recent scholars. This contrasts with al-Baḥrānī's willingness to criticise any scholar from any era if he feels their views deviate from orthodoxy.

[116] See Gleave, "Ijaza". The *isnād*s in this *ijāza* (*Marʿashī Fihrist*, v. 15, p. 9, MS#5605) are almost identical to those found in *Lu'lu'at al-Baḥrayn*. See also on the *Lu'lu'a*, Salati, "La Lu'lu'a al-Bahrayn".

arrangement is disrupted, firstly by the fact that some teachers outlived their pupils and hence continued to have academic effects after their pupil's death. More often, the chronology is disrupted by al-Baḥrānī breaking off from the transmission chain at a point (signified by the term *ḥaylūla* in the text), and returning to an earlier node in the chain to trace a second *isnād*, contemporaneous with the one abandoned. The new branch to the *isnād* complex provides al-Baḥrānī with additional names, and an opportunity to include additional biographical details concerning the scholars mentioned there. In this way a network of scholars (rather than a single *isnād*) is created, achieving a unified and connected portrayal of the scholarly class. The message is clear: without a teacher of note, and pupils who contribute to the academic activities of the *ʿulamāʾ*, one cannot be included in al-Baḥrānī's conception of the scholarly class. The *Luʾluʾat al-Baḥrayn* is, then, an accomplished piece of scholarship blending the *ijāza* and *tabaqāt* literary forms. The entry on Astarābādī, for example, is placed within the *isnād* extract:

> al-Ḥurr al-ʿĀmilī>Zayn al-dīn b. Muḥammad b. Ḥasan b. al-Shahīd II[117]>al-Astarābādī>Ṣāḥib al-Rijāl (that is, Muḥammad b. ʿAlī al-Astarābādī).[118]

The last two names in this *isnād* are given entries at this point, the first two having been dealt with in a previous *isnād*.

The established model, taken over from al-Afandī and before him al-Ḥurr, is also in evidence. Al-Baḥrānī cites their works and others, fulfilling the standard requirements of a biographical entry outlined above. At times he adds details not found in previous sources, but (presumably) derived from oral tradition. Where his presentation deviates from preceding *ṭabaqāt* works is in his frequent criticism of past scholars. He lists Astarābādī's scholarly characteristics in panegyric terms at the outset of his entry (*kāna fāḍil*^{an} *muḥaqqiq*^{an} *mudaqqiq*^{an} *māhir*^{an} *fīʾl-uṣūlayn*), but he goes on to say that Astarābādī strayed from the path when he slandered the *mujtahids*, accusing them of "ruining religion" (*takhrīb al-dīn*). Al-Baḥrānī also makes reference to his own refutations of Astarābādī's position in his other works. Astarābādī is not alone in receiving such criticism. Reading through

[117] Author of *al-Durr al-Manthūr*, on whom, see below, pp. 156–157.
[118] Baḥrānī, *Luʾluʾa*, pp. 117–120.

the *Lu'lu'at al-Bahrayn* a number of scholars are subjected to criticism from extremely mild comment (for example, "He used to mix with the nobility and sultans and those who came after him criticised him for this)."[119] to withering rebuke (for example, "His opinion was excessive and pure extremism, and hence in some of his works the views of Sufis and philosophers are expressed such that he is almost an unbeliever)."[120] The scholars singled out for particular criticism are Sufis, philosophers and Akhbārīs. Whilst the former two are often targets for scholastic criticism, the last category is surprising since, as I have shown elsewhere, al-Baḥrānī's own methodology is a quite sophisticated application of Akhbārī principles to the questions of *uṣūl al-fiqh*.[121] It is possible that al-Baḥrānī senses the demise of Akhbarism (which was to occur fully a generation later), and so, in the *Lu'lu'a* (one of his last works), he recognises that the Akhbarism of Astarābādī and others did great harm to the community.[122] It is not so much the objects of criticism, however, that are remarkable in al-Baḥrānī's work, but the fact that he makes such comments at all in a work of *ṭabaqāt*. As we have seen, the standard mode of presentation in late classical Imāmī *ṭabaqāt* works was to depict a united and harmonious picture of the scholarly elite. With al-Baḥrānī, a polemic use of the *ṭabaqāt* genre emerges. The standard elements of the established genre remain (that is, the "six point" model mentioned above), and extensive use is made of previous works in the form of quotations and references. However, this is then supplemented by explicit doctrinal comments, condemning certain tendencies within Imāmī intellectual history. In this al-Baḥrānī's *Lu'lu'a* can be considered an adaptation of the Safavid *ṭabaqāt* tradition for use in an intra-*ʿulamā'* polemic.

Polemic employment of the *ṭabaqāt* genre in the Qajar era established itself in the wake of al-Baḥrānī's innovative work. Analyses of, for example, al-Tanukābunī's *Qiṣaṣ al-ʿulamā'*[123] and al-Khwānsārī's

[119] The comment is made in relation to Niʿmat Allāh al-Jazā'irī (Baḥrānī, *Lu'lu'a*, p. 111, #42), on whom, see below, pp. 169–170.

[120] The comment is made in relation to Muḥsin Fayḍ al-Kāshānī (Baḥrānī, *Lu'lu'a*, p. 121, #46), on whom, see below, pp. 153–154.

[121] Gleave, *Inevitable Doubt*, passim.

[122] Al-Bahrānī, himself, often agrees with Astarābādī's position, though he criticises the manner in which he expresses it. See, for example, Baḥrānī, *al-Ḥadā'iq*, v. 1, p. 55 and v. 1, p. 180.

[123] Gleave, "Biography and Hagiography". Interestingly, al-Tanukābunī wrote two

Rawḍāt al-Jannāt[124] demonstrate the almost ubiquitous intrusion of doctrinal criticisms into scholarly biographical compendia. The entries on Astarābādī (and other Akhbārīs) exemplify this trend. Akhbārīs were not the only ones to be subject to criticism in *ṭabaqāt* works. Ṣūfīs, philosophers, astrologists, Shaykhīs (a category which sometimes incorporated Bābīs and Bahāʾīs) and those who asserted self-aggrandisement through magical practices were all criticised. The trend towards a more expressive (and entertaining) style in *ṭabaqāt* writings requires a full explanation and a detailed separate study. However, a few explanatory comments can be made here. Firstly, the establishment of a distinct financial base for the *ʿulamāʾ*, independent from state sponsorship, is said by some to have occurred in the wake of the fall of the Safavid dynasty in 1722 CE.[125] As this independent base emerged, the *ʿulamāʾ* lost any residual authority from association with the Shah, and this relationship did not re-emerge in the same form after the consolidation of Qajar rule. The heightened need for the scholarly class to engage with the population at large in the search for legitimacy may have led to a greater level of competition between scholars for popular support. Hence there was a new willingness on the part of particular authors to vilify scholars of the past and present. Furthermore, there may have been a need to represent scholars as personalities who might inspire popular loyalty, and an enthusiasm for recording and exemplifying a scholar's characteristics through anecdotes and popular lore emerged. The result was biographical entries in *ṭabaqāt* works which explicitly link a scholar's personality to his doctrinal position. I have argued that works such as al-Tanukābunī's *Qiṣaṣ al-ʿUlamāʾ* were attempts to hijack the popular charisma normally reserved for mystical thinkers and place it upon the more traditional scholarly class.[126] The increased use of Persian as the medium for *ṭabaqāt* works, together with a less bookish style, also indicate an audience outside of the (Arabic literate) scholarly class.

This polemic employment of *ṭabaqāt* became accepted and continued into the Twentieth Century. Eventually, Western conceptions of

---

works of *ṭabaqāt* (both in Persian). One is more traditional in style (*Tadhkirat al-ʿulamāʾ*) and contains no entry on Astarābādī. The other is more polemic in tone (the better known *Qiṣaṣ al-ʿulamāʾ*).

[124] Newman, "Anti-Akhbārī Sentiments".

[125] See Floor, "Economic Role of the *Ulama*", pp. 60–67.

[126] See Gleave "Biography and Hagiography".

scholarly neutrality began to affect the collection and presentation of bio-bibliographical data in Imāmī scholarship with the biographical work of Muḥsin Amīn and Āghā Buzūrg al-Ṭihrānī. There we have a re-emergence of the traditional formalism, though with an increased emphasis on the collation and referencing of sources, and an awareness of a strictly academic as well as confessional readership. Of course, polemic use of *ṭabaqāt* has not disappeared completely, and the assertion of scholarly prerogative after the Islamic Revolution in Iran has inevitably changed the character of recent *ṭabaqāt* works.[127] The ways in which the scholarly biographical compendium might be employed, and the different societal and religious settings which influence that employment, have made possible a variety of different presentations of the history of the Shīʿī *ʿulamāʾ*. My argument is that, with respect to *ṭabaqāt* works, elements can be traced back to the Eighteenth and Nineteenth Centuries and the need for the *ʿulamāʾ* to engage more directly with the Shīʿī community (particularly in Iran) after the collapse of the Safavids.

All of the above analysis serves to illustrate and explain the lack of connection between Astarābādī and Akhbarism in works of this genre. It does not demonstrate that there was no linkage, nor that there was no perception of a linkage amongst the *ʿulamāʾ*. It does, however, show that the general picture of late classical Muslim religious literature was, to a large extent, controlled by the demands of genre. Innovation (a distinctly "modern" normative assessment) played a much reduced role when compared with the usual exaltation of originality in analyses of the genre of biography outside of the Muslim tradition. The type of source one is consulting (and by this I mean, not just authorial bias, but also structure and composition) inevitably affects the character of the information historians can extract. The vilification of Astarābādī as the founder of an iniquitous juristic school of Imāmī Shiʿism is certainly the result of an increased confidence on the part of (Uṣūlī) *ṭabaqāt* writers in the Nineteenth Century CE. However, the apparently anodyne references in earlier works do not indicate an ambivalence concerning his role as founder of the Akhbārī movement. Evidence for a widespread perception of Astarābādī's position as reviver of the Akhbārī "way", founder (or leader) of the Akhbārī "trend" (or even "school") is to be found

[127] For example, Jarfadāqānī, *Az Kulaynī tā Khumaynī*.

in other genres of literature, including Astarābādī's own writings. He viewed himself as the reviver of Akhbarism and his supporters, within thirty years, also considered him as such. His detractors also considered him the instigator of a dangerous new threat to juristic methodology within the same time frame. They wrote refutations of his work, portraying him as an innovating heretic (*mubtadiʿ*). This view of Astarābādī maintained its currency throughout the Safavid period. In the Nineteenth Century, independent developments within *ṭabaqāt* literature produced some highly vitriolic representations of both Astarābādī and the school which he is said to have founded. It is against this background that we understand, not only Astarābādī's scholarly contribution toward the history of Shīʿī Islam, but also the role he has been given in the development of Imāmī Shīʿī thought.

CHAPTER THREE

ASTARĀBĀDĪ'S LEGAL THOUGHT

In any examination of the highly formalised tradition of pre-modern Muslim legal scholarship, a jurist's legal thought is most usefully derived from his (or occasionally, her) output in a number of reasonably standard genres of legal literature. Foremost amongst these are works of legal theory (*uṣūl al-fiqh*), substantive law (*furūʿ al-fiqh*) and legal responsae (*fatāwā*). The generic names for the first two forms are derived from a horticultural metaphor in which the law's roots (*uṣūl*) provide sustenance for the branches (*furūʿ*), the latter being entirely dependent on the former. That is, the legal rulings that a jurist gives on particular occasions are conceived of as a natural outgrowth of his legal theory. The jurist adopts a theoretical position regarding how texts become legal sources and how these legal sources are to be interpreted. He then applies this theory to the texts deemed to be relevant to a particular issue. The resultant ruling is, supposedly, determined by the previously adopted stance.[1] Theoretically, then, the *uṣūl* and *furūʿ* are in perfect harmony. A large proportion of the controversial material generated by opposing parties of Muslim jurists (particularly those belonging to the same school or *madhhab*) concerns the correct means of applying agreed theoretical procedures to particular legal issues. More seriously disruptive to the intellectual coherence of the tradition are those occasions on which the dispute concerns the correct theoretical procedure to adopt, and not merely the manner of its application. The dispute between the Akhbārīs and the Uṣūlīs was a dispute of this kind. The Akhbārīs and Uṣūlīs developed distinct theories concerning both the manner in which scriptural texts become legal sources (that is, how texts assume probative force—*ḥujjiyya*—in legal argumentation), and the range of

---

[1] That the relationship here described between *uṣūl* and *furūʿ* is somewhat idealistic is a common element in Western accounts of Islamic Law. That substantive law (in the form of *fiqh* works and a jurist's *fatāwā*) was actually closely related to legal theory (i.e. *uṣūl al-fiqh*) has been argued most vehemently in the recent past by Hallaq (see "*Uṣūl al-fiqh*" and "Murder in Cordoba"). The converse has been argued by Calder ("al-Nawawī"), and (in modified form) by myself ("Marrying Fatimid Women").

possible legal meanings which can be derived from them (that is, how these sources are to be interpreted once they have acquired probative force). The two theories were mutually exclusive, though there were those of both schools who attempted to develop compromise positions which bridged the gap.

Works of *uṣūl al-fiqh* outline, in an abstract manner, the resources available to the jurist in his attempt to attain knowledge of God's ruling on a particular legal question. Issues examined in such works include the probative force of particular textual sources, the manner in which these sources should be interpreted and the means whereby the jurist's ruling becomes (or fails to become) authoritative, both for himself and for others. Classical works of *uṣūl al-fiqh* demonstrate remarkable structural stability. In both the Sunni and Shīʿī traditions, the structures of works of *uṣūl al-fiqh* remained, on the whole, fixed until the Thirteenth Century AH/Nineteenth Century CE. This is not to say that the content of these works remained unchanged. Development and innovation certainly occurred, most commonly in the content of the works, but also occasionally in structure. As in all highly formalised traditions of literature, some innovations were incorporated into the tradition and became accepted elements of a more refined tradition; others were rejected, and the main tradition was not diverted or significantly altered. The stability of the classical *uṣūl* genre was such that a work dealing with issues of *uṣūl al-fiqh* which did not follow the established structure invariably also presented a challenge to the theoretical stance of the tradition in which the writer was working.[2] *Al-Fawāʾid al-Madaniyya*, Astarābādī's work of *uṣūl al-fiqh* can be characterised in this way—it deviated from the established Shīʿī *uṣūl* tradition, both in terms of its structure, and also (and perhaps more significantly) in the theoretical position for which Astarābādī argues. The deviation is such that it could be questioned whether *al-Fawāʾid* counts as a work of *uṣūl al-fiqh*.[3] An analysis of this work forms the first part of this chapter.

[2] For example, al-Shāṭibī's (d. 790/1388) introduction of a chapter entitled *kitāb al-maqāṣid* which comprises the second part of his *al-Muwāfaqāt* (see Shāṭibī, *al-Muwāfaqāt*, pt. 2 in vol. 1) signals his introduction of this topic as a central element of his innovative theory of law.

[3] See, for comparison, the discussion over whether or not al-Qāḍī al-Nuʿmān's *Ikhitlāf uṣūl al-madhāhib* can be considered a work of *uṣūl al-fiqh* in "Alta Discussion", pp. 419–420 in "Alta Discussion" in Weiss, *Studies in Islamic Legal Theory*, pp. 385–429.

A medieval Muslim jurist's legal views on specific topics can be found in works of *furūʿ al-fiqh* (usually just termed *fiqh*). Works of *fiqh* also follow a relatively stable format in which the author's views on specific areas of the law are stated in an established order. Works of *fiqh* begin with an examination of the laws of ritual purity (*ṭahāra*), followed by sections concerning the four personal duties of a Muslim (prayer, fasting, payment of alms tax and pilgrimage). The order of these first sections in *fiqh* works is relatively predictable. The structure of the remainder of a *fiqh* work is less predictable. The order of subsequent chapters (concerning issues such as marriage and divorce, inheritance, crime and punishment, contracts, slavery and manumission, court procedure, the operations of the state—including the regulations of legitimate military action) is not so fixed. Though the chapter order varies, the legal matters dealt with form a canon of subjects. Conservatism, which is not to be confused with lack of originality or atrophy, characterised both genres of *fiqh* and *uṣūl al-fiqh*, and this is borne out by the proliferation of commentaries upon established, authoritative works of the past. Much of the literary production of pre-modern Muslim jurists comprised commentaries (*shurūḥ*) or marginalia (*hawāshī*) on works of *uṣūl* or *furūʿ* composed by great scholars of the past. An account of Astarābādī's legal views (and hence an analysis of the manner in which he applied his legal theory to specific issues) is hampered by his explicitly stated refusal to write a work of *fiqh*.[4] He did compose marginal comments (*ḥāshiya*) upon the great Safavid *fiqh* work, *Madārik al-aḥkam*, but these have not survived.[5] In any case, this work was most likely composed before his conversion to Akhbarism, and hence would probably not provide a useful guide to how his legal theory (as found in *al-Fawāʾid al-Madaniyya*) might be applied to *fiqh* issues. As a poor substitute for a comprehensive work of *fiqh*, we have some responses to *fiqh* questions,[6] a *risāla* (and accompanying *fatwā*) on

---

[4] See above, p. 158. Astarābādī refused to write a work of *furūʿ* because he followed the "way of the ancients". The "ancients" considered the *akhbār* sufficient in themselves, so he did also. *Fiqh* was not a particularly popular genre amongst the post-Astarābādī Akhbārīs. There are, of course, Muḥsin Fayḍ's *Mafātīḥ al-Sharāʾiʿ* and al-Baḥrānī's *al-Ḥadāʾiq al-Nāḍira*, but both of these works are distinctive for different reasons (see Gleave, "The Qāḍī and the Muftī"). More common were comments (*shurūḥ* and *ḥawāshī*) upon the *akhbār* themselves, and *risālas* upon specific legal issues.

[5] See above, p. 39.

[6] Found in his *Jawāb*, edited and translated in Gleave, "Questions and Answers".

the purity of wine[7] and scattered comments on the *akhbār* dealing with matters of legal import.[8] An analysis of these works, and their relationship to the ideas laid out in *al-Fawā'id al-Madaniyya* form the second part of this chapter.

Astarābādī's al-Fawā'id al-Madaniyya

Al-Fawā'id al-Madaniyya (hereon *al-Fawā'id*) was viewed within the Shi'ite tradition (both by Akhbārīs and their opponents) as the first (and most authoritative) expression of the later Akhbārī position.[9] It also forms the starting point for almost all modern analyses of Akhbārī ideas, both within the Shi'ite tradition and outside of it.[10] The ideas presented within it, then, form an element of any account of the central tenets of Akhbarism, at least in the century following Astarābādī's death when his approach began to gain popularity amongst the Shi'ite *'ulamā'*. The work itself is rather haphazardly structured, and though much of it concerns issues of *uṣūl al-fiqh*, there are also chapters on the shortcomings of the theologians and philosophers. Here I am primarily interested in the legal theory proposed by Astarābādī in *al-Fawā'id*.[11]

Whilst issues of legal methodology and hermeneutics dominate the work, its structure is not that of a work of *uṣūl al-fiqh*. The deviation from the established format of *uṣūl* works indicates Astarābādī's dis-

---

[7] *Al-Risāla fī ṭahārat al-khamr*, edited and translated in Gleave, "The Purity of Wine". The *fatwā* is edited and translated in Appendix 3, below.

[8] See above, pp. 93–97, for the editions and manuscripts containing Astarābādī's *ḥawāshī* and *shurūḥ* on the *akhbār* collections.

[9] See the references outlined in the previous chapter, pp. 40–60.

[10] Recent Western accounts include Kohlberg, "Akhbari" (a summary of the ideas in *al-Fawā'id* is found on pp. 134–137); Stewart, *Islamic Legal Orthodoxy*, pp. 175–208 (the analysis of *al-Fawā'id* is combined with citations from other Akhbārī works, pp. 184–202); Abisaab, *Converting Persia*, p. 106 (where *al-Fawā'id* is erroneously named *Safīnat al-najāt*, the latter being the title of a work by Muḥsin Fayḍ al-Kāshānī). Jābirī's analysis (*al-Fikr al-Salafī*, pp. 280–284) begins with an exposition of Astarābādī's views as expressed in *al-Fawā'id al-Madaniyya*. Al-Gharāwī's analysis (Gharāwī, *Maṣādir al-Istinbāṭ*, p. 73 onwards) is more systematic, though is embedded within the Shī'ī *uṣūl* tradition. Sefatgol's analysis of Akhbarism also begins with such an analysis (*Sākhtār-i Nihād*, pp. 520–526), as do those of Dhākirī ("Akhbārīgirī", pp. 320–322) and Sharīfī ("'Aql az Dīdgāh-i Akhbāriyān", pp. 28–36, and as the prime source for subsequent sections of the article).

[11] Reference is made to the theological and philosophical sections of *al-Fawā'id* in the next chapter, see below, p. 103.

satisfaction with the study of *uṣūl* generally.[12] After a brief account of his reasons for writing the work and its contents,[13] the work comprises of a lengthy introduction,[14] twelve subsequent chapters of varying length (the first ten of which are explicitly concerned with matters of legal theory)[15] and a conclusion (*khātima*). The conclusion does not summarise the work as such, but adds new information concerning the hermeneutic method of the "early" Akhbārīs and some interesting anecdotes which aim to further legitimise the way of the Akhbāriyya.[16] The work as a whole is best characterised, not as an exposition of Akhbārī legal methodology,[17] but as a refutation of current trends in Shiʿite *uṣūl al-fiqh*. Only towards the end of the work[18] does the construction of an Akhbārī alternative (rather than a destruction of other opinions) dictate Astarābādī's presentation. Of course, in his criticism of the Sunnis, Shīʿī *mujtahids*, theologians and philosophers, Astarābādī frequently uses "Akhbārī" principles, and there is even occasional justification of Akhbarism from first principles. However, the bulk of the book comprises reactive and polemical objections to the *mujtahids* (and others), and hence the (perceived) inadequacies of his opponents' positions control the discussion. Given the nature of *al-Fawāʾid al-Madaniyya*, an account of its contents will inevitably

---

[12] According to al-Afandī, *al-Fawāʾid al-Madaniyya* began life as a commentary on al-Shahīd al-Thānī's *Tamhīd al-qawāʾid*. It soon, however, became a separate work, incorporating criticisms of al-Shaykh al-Bahāʾī (see Afandī, *Riyāḍ*, v. 5, p. 36 and *Taʿlīqāt Amal al-Āmil*, p. 246). Al-Afandī claims to have seen the original manuscript (ie the refutation of Shahīd II's *Tamhīd*) in Bārfurūsh. Whilst there are regular citations from Shahīd II's *Tamhīd* in *al-Fawāʾid*, a comparison of the two works does not reveal that the latter can be characterised as a commentary on the former.

[13] Astarābādī, *al-Fawāʾid*, pp. 27–41.

[14] Ibid., pp. 41–180.

[15] For example, the first chapter comprises of 70 pages in the printed edition, whilst the third, fourth and fifth chapters (pp. 261–264) are a page or less each.

[16] Ibid., pp. 516–544.

[17] There were, after Astarābādī, to be expositions of Akhbārī methodology, which (of course) make mention of the opponents' opinions but do not allow the polemic to control the structure of the work. An example of this in its early form is Ḥusayn al-Karakī's *Hidāyat al-Abrār* (analysed below, p. 166). Akhbārī methodology is more comprehensively put forward by Yūsuf al-Baḥrānī in his "introductions" (*muqaddimāt*) to his work of *furūʿ*, *al-Ḥadāʾiq al-Nāḍira* (analysed in Gleave, *Inevitable Doubt*, though see also below, p. 289). Al-Baḥrānī only turns to the dispute between the Akhbārīs and Uṣūlīs in the last *muqaddima* of *al-Ḥadāʾiq*, though naturally the dispute between the perspectives permeates much of his presentation of matters of *uṣūl al-fiqh* in the preceding *muqaddimāt*.

[18] Specifically, chapters 9 and 10 (Astarābādī, *al-Fawāʾid*, pp. 371–405) and the conclusion.

involve an examination of what Astarābādī considers mistaken in his opponents' positions. The following analysis is an attempt to extract from *al-Fawāʾid al-Madaniyya* an account of Astarābādī's legal theory itself, rather than an exposition of his opposition to the *mujtahids*. Furthermore, *al-Fawāʾid* does not contain justifications for certain assumptions which underlie Islam generally and Shīʿī Islam in particular. These include, not only theological doctrines (the existence and nature of God, his communication with humanity through revelation to Prophets and his demand from humanity to follow his law),[19] but also certain elements of a mature Shīʿī legal theory (such as the role of the Imams as interpreters of revelation). It is clear, therefore, that Astarābādī's target audience are those who are already committed to these doctrines; he therefore perceives little need to reiterate any justification for them in *al-Fawāʾid*.

Astarābādī's legal theory is based upon a stratified epistemology of the law, in which certainty (indicated by the terms *al-ʿilm* or *al-yaqīn*) plays a central role. Believers can, he argues, attain knowledge regarding the actions with which the believer is charged (*taklīf*). Knowledge of these requisite actions is to be distinguished from knowledge of God's law itself.[20] At first blush, there would seem to be little difference between these two types of knowledge. Knowing the law, it might be argued, is indistinguishable from knowing what it is Muslims are charged to perform. However, Astarābādī wishes to drive a wedge between the two for theological reasons. Essential (*dhātī*) knowledge of God's will (that is, the law) is only available to God, since he is the origin of that will. If it is, on occasions, available to believers, then (philosophically speaking) it is so accidentally (*ʿaraḍiyy^{an}*) rather that essentially. In other words, God's knowledge of his will springs from his role as its originator. He knows the demands he makes upon humanity because he is the creator of those demands. For the rest of humanity, any knowledge that one might attain of

---

19 Weiss, in his study of the Sunni writer Sayf al-din al-Āmidī (d. 631/1233), terms these "theological postulates" (Weiss, *God's Law*, pp. 33–80). Their presence in a work such as Āmidī's *al-Iḥkām*, when contrasted with their absence in *al-Fawāʾid*, demonstrates the quite different purposes of the two works. Some of the "theological postulates" which underpin Astarābādī's legal theory are discussed in the next chapter (pp. 110–117).

20 Much of this "meta-theory" (if you will) of the law is only hinted at in *al-Fawāʾid*, and is given fuller expression in Astarābādī's theological works, discussed in the next chapter, pp. 117–137.

God's demands is mediated through contingent statements made by God himself (in the form of direct revelation, such as the Qurʾān) or his representatives (such as the Prophets). Knowledge of any created thing gained by a report (even a report of the thing's creator) has the possibility of error inherent within it. This is not a comment about whether or not the Qurʾān or the reports of the Prophet's actions and words accurately reflects God's will. It is a comment concerning the ontological status (essential or accidental) of one's knowledge of the law. The possibility of error in a report makes such knowledge distinct from knowledge gained through being its creator. The truth or otherwise of the knowledge gained is not being questioned here; the difference concerns the nature of the knowledge. This distinction recognises that God's knowledge of himself (an element of which is, of course, his knowledge of his own law) must be distinguished from one's knowledge of him gained through reports (whether they are verbalised by himself, or by his representative). The latter have the inescapable possibility of error (*iḥtimāl al-sahw*—even if they actually contain no error themselves). This possibility of error makes such knowledge accidental rather than essential. God does not expect humanity to know the law in the same manner as he knows it himself. He knows it through being its creator; humanity knows it through reports concerning the law's content.[21]

If the transmission of knowledge works perfectly, knowledge flows from the law into one's minds without hindrance and one gains knowledge of the law itself. However, any breakdown in the transmission process will prevent the hearers gaining even this contingent knowledge of the law. One question which appears in works of *uṣūl al-fiqh* is how one is to judge the effectiveness of the knowledge transmission process. For Astarābādī, a full judgement on the effectiveness of the process is not possible. Knowledge of God's law can never be identical with God's self-knowledge. One can (perhaps) view the methods whereby one came to this knowledge as insignificant. That is, although God's self knowledge and humanity's empirical (report-based) knowledge have come about through different means

[21] It is for this reason that Astarābādī condemns the Sunni juristic mechanism known as *qiyās* (analogy) in which it is necessary to know the reason behind a ruling so that this reason (*ʿilla*) can be detected in a novel case, and the law extended. To know the reason behind a ruling is to know it "as God knows it", and this is impossible. For Astarābādī's refutation of *qiyās* generally, see Astarābādī, *al-Fawāʾid*, p. 269.

and have different ontological statuses (the former necessary, the latter contingent), one could argue that at least the content of one's own and God's knowledge is identical. But Astarābādī argues that such a judgement is dependent upon knowing (with absolute certainty) that the transmission process has occurred perfectly. One cannot have knowledge that the transmission process has been totally effective, since to do so would require already knowing the law "in itself" (*fī nafsihi*) and comparing it with one's own knowledge, gained through reports, and our inability to gain the former is precisely the point at issue here. For this reason humanity is not charged with knowing the law itself. Instead, it is charged (*mukallafūn*) with knowing those actions which justify its status as obedient servants. Whilst such a distinction is a subtle one, it underpins Astarābādī's typology of knowledge.[22]

We begin, then, with two sorts of knowledge of the law: God's knowledge of his law, and human knowledge of his law (gained from reports). Since the former is unobtainable unless one seeks some sort of unity with God's essence (a philosophical position considered untenable and ultimately heretical by Astarābādī),[23] it is the latter which occupies Astarābādī in *al-Fawā'id*. The only evidence one has of the law of God is the indicators he has given humanity, and hence the question which concerns Astarābādī above all others is how to gain knowledge of what humanity is charged to perform in order that believers might call themselves obedient servants. This knowledge, gained from the indicators provided by God, is not of the highest level of certainty, according to Astarābādī.[24] Though it may not be of the highest level, this does not mean it is to be disregarded. Astarābādī's point is that knowledge of the "charged" actions may

[22] That such a distinction owes much to the Avicennan tradition is clear. Astarābādī's thoughts on the particular characteristics of God's knowledge are discussed further in the next chapter (see below, pp. 117–137).

[23] The position is associated with the Sufis, who are criticised on numerous occasions (Astarābādī, *al-Fawā'id*, p. 407, p. 543).

[24] When referring to the possibility of error in the reports from the Imams, Astarābādī says, one can defend oneself against this accusation by referring to various pieces of evidence which might establish the authenticity of any *ḥadīth* under examination. "If this fails, then we declare openly that we have *customary certainty* that these issues, recorded in our books of *ḥadīth*, were posed to the Imams, and they were asked about them, and they gave answers, and the answers are found in those *ḥadīths* which were current amongst their companions." (Astarābādī, *al-Fawā'id*, p. 378, emphasis added).

not be of an indubitable kind. However, it is sufficient to establish duties to perform actions.[25]

That there is a distinction between the law itself and what believers are charged to perform in order to be considered obedient to the law is most clearly exemplified by the specifically Shīʿī problem of dissimulation (*taqiyya*). As is well-known, the Imams (on whom humanity relies for knowledge of God's law) at times were unable to declare the law openly to their followers for fear of oppression from their enemies (both with regard to themselves and with regard to the Shīʿa in general). At times, they dissimulated, describing the content of the law in deliberately misleading ways. Later Akhbārīs were to struggle with the means whereby this deliberate sabotage of the transmission process by the Imams might be identified, and some devised mechanisms to this effect.[26] However, Astarābādī is quite clear that identifying a *taqiyya*-generated report is not possible, and that the Shīʿa are justified if they follow such reports even though they do not reflect the law:

> Amongst the blessings of God for this sect [that is, the Shīʿa] is that He, may He be glorified, permits [the Shīʿa] to act upon all [reports] that come from the Possessors of Sinlessness [that is, the Imams], even if they were issued under dissimulation.[27]

Even if one's knowledge of the law is deficient in some way and the reports on which one bases one's knowledge do not reflect the law itself, one is justified in acting on the basis of these reports. One is, not charged with following the law in reality (*al-ḥukm al-wāqiʿī*), but with following the reports which have reached the community concerning the law's content.[28]

Although Astarābādī never outlines his conception of the different "levels" of knowledge in an unambiguous way, it is clear that, for Astarābādī, the category of knowledge termed "customary" or

[25] The argument here is based upon the theological principle that God would not ask the impossible of his servants (*taklīf mā lā yuṭāq*). Though Astarābādī does not explicitly refer to this principle, it is clearly the underlying premise of his argumentation, and a principle he shares with his opponents.

[26] Al-Baḥrānī, for example, outlines procedures but has little faith in their potential, and in the end recommends suspension of judgement and caution. See Gleave, *Inevitable Doubt*, pp. 112–121. Other Akhbārī means of detecting *taqiyya*-generated *ḥadīths* can be found detailed below, pp. 291–292.

[27] Astarābādī, *al-Fawāʾid*, p. 390.

[28] See below, p. 84, p. 86, for a fuller exposition of Astarābādī's views on *taqiyya*.

"ordinary" knowledge (*al-ʿilm al-ʿādī, al-yaqīn al-ʿādī, al-qaṭʿ al-ʿādī*) concerns him most. "Customary knowledge" is sufficient to establish the nature of the act with which the believer is charged. This, it seems, is the level of certainty attained by ordinary people (that is, when a person says "I know such and such" or "I am certain of such and such"). It is not indubitable (that is, uncontestable) knowledge. As Astarābādī puts it:

> What is meant by certainty (*al-yaqīn*) in the two topics [to be discussed][29] comprises "customary certainty" (*al-yaqīn al-ʿādī*). Obtaining any of the types of certainty (*afrād al-yaqīn*) stronger than this is not required.[30]

There are then "stronger" (*aqwā*) types of certainty than customary certainty, but they are not required in the study of either the hermeneutics of the law (*uṣūl al-fiqh*) or the actual derivation of the law (*furūʿ al-fiqh*). "Customary certainty" is all that is required here.[31]

An example of the implications of this epistemology in the context of a discussion of *uṣūl al-fiqh* can be found in Astarābādī's account of the authenticity of *ḥadīths* reported to have come from the Imams and recorded in the early Shīʿī collections. He argues that the authenticity of these reports is established by pieces of evidence (*qarāʾin*) which lead one to attain "customary certainty" of the *ḥadīths*' authenticity. Astarābādī lists seven *qarāʾin*:

> [1] Amongst [the *qarāʾin*] is that, on the whole (*kathīr*[an]), we are certain, due to circumstantial and spoken evidence, that a transmitter who is truthful in his transmission does not approve of fabrication, nor of transmitting something which has no clear evidence in his opinion. [This is the case,] even if he is someone with corrupt beliefs or someone who performs sinful practices. This type of evidence is common amongst the *ḥadīths* in our colleagues' books.
>
> [2] Amongst [the *qarāʾin*] is that the [*ḥadīths* in different sources] support each other.
>
> [3] Amongst [the *qarāʾin*] is that a learned, truthful and pious individual—in any book he might write as a guide for the people, and

[29] By which he means the disciplines of *uṣūl* and *furūʿ*.
[30] Astarābādī, *al-Fawāʾid*, pp. 106–107.
[31] One is reminded here of G.E. Moore's defence of "common sense" certainty in the possibly apocryphal statement he made about Bertrand Russell: Russell may have been a sceptic, but he was, on thousands of occasions, certain that he was sitting down. See Moore, "Four Forms of Scepticism".

as a source for the Shīʿa[32]—always reports a transmitter's source;[33] or [he reports] what the transmitter relates, having the chance to reveal the character [of the transmitter, that is, whether he is truthful or not] or the transmission [whether it is acceptable or not] and whether one can deduce the Imams' rulings with certainty [from the *ḥadīth*].

[4] Amongst [the *qarāʾin*] is that [al-Kulaynī] always clings to the *ḥadīths* from [a transmitter's] source or in that [transmitter's] version [of the source], even though he had the opportunity to turn to other sound reports.

[5] Amongst [the *qarāʾin*] is that *ḥadīths* are found in the two books of the Shaykh, *al-Kāfī* of al-Kulaynī and *Man lā Yaḥḍuruhu al-Faqīh*,[34] and they all bear witness together that the *ḥadīths* found in their books are sound, or that they are taken from those sources which are agreed to be sound.

[6] Amongst [the *qarāʾin*] is that the transmitter [of a source] may be one of the group who, as is agreed [by everybody], only transmit sound material.

[7] Amongst [the *qarāʾin*] is that the transmitter is one of the group who are described by the Imams as "trustworthy and dependable", or [that the Imams said,] "Take from the sign posts of your religion!", or [that the Imams said,] "They are God's dependable ones on his earth" and such like.[35]

Astarābādī's point here is that these seven pieces of evidence, both individually and collectively, do not establish with indubitable certainty that the *ḥadīths* in the sources come from the Imams. Instead, they (collectively, and perhaps individually also) establish a lower level of "customary certainty" (*al-yaqīn al-ʿādī*) that this is the case. That a truthful individual does not usually lie is powerful evidence for the accuracy of his transmission. This is the case, even if he holds deviant theological opinions (point [1] above). That a report re-occurs in different collections also lends weight to one's belief in

---

[32] Meaning here al-Kulaynī and his *ḥadīth* collection *al-Kāfī*.

[33] *aṣl rajul*[in]—by which Astarābādī means the source from the so-called four hundred sources which recorded the Imams' words, and which were rendered redundant by the composition of the early Imāmī *ḥadīth* collections.

[34] This is a reference to the canonical nature of the so called "Four Books", on which see Gleave, "Between *ḥadīth* and *Fiqh*".

[35] Astarābādī, *al-Fawāʾid*, pp. 176–178. Majlis MS#2706, f.63a is missing [7] in this list.

its accuracy (point [2] above).[36] That an eminent collector of *ḥadīth* would not record a dubious *ḥadīth* without alerting his readers to the fact (points [4] and [5]) is also evidence of this (and so on). Although these observations would not persuade a determined sceptic, this does not concern Astarābādī. He is only hoping to establish customary certainty—that is, certainty on which it is reasonable to base subsequent action.

With the bar of *ʿilm, yaqīn* and *qaṭʿ* (I recognise no distinction in Astarābādī's use of these terms) set relatively low, Astarābādī lays out what it is one can know of the law of God. One comes to know the law of God through reports of its content found in sources. Principal amongst these sources are, of course, *Kitāb* (that is, the Qurʾān) and *Sunna* (found in the *aḥādīth*). However, according to Astarābādī, one cannot know these sources directly—that is, one cannot simply read them and understand the law of God:

> It is known that a wise person, when explaining or elucidating [an idea] does not talk in a manner which deviates from the apparent meaning of his speech (*khilāf ẓāhirihi*) unless there also be a directed, clarifying piece of evidence (*min ghayri qarīnat*^in^ *ṣarīfat*^in^ *bayyinat*^in^), this is especially the case with one in whom the extent of knowledge is such that he is sinless.
>
> This is not the case, however, with most of the speech of God, nor most of the speech of the Prophet of God in relation to us. The [Imams] themselves said, "Only the person to whom it is addressed can understand the Qurʾān".[37] They also said, "The Prophet's speech is like the speech of God. It is liable to abrogate or be abrogated. Perhaps it is of general reference, perhaps it is of particular reference. Perhaps it should be interpreted. This is not known except[38] through us because we are the ones to whom they [that is, Qurʾān and *Sunna*] are addressed, and we know what is intended by them both."[39] Furthermore, the explicit statement of the Imams concerning the difference between their own speech and that of God and the Prophet leads one to the conclusion that [Qurʾān

---

[36] Astarābādī is indicating here that, although the report may be recorded in a number of places, that number does not establish the level of certainty accorded to *mutawātir ḥadīths*. On *tawātur* generally, see Weiss, "Knowledge of the Past".

[37] This *ḥadīth* is found in Kulaynī, *al-Kāfī*, v. 8, p. 312.

[38] The editors of the most recent edition of *al-Fawāʾid* totally misrepresent Astarābādī's point here (p. 179, ln. 2), copying mistakes made in the lithograph. They insert brackets inappropriately, and omit the crucial word *illā* (except) which is found in most manuscripts (see, for example, Majlis MS#2706, f.64a, l.2).

[39] A *ḥadīth* with this wording is not to be found in the collections (though the editors give the impression that it is a direct quote from the Imams).

> and *Sunna*] have various different aspects; that they could be abrogating or abrogated; that they were, mostly, revealed in a way such that their meaning was hidden to the minds of the ordinary folk...[40]

Wise persons (in particular those who are so wise one might call them sinless) say what they mean, and when they mean something other than what they say, they provide indicators that their intended meaning is other than the clear (or perhaps "literal") meaning of their words. The indicators of a meaning other than the literal are known to the addressee of the speech. In the case of God's revelation (both Qur'ān and *Sunna*), the addressee of the speech is not the people generally, but the Imams specifically. Hence only the Imams know whether the intended meaning of the authors (in this case, God and his Prophet) are the literal meaning of the words, or a diverted meaning. Just as one might misunderstand an overheard conversation between two strangers, so one might misunderstand the Qur'ān. The Imams, on the other hand, know the meaning of these texts because they have access to the possible indicators which might divert the meaning. The ordinary folk (*al-raʿiyya*) have no such access, and hence are unable to determine the meaning of God's revelation directly. They must rely on the Imams' words which are "not liable to be abrogated, and which are verbalised in a manner the ordinary people can understand. They are the ones addressed by [the Imams' speech]."[41]

Of course, Astarābādī's proof for this position contains a *petitio principii*. He argues that the Imams themselves have declared that only they can understand the Qur'ān. Such an argument will only appeal to those who have already accepted that the Imams' words are the decisive proofs. What is required to convince one who believes one can understand the Qur'ān and *Sunna* directly is external evidence that the Imams are necessary elements in the process of understanding God's revelation. Traditionally within Imāmī Shiʿism, this evidence consists of proofs, both rational and transmission-based, that an Imam is necessary at all times as a guide for the Muslim community, that this Imam must be sinless and that the only candidates for this position are the Twelve Sinless Imams. The fact that Astarābādī does not feel the need to replicate these arguments, or indeed even hint at their necessity in *al-Fawāʾid*, demonstrates that his principal audience are

[40] Astarābādī, *al-Fawāʾid*, pp. 178–179.
[41] Astarābādī, *al-Fawāʾid*, p. 179.

his opponents within Imāmī Shiʿism (that is, *mujtahids* and the like) rather than the Muslim community more broadly.

It is, perhaps, worth noting here that Astarābādī speaks of the fact that the Qurʾān and *Sunna* can "mostly" (*fīʾl-akthar*) only be understood with reference to the speech of the Imams. By implication, then, there are truths which can be understood directly from the revelatory texts. Their existence indicates that the necessity of the Imams' interpretive presence is not total. That is, humanity is not entirely dependent upon the Imams for its understanding of the message of God. It is possible that Astarābādī is indicating here doctrines which the Shīʿa share with the other (non-Imāmī) Muslim groups (the unity of God, the finality of Muhammad's prophecy etc.).[42]

Astarābādī demonstrates to his satisfaction that the sayings of the Imams are, in legal terms, the only legitimate mechanism for understanding the Qurʾān and Sunna. He spends a large proportion of *al-Fawāʾid* explaining where these sayings might be found, and how the sources in which they are found can be trusted as authentic, bringing knowledge[43] of the Imams' meaning (which is, in fact, identical with the meaning of the Qurʾān and *Sunna* to which one does not have access). His central argument is that the manner in which the sayings of the Imams (sing. *khabar*, pl. *akhbār*) were recorded guarantee their authenticity, and hence enable them to be the basis for a believer's valid legal action. In order to argue this position, Astarābādī has to demonstrate that the transmission process is sufficiently well-

[42] The later Akhbārī, Yūsuf al-Baḥrānī, considers Astarābādī's position to be that it is not permitted to base one's action upon the text of the Qurʾān without the *tafsīr* of the Imams (see Baḥrānī, *al-Ḥadāʾiq*, v. 1, p. 169). Al-Baḥrānī's characterisation would seem to contradict Astarābādī's position as found in *al-Fawāʾid*: Astarābādī talks of understanding the Qurʾān *and* Prophetic *Sunna* as being *mostly* dependent on the *tafsīr* of the Imams. It is possible, with some exegetical effort on my part, to marry the two: Astarābādī asserts that one cannot act on the basis of the Qurʾān without the *tafsīr* of the Imams (i.e. legally speaking, the Imams' *tafsīr* is indispensable), but one can understand (directly) verses which relate to religious doctrine, providing they do not become the basis for subsequent action. Astarābādī's position is, in fact, not so different from that of al-Baḥrānī himself, notwithstanding al-Baḥrānī's criticism of Astarābādī (see Gleave, *Inevitable Doubt*, pp. 48–55). See above, p. 170.

[43] The terms used here are *al-yaqīn, al-ʿilm* and *al-qaṭʿ* (certainty, knowledge and surety respectively, though he uses these terms interchangeably). They are not normally used with the modifier of *al-ʿādī* ("customary"), though it is clear this is what is meant and not other (stronger) types of certainty. As he says, "What is meant by certainty… comprises customary certainty" (Astarābādī, *al-Fawāʾid*, p. 105), by which I take it that he means that when he describes something as "certain" (*maʿlūm, qaṭʿī, yaqīnī*) he means that it (at least) reaches the grade of customary certainty outlined above.

established to engender at least customary certainty in the mind of the believer. He describes this process, beginning with the statement of the Imam on a particular occasion and ending with it being recorded in a currently accessible text. The transmission process he describes is, firstly, the collection of the Imams' statements by companions in works known as *uṣūl* (sources), many of which have not survived in themselves, but most of which acted as the sources for the authors of extant collections of the Imams' sayings. There are, then, at least two stages in the transmission process—from Imam to *uṣūl*, and from *uṣūl* to extant collections. The arguments are similar, though discrete, for each of the two stages. With respect to the first stage, Astarābādī records a series of *akhbār* from the Imams in which writing and recording are viewed as a religious duty in order that knowledge might not be lost. Amongst these are sayings such as "The heart relies upon writing." (from al-Imām al-Riḍā) and "Write! For you will not remember until you write." (from al-Imām al-Ṣādiq).[44] These establish for his audience the importance the Imams placed upon the recording of their words. This argumentation aims to establish that the companions of the Imams recorded the Imams' words in the *uṣūl*. By Astarābādī's time, it was established doctrine that there were four hundred such collections (*al-uṣūl al-arbaʿumiʾa*) which were rendered superfluous (and hence no longer recorded and transmitted) by the canonical four collections of *akhbār*.[45] As a challenge to this doctrine, Astarābādī expands the amount of material available to the early scholars, beyond the four hundred *uṣūl*. There were, he states, four hundred *uṣūl* from al-Imām al-Ṣādiq alone:

> These four hundred were taken from a single Imam. Our early scholars had other *uṣūl* apart from the four hundred. Whoever consults *al-Fihrist* of al-Shaykh al-Ṭūsī, or *al-Fihrist* of al-Najāshī or *al-Fihrist* of Muḥammad b. Shahrāshūb will testify to this.[46]

Astarābādī is clearly concerned to maximise the amount of written material recorded during the time of the Imams, even if only a small proportion of it has survived (to his time) in its original form

---

[44] These, and the other *ḥadīths*, are found in Astarābādī, *al-Fawāʾid*, pp. 138–139, with references to the relevant *akhbār* collections.

[45] See generally, Kohlberg, "al-Uṣūl al-Arbaʿumiʾa". The symmetry of four hundred *uṣūl* being reduced to Four Books is, of course, a little too convenient to be entirely trusted historically.

[46] Astarābādī, *al-Fawāʾid*, p. 131.

(that is, in works called the *Aṣl* of so-and-so). The more material which can be established as extant at the time of the compilation of the Four Books, the stronger Astarābādī's case for the authentic preservation of the Imams' words and deeds in the early collections of Shīʿī *akhbār*.

In order to establish this position, Astarābādī needs not only to demonstrate that the *uṣūl* existed, but that they survived and were employed as sources by the authors of the extant *akhbār* collections. Evidence for this is taken from early Shīʿī authors in which they record that they "rely upon the *uṣūl* for their beliefs and in action". The *uṣūl*, it seems, fulfilled this function not only for the earliest collectors (al-Kulaynī, Ibn Bābūya and al-Ṭūsī), but also for later scholars (Ibn Idrīs and al-Muḥaqqiq al-Ḥillī). They all testified (and are quoted to this effect) to the authenticity (*ṣiḥḥa*) of the *aḥādīth* they found within the *uṣūl*. Astarābādī is also keen to demonstrate that *ṣiḥḥa* means that one can be certain (to the customary level) that the reports are an accurate depiction of the Imams' words and actions. In technical terminology, the meaning of *ṣiḥḥa* for these early scholars "is that which is known to come from the sinless one."[47] Other early writers are cited who claim that all the *akhbār* they record are "known" to be authentic (*ṣaḥīḥ*), and are therefore reliable bases for action. Of particular note is an argument which was to re-occur frequently in Akhbārī writings, and is probably expressed here for the first time: al-Kulaynī, in collecting *al-Kāfī*, states in the introduction to the book that he is writing the work to dispel the confusion (*al-ishkāl wa'l-ḥayra*) within the Shīʿī community concerning what is authentic (*ṣaḥīḥ*) and what is not. To include both sound and unsound *akhbār* in his collection would frustrate this aim, therefore, all the reports mentioned there must be sound. Furthermore, al-Kulaynī does not mention any principle by which to distinguish between authentic and inauthentic reports, therefore, it must be the case (Astarābādī argues) that all the *akhbār* in his collection are *ṣaḥīḥ*.[48] *Ṣaḥīḥ* (authentic) and *ṣiḥḥa* (authenticity), when used by the early writers, merely refer to their (customary) certainty that these *akhbār* accurately reflect the Imams' words and deeds.[49]

---

[47] Astarābādī, *al-Fawā'id*, p. 109.

[48] Astarābādī, *al-Fawā'id*, p. 112.

[49] It therefore differs from the meaning of *ṣaḥīḥ* proposed by later Shīʿī scholars in which *ṣaḥīḥ* is a judgement on the "sound" character of the transmission chain

Astarābādī quotes from al-Muḥaqqiq al-Ḥillī as evidence of this definition of *ṣiḥḥa*:

> Most of the *akhbār* recorded in our books are known with certainty to be authentic (*ṣaḥīḥ*). [This is] either by the fact that they are well-attested and widely transmitted, or that there are [other] signs and indications which prove their authenticity and the accuracy of their transmission. These [signs and indications] give rise to knowledge and engender surety even if we find them appearing with only one transmission chain in the books.[50]

There is, of course, a mismatch here between Astarābādī's use of the terms associated with certainty (*al-yaqīn, al-qaṭʿ, al-ʿilm*) meaning *al-ʿilm al-ʿādī* and the early scholars' use of these terms. As I have argued elsewhere, the earlier scholars are, in fact, arguing for a stronger (and certainly less nuanced) epistemology than that proposed by Astarābādī. For them, *ʿilm* is always an indubitable type of knowledge,[51] whilst for Astarābādī, *al-ʿilm al-ʿādī*, which falls short of these stringent demands, is not indubitable, but sufficient to establish a religious duty. Astarābādī, though, is concerned with establishing a pedigree for his position, and finds a useful prefiguration in the stringent rejection of mere opinion (*al-ẓann*) and the forceful affirmation of certainty (*al-ʿilm*, unmodified by any adjective and seen as undifferentiated) amongst the early Shīʿī jurists.[52]

The reason why the three early collectors of *ḥadīth* (namely, al-Kulaynī, Ibn Bābūya and al-Shaykh al-Ṭūsī) relied on the "sources and books" (*al-uṣūl wa'l-kutub*) of the companions of the Imams was due to a number of factors:

> Know then that the reason why they [the collectors] relied upon these *uṣūl* and *kutub* comprises different points:
>
> 1. They were certain that the transmitter [of the source] was sound (*thiqa*) in his transmission.
> 2. [The transmitter himself] declared that his book was from one or other of the Possessors of Sinlessness [that is, the Imams]

(*sanad*, *isnād*), but does not, in itself, bring certainty of authenticity. On this, see below, pp. 201–202.

[50] Astarābādī, *al-Fawā'id*, p. 109.

[51] See Gleave, "Qiyās", pp. 272–273. This was the primary objection of the early scholars to the Sunni juristic mechanism of *qiyās*.

[52] I return to this epistemological discussion below, pp. 87–88.

3. [The collectors] compared his book with other books with certain (*maqṭūʿ*) authenticity.
So the three imams [that is, the three collectors] took the reports in their books from these *kutub* and *uṣūl*.[53]

According to Astarābādī, then, one knows that both the Imams' companions and other early Shīʿī scholars recorded the Imams' words in *kutub* and *uṣūl*. One also knows that the collectors of the Four Books relied on these *kutub* and *uṣūl* when composing their own *ḥadīth* collections (which went on to become the canonical "Four Books"). These "facts" should assure the reader that the transmission process was sufficiently robust to establish (at least to the level of "customary certainty") that when one reads the early *ḥadīth* collections (such as the "Four Books") one is actually reading the Imams' words or descriptions of their deeds.[54] Of course, it is these words and deeds

---

[53] Astarābādī, *al-Fawāʾid*, p. 146.

[54] In the rather repetitive structure of *al-Fawāʾid*, Astarābādī makes these points in his introduction, and then repeats them in the ninth chapter (pp. 371–377), listing twelve reasons why we have "customary certainty" (*qaṭʿ ʿādī*) that the reports in the Four Books, "for example" (*mathal[an]*) are authentic. They are:

1. We have customary certainty that a group of the Imams' companions for a period of 300 years or more asked the Imams their opinions and then wrote them down.
2. We have customary certainty that these sources formed the basis of belief and action (*ʿaqāʾiduhum wa-aʿmāluhum*) during the time before the collection of the Four Books.
3. Such is the wisdom of God and the kindness of the Prophet and the Imams to the Shīʿa, that they would not allow the *ḥadīth* to be lost, and the Shīʿa to be without a source on which to base their action.
4. There are numerous reports that the Imams told their companions to write and publicise the Imams' legal decisions so that they might become the basis for the action of the Shīʿa.
5. We know also from the earliest works of transmitters (*rijāl*), such as *Rijāl al-Kashshī*, that whole cohorts of companions were declared sound by the Imams themselves, and by the early transmitters of *ḥadīth*.
6. The authors of the Four Books all declare that they only record reports of declared authenticity (*ṣiḥḥa*).
7. If the *ḥadīths* in the collections did not come from the *uṣūl*, then this would mean our *ḥadīth* are not sound, and hence any action based on them would be invalid.
8. Most of the reports which al-Ṭūsī rejects would be considered *ṣaḥīḥ* (of sound *isnād*) by modern scholars, and most of what he acts on would be considered *ḍaʿīf* (weak). Therefore, he must have known something about their authenticity we do not.
9. Similarly al-Ṭūsī relies on a report with a weak *isnād*, when a "sounder" *isnād* (according to the categorisation of modern scholars) was available. Therefore, he must have known something about their authenticity we do not.

which establish (again, to the level of customary certainty) that performing actions in compliance with these words and deeds is a duty for the individual believer.

After establishing the authentic status of the *akhbār* collections, there remains a description of the means whereby they are to be interpreted. Interpreting these texts is of particular importance to judges and *muftīs*. These are, for Astarābādī, the transmitters of *ḥadīth* (*muḥaddithūn*, though this could equally be a reference to Akhbārī jurists generally). He devotes a chapter to demonstrating that only the *muḥaddithūn* should be judges and *muftīs*, refuting the position that it is the *mujtahids* who should take on these community roles. It is clear that, for Astarābādī, scholars (specifically, *ḥadīth* experts) are to take on the role of implementing the relevant sections of the law within the Shīʿa community.[55] Astarābādī's hermeneutics, outlined below, are primarily designed to aid the judge and the *muftī* in their task of interpreting the law. As mentioned earlier, Astarābādī considers the Qurʾān and *Sunna* of the Prophet to be (in the main) unavailable for direct interpretation. Rather, it is the *akhbār* of the Imams which provide the community with substantive legal knowledge. Interpreting these reports is not portrayed as particularly problematic. The Imams' speech "came in such a way that the ordinary folk can understand them. They are the ones addressed by the speech."[56] The *akhbār* are "devoid" (*khāliyy*[an]) of both abrogation (or, more accurately, the potential to be abrogated) and having an obscure meaning.[57] The interpretive process appears unproblematic. However, this is not the whole story. Whilst reports cannot be totally obscure in meaning, they can be open to more than one interpretation. The principal example

10. We know that al-Ṭūsī does not normally lie, and he says that he took his reports from the *uṣūl*.
11. Ibn Bābūya and al-Kulaynī say this also, and we know that they do not lie.
12. "We are certain in a customary way" that most of the transmitters of our *ḥadīths* are reliable. We know this because reports have reached us which describe how unhappy they are with fabrication in *ḥadīths*.

[55] Astarābādī, *al-Fawāʾid*, pp. 301–311.

[56] Astarābādī, *al-Fawāʾid*, p. 179.

[57] Unfortunately, Astarābādī does not describe the interpretive process (or rather, the supposed lack of an interpretive process) involved in drawing meaning out of the *akhbār*. Later Akhbārīs presented more sophisticated approaches (see below, pp. 000–000, and with respect to al-Baḥrānī, Gleave, *Inevitable Doubt*, pp. 96–102 and pp. 147–162). The impression gained is that, for Astarābādī, words have meanings which are unambiguous and which common people can understand immediately on hearing. In this sense, Akhbārī linguistics might be considered a form of literalism.

of this ambiguity[58] is an order (*amr*) which could be interpreted in one of two (or possibly more than two) ways. Although Astarābādī does not explore the presumptions underlying his view of language, it seems that he is committed to the conception of language found in most works of *uṣūl al-fiqh*. Words, phrases and grammatical constructions have "literal" (or "given" or "apparent" or "inherent") meanings. In the absence of any indication that the intended meaning is other than these, these form the default interpretive assumption as to intended meaning.[59] Astarābādī is asked his view on ambiguous reports in the eighth chapter of *al-Fawā'id al-Madaniyya*. For example:

> Question 7: What is your procedure for an authentic *ḥadīth* which could be interpreted as implying either an obligation or a recommendation?
>
> Answer: It is obligatory for us to suspend judgement as to which of the two possibilities is correct. Then we say: if the literal meaning [of the report] is obligation, then one performs the action, but with an intention to perform it cautiously (*bi-niyyat[in] iḥtiyāṭ[an]*). This is also the case if the two possible interpretations [obligations and recommendations] are equal. If the literal meaning is recommendation, but its inner meaning (*bāṭinuhu*) is obligation, then we have been freed from the burden of its obligation.[60]

There are a number of points to be made about Astarābādī's position here. Firstly, he wishes to maintain a commitment to the literal meaning of a command, be it obligation or recommendation. The question, as it is phrased, glides over two possible causes of the uncertainty:

1. Uncertainty could arise from there being two possible meanings—one literal and one non-literal (the "inner" meaning in the above passage) without a way of deciding between them. Astarābādī's answer is that the literal meaning always predominates here. However, it is important to recognise that in opting for the literal meaning, the believer is not deciding which interpretation is correct, and there-

[58] I use the word "ambiguity" here in its most basic sense—that is, that the linguistic phenomenon under consideration can have two (or possibly more) meanings. This should be distinguished from "obscurity" in which no meaning at all can be derived from the text.

[59] Which of the terms is the best translation of the various Arabic terms here (*ẓāhir, ḥaqīqī, mawḍū'* etc.) depends on context. Hence I will refer to them the "given" or "owned" meaning of words, phrases and grammatical constructions as the "literal" meaning.

[60] Astarābādī, *al-Fawā'id*, p. 334.

fore determining the meaning of the report. He is acting cautiously (*iḥtiyāṭ^an*), and his intention in performing the act must be an intention to act cautiously.

2. Uncertainty could arise because there is no means of determining which of two possible literal meanings is the intended one (that is, they are equally likely—*maʿa tasāwī al-iḥtimālayn*). This is a case of the literal meaning itself being ambiguous (that is, homonymy), and not a potential conflict between the literal and non-literal meanings. When such cases involve obligation and recommendation, one treats the order as if it is obligatory. Once again, however, this course of action does not represent a decision about the meaning of the report. The report remains irreducibly ambiguous. It is merely that action is necessary, and caution establishes the right course of action.

Astarābādī's answer fits well within his general epistemology of the law, outlined above. Here he is describing not the law itself, but actions which are justified (that is, legally valid) in response to the reports provided by the Imams. It is important to note here that observing the regulations concerning caution is, according to Astarābādī, sanctioned by the Imams themselves.[61] It is also interesting to note that his commitment to the literal meaning of a report is such that, on occasions, it contradicts the common sense assessment of the most cautious course of action. In the above excerpt, when the literal meaning is recommendation and the "inner" meaning is obligation, the most cautious course of action would (surely) be to treat the action concerned as obligatory. Yet, Astarābādī says that this burden is "lifted from us" (*mawḍūʿ ʿannā*). In cases where there is a potential conflict between a literal and non-literal meaning, caution always dictates treating the literal meaning as the meaning of the report, though at the same time recognising that the report is irreducibly ambiguous. This is the case even when (in abstract terms) the more cautious course of action might be to follow the non-literal meaning.

Astarābādī fleshes out the rules concerning the interpretation of ambiguous reports in his answers to a number of questions set by his interlocutor. In all cases of ambiguity, the individual suspends judgement (*tawaqquf*) as to the true meaning of the report. However,

[61] Amongst the reports cited to demonstrate the validity of caution as a juristic principle is the statement from al-Imām al-Kāẓim who was asked by one of his companions about cases for which there was no clear indicator from the Imam. He said, "When you come across cases like this and you do not know [the answer], then you must perform caution, until you ask [the Imam] about it, and then you will know." Cited in Astarābādī, *al-Fawāʾid*, p. 335, found in Kulaynī, *al-Kāfī*, v. 4, p. 391 #1.

this does not mean a suspension of possible action. A ruling based on caution can act as a substitute (*miṣdāq*) for the true ruling, but it must not claim to be the meaning of the report (and hence the true ruling contained within the report). Furthermore, whilst following the literal meaning of the ambiguous report is taken to be the most cautious course of action (*al-aḥwaṭ*), this does not extend to cases where there is an ambiguity between prohibition and another classification. In all such cases, the most cautious course of action is to avoid performance of the action: according to Astarābādī, one should always "avoid performing actions when we are not certain that they are permitted."[62] It is, it seems, a more serious transgression of the law to perform a forbidden action than to fail to perform an obligatory one.[63] Hence in cases where there is ambiguity in a report (it could indicate either an obligation or prohibition), one should never perform the action.[64] However, in such circumstances, one should not only avoid performing the action. One should also not criticise those who do perform the action (*tark tafsīq fāʿilihi*).[65] Caution dictates one course of action, but Astarābādī here does allow difference (*ikhtilāf*) in terms of action. At another point, Astarābādī states:

> If someone else does not avoid a thing which must be avoided in our opinion, because, for us, there is uncertainty (*shubha*, pertaining to its ruling), then it is not permitted for us to forbid it [to him]. This is because forbidding vice can only occur when the one forbidding the action knows [the action in question] to be forbidden.
>
> Let it not be said that suspending [judgement] when there is uncertainty is obligatory here also, for we say, "Perhaps he is ignoring the [uncertainty] or perhaps he knows [and therefore has no uncertainty]." This answer is based on the [principle] that it is obligatory for the

[62] Astarābādī, *al-Fawā'id*, p. 320.

[63] By such a position, Astarābādī is ruling out the view found in some *uṣūl* works that any order necessarily entails an opposite prohibition. This was explicitly stated by later Akhbārī authors (see, for example, Baḥrānī, *al-Ḥadā'iq*, v. 1, p. 59).

[64] Astarābādī, *al-Fawā'id*, p. 333.

[65] In most cases, caution dictates taking the literal meaning, and there is little tolerance of those who ignore this rule in the face of the Imams' decrees. However, when there is a radical ambiguity between obligation and prohibition, Astarābādī accepts that caution dictates avoiding the action, but he permits those who perform it to remain with sound faith (that is, they are not subject to *tafsīq*). The case is similar to that of two contradictory reports, one of which is *taqiyya*-generated and the other not (see below, pp. 291–292). The Shīʿa are justified in following the rulings contained in either report.

> learned person to disseminate his knowledge, but it is not obligatory for him to disseminate things about which he is uncertain.[66]

How does this marry with Astarābādī's encouragement of the judge and the *muftī* to give out *fatwās* and to make rulings based on caution? Though not spelled out, it would seem that, for Astarābādī, the judge and *muftī* can make cautionary *fatwās* and rulings when asked. However, they cannot, in areas of uncertainty (*shubha*), perform the public (proactive) duty of "forbidding vice" (*al-nahy ʿan al-munkar*). They cannot force others to act cautiously. On issues of *shubha*, there may be different norms within the community based on different interpretations of the same report (perhaps on different reports also). There is, then, a tolerance of diversity in areas of *shubha*. However, it should be noted that here the difference is not over the ruling itself, but over the most cautious course of action in these circumstances. Other cases of ambiguity are more easily dealt with by Astarābādī. For example, caution dictates that an indicator in a report which is ambiguous (namely, it could indicate either a prohibition or discouragement) leads to avoidance.

All of these rules concerning caution refer to cases where there are two possible interpretations of a single report. There is, however, another possible source of ambiguity in the law; that is, when two reports have incommensurate literal meanings. Astarābādī's solution to this problem is laid out in the ninth chapter of *al-Fawāʾid*. He begins by listing twenty-six reports which relate to the problem of deciding between contradictory reports, both in the area of issuing advisory decrees (*fatāwā*) and making legal decisions (*qaḍāʾ*).[67] These reports are themselves contradictory, some of them recommending that the believer choose (freely) between the two reports, and others recommending suspending any decision until one meets the Imam. Astarābādī states his own understanding of the reports:

> What I understand from [the Imams'] words is that when the subject of the two conflicting reports is a matter of personal devotion alone, such as prayer, then we can choose which to act upon. If it is a matter of the rights of men, such as a debt or inheritance, or a bequest to particular people, or sexual intercourse, or alms tax (*zakāt*) or the

[66] Astarābādī, *al-Fawāʾid*, p. 352.

[67] Most of these reports are those listed by al-Baḥrānī, and described in Gleave, *Inevitable Doubt*, p. 117.

> fifth tithe (*khums*), then on these occasions it is obligatory to suspend (*tawaqquf*) the performance of any act which is based upon identifying one of the two [reports as sound].[68]

Astarābādī's solution to the problem of conflicting reports brings to the fore the division between personal devotion and more public matters. The difference is between those areas a judge might act upon, and those with which he would not concern himself. His proposal (choice in personal devotion, suspension of judgement in public affairs) would appear to place a judge in a position whereby he is barred from making a decision in all matters of the law (other than personal devotion) when there is conflicting evidence from the *akhbār*. This is not, however, Astarābādī's position. What he wishes to prevent is judges (and *muftīs* when asked to provide *fatwās* by the community)[69] making decisions in cases based upon a judgment that one of the two reports is authentic and the other not (or one more likely to be authentic than the other). For Astarābādī, both reports are authentic (that is, their *ṣiḥḥa* is guaranteed). However, one of them (and it is not known which) was issued under conditions of *taqiyya*.[70] In such cases, Astarābādī states, the Imams have allowed the Shīʿa to follow the rulings in either report. If the issue concerns more than one party and it is brought before a judge, then the judge is the one who chooses which report to follow. Here the distinction between the law (in reality) and valid rulings (which may or may not reflect the law) is emphasised. A judge may rule on the basis of a *taqiyya* report, just as he may rule on the basis of two just witnesses. Two just witnesses do not establish the truth of a party's case. Their testimony merely establishes the validity of the judge's ruling. Similarly, when he rules on the basis of a *taqiyya* ruling, he may not be following the law, but he is justified, and he is acting as an obedient servant of God.[71]

---

[68] Astarābādī, *al-Fawāʾid*, p. 390.

[69] Astarābādī sees the task of the *muftī* as more demanding, and of greater seriousness than the task of the judge. In the first, the *muftī* is declaring the law "until the day of resurrection". In the second, the judge is merely declaring a valid legal decision in a particular case. See Astarābādī, *al-Fawāʾid*, p. 388.

[70] There are two types of *ikhtilāf*, Astarābādī states. One is based on the foolish interpretive principles of the Uṣūlīs (*al-istinbāṭāt al-ẓanniyya*, on which see below, pp. 87–88), and the other is based on the different reports from the Imams. See Astarābādī, *al-Fawāʾid*, p. 321.

[71] The reasoning for this position is laid out in Astarābādī, *al-Fawāʾid*, pp. 315–318.

The above instances of ambiguity are related to cases where the texts are unclear. Astarābādī also has some comments concerning those areas where the texts are entirely silent. He is asked:

> Question 17: What do you say concerning an animal which comes from the sea when we do not know God's ruling on it?
>
> Answer: Suspension [in making a ruling], and the substitution [*miṣdāq*, for a ruling] means one should avoid eating it; not giving any *fatwās* concerning its permissibility or its prohibition; and not denying anyone else from eating it if [one judges that] that he could know it to be permitted.[72]

When a *muftī* encounters a case for which he can find no indicators in the revelatory texts, this does not mean that there are no such indicators. For every situation "there is a certain indicator (*dalīl*[an] *qaṭʿiyy*[an]). The people are ordered to seek [the indicator] from the Preservers of Religion, the People of Remembrance [that is, the Imams]."[73] The fact that the *muftī* does not recognise the indicator as an indicator is evidence of his personal ignorance, not a deficiency in the law.[74] His ignorance here does not lead him to avoid giving a *fatwā* concerning the case, but instead to give a *fatwā* concerning the most cautious course of action. His recommendation is based both on the principles of Akhbārī hermeneutics outlined above, and also (and fundamentally) on the principle that one should always avoid actions which are potentially forbidden. Astarābādī's view at this point accords with his more general position that it is more serious to perform a forbidden act than it is to neglect an obligatory (or permitted) one. It is clear, however, that the *muftī's* ignorance here is personal, and the *muftī* cannot condemn another *muftī* for declaring the unknown animal permitted. It is possible that the other *muftī* knows of an indicator that he does not. Two important points emerge from Astarābādī's rather scattered comments here.

Firstly, it is clear that Astarābādī, whilst highly critical of juristic difference (*al-ikhtilāf*) within the community, is forced to permit it in certain circumstances. The juristic difference he permits does not,

---

72 Astarābādī, *al-Fawāʾid*, p. 351.

73 Astarābādī, *al-Fawāʾid*, p. 75.

74 *lillāh fī kull wāqiʿa taḥtaju ilayhā al-ummu ilā yawm ul-qiyāma ḥukm*[an] *muʿayyan*[an] (Astarābādī, *al-Fawāʾid*, p. 75): "God has a discrete ruling for every situation which the community will encounter until the day of resurrection." This dictum is supported by citations from the *akhbār*.

however, concern the content of God's ruling in reality (*fī'l-wāqiʿ*). Rather, difference can come about in a number of circumstances due to the failure of jurists, *muftīs* and *ḥadīth* experts to recognise the indicators in the revelatory texts. The permitted occasions for *ikhtilāf* that I have found mentioned in *al-Fawā'id* are as follows:

i) when there are conflicting reports from the Imams, giving different rulings. This, as has already been stated, occurs because one or more of the reports has been *taqiyya* generated.[75] In such circumstances, the Imams have permitted the Shīʿa to follow the ruling contained in any one of the reports. Different actions performed by different groupings of the Shīʿa do not reflect contradictions in God's law. All groups have to agree that this is unknowable. Rather, the different actions reflect preference for different indicators in the law, each of which make an action legally valid. On such occasions, the Shīʿa are commanded to follow the indicators of God's law, rather than God's law itself, and all parties fulfil this command (even though their actions diverge).

ii) when a report is ambiguous, in that it could be ordering an action or prohibiting it. In such cases, Astarābādī argues that one should not perform the action (because of caution), but also not condemn those who do perform it. Here, the difference of opinion is not over which report to follow, but over whether or not an individual report is clear in its indication of God's ruling.[76]

iii) when one perceives there to be no report pertinent to the case under consideration (though one knows, of course, that there must be a "certain indicator"). On such occasions (as with ii) above), one cannot sanction performance oneself, but equally one cannot condemn others for performing the action.

The second important point to note here is that these regulations apply only to the *muftī*, and that *ikhtilāf* is only permitted in the community if it stems from *ikhtilāf* amongst the *muftīs*. The phrase in the above quotation, in which another is not to be condemned if "it is *possible* that he knows [the action] to be permitted," (emphasis added) is an indication of this.[77] The only person for whom it is *pos-*

[75] Astarābādī, *al-Fawā'id*, p. 321. "This type of *ikhtilāf* does not lead to a contradiction (*tanāquḍ*)" since all *muftīs* can say "this *fatwā* is proven to have come from [the Imams], and it is not clear to me yet whether or not it is *taqiyya*-generated." "Both are justified (*kull wāḥida minhumā ḥaqq*)—one due to choice and the other due to the exigencies of *taqiyya*." (Astarābādī, *al-Fawā'id*, pp. 96–97).

[76] See above, pp. 272–273.

[77] There is a certain lack of clarity in Astarābādī's thought here (at least as it is

sible to know whether there is an indicator here is one who knows the *ḥadīth*. An ordinary believer would not be able to ascertain this, and hence (by implication) his actions are to be condemned unless they are supported by a *fatwā*. There is an elitist thread of thought running through Astarābādī's work. The task of announcing the law of God falls to *muftīs*; these are to be drawn from those who are experts in *ḥadīth*. The task of implementing the law of God falls to the judges; they too are to be drawn from the *ḥadīth* experts. Those who differ over the most cautious course of action (i) above) do so because of "their limited abilities in the discipline of *ḥadīth* study. Hence it is necessary for the one who is confused about the most cautious course of action to turn to one who is more learned than he."[78] The idea that Astarābādī is promoting some sort of egalitarian approach to the scriptural texts should, therefore, be discarded.[79] Whilst he does not outline the precise terms of his hierarchy, it is clear he assumes there to be one.

As is well known, Astarābādī's juristic theory was entirely at odds with the dominant "*ijtihādī*" Shīʿī legal theory. For the *mujtahids* (or Uṣūlīs), knowledge of the law (*al-ʿilm, al-yaqīn, al-qaṭʿ*) is only available on rare occasions. Most of the law has to be deduced from the sources, and these sources are not always established (with absolute certainty) to be authentic. Furthermore, the sources are regularly ambiguous and hence the jurist is required to use fallible interpretive mechanisms in order to understand the sources. These mechanisms will inevitably be applied differently by different scholars, and hence the results are inevitably uncertain. The result is a theory in which much of the law is open to doubt. Juristic anarchy is prevented by

portrayed in *al-Fawāʾid*). Major issues are glossed over: What is the relationship between *muftīs*? (What happens when one *muftī* is certain that one *dalīl* is effective and another ineffective whilst his opponent insists the opposite?) How do the "ordinary folk" recognise a true expert in *ḥadīth* (and hence a legitimate *muftī*)? Will even limited *ikhtilāf* entail the Shīʿa's practices being diverse?

[78] Astarābādī, *al-Fawāʾid*, p. 350.

[79] Stewart has already noted this (*Islamic Legal Orthodoxy*, p. 196). However, Stewart's assertion that Astarābādī makes a distinction between *ḥadīth* experts and jurists is not obvious. His main distinction is not between *muḥaddithūn* and *fuqahāʾ*, but between *akhbāriyyūn* and *mujtahidūn*. The *ḥadīth* experts who are to determine the most cautious course of action and give *fatwās* to that effect utilise juristic skills of interpretation. As has been seen, Astarābādī was not devoid of hermeneutic awareness, and required the scholar (*al-ʿālim*) to be able, not only to know *ḥadīth*, but also to understand the procedures relating to *iḥtiyāṭ*.

the establishment of qualifications required of a jurist, before he can make a judgement on the authenticity of the sources. These qualifications (*sharāʾiṭ al-ijtihād*), once attained, enable the jurist to make authoritative rulings (*fatāwā*) on the basis of his personal opinion (*ẓann*) of the sources' status and meaning, after doing a thorough search of the available evidence (*ijtihād*).[80] Sources then have variable levels of reliability, and the application of interpretive techniques by qualified persons leads to different (but equally valid) opinions of the law.

Whilst there are clearly significant differences between Astarābādī's theory and that proposed by the *mujtahids*, there are also similarities. Epistemological flexibility characterises both approaches, in that, for Astarābādī, not only is *ikhtilāf* inevitable in certain areas, there is also a variety of types of knowledge of God's law: knowledge of the law "in itself", knowledge gained from reports, knowledge which is indubitable, "customary knowledge", knowledge not of the law but of what is required of obedient servants, knowledge of the most cautious course of action (as distinct from knowledge of God's ruling) and so on. This flexibility is not so different from the *mujtahid* assertion that few legal matters are established to an indubitable level of knowledge, and that the community is forced to content itself with the opinion of the qualified jurist. The epistemological difference between the two approaches is, in part, terminological. Astarābādī rejects the category of *ẓann*, but admits plurality in the category of *ʿilm*; the *mujtahid* reserves *ʿilm* for indubitable knowledge, and argues for plurality in the category of *ẓann*. Furthermore, both epistemologies lead to hierarchical conceptions of the scholarly class. For Astarābādī, *ʿilm* is only available to the experts in *ḥadīth*; for the *mujtahids*, only a qualified jurist can issue a *fatwā* based on his own opinion. These similarities do not, however, lessen the significance of the difference in hermeneutics between the two approaches, and it is the effect of this difference upon the derivation of the law to which I now turn.

---

[80] I have already outlined the theory of Muḥammad Bāqir al-Bihbihānī, a representative member of the *mujtahid* school (Gleave, *Inevitable Doubt*, passim). My intention here is to outline Akhbārī juristic thought as found in the works of Astarābādī, and hence, in the above, I have restricted *mujtahid* criticisms of his ideas to footnotes.

Astarābādī's Positive Law

As mentioned earlier, Astarābādī did not compose a work of *fiqh*, and his marginalia (*ḥawāshī*) on Muḥammad b. ʿAlī al-ʿĀmilī's *Madārik al-aḥkam*, probably written before his conversion to Akhbarism, have not (yet) come to light. However, a number of sources for Astarābādī's *fiqh* have survived, and although they do not cover all areas of law, they do offer some insight into his method. These include a *fatwā* and a *risāla* on the purity of wine, comments on collections of *akhbār* relating to matters of *fiqh* and answers to specific questions of substantive law. A reading of this material reveals that Astarābādī's hermeneutic principles were often faithfully put into practice in his exposition of *fiqh* topics. A pertinent example of this is found in *al-Fawāʾid al-Madaniyya* when he is asked about the major ritual purification on a Friday (*ghusl yawm al-jumʿa*, or simply *ghusl al-jumʿa*) by an anonymous interlocutor.[81] The major Friday purification is a disputed element of the law. It is (according to most Shīʿīs) obligatory even if it is not ritually required.[82] It is, it seems, required due to the significance of Friday within the Muslim week and not because of any purity infraction. The question set to Astarābādī concerns a man who performs *ghusl al-jumʿa*, although he has no need to perform even minor ritual purification (*wuḍūʾ*). He has, then, doubt as to whether he should also perform *wuḍūʾ* after having performed *ghusl al-jumʿa* and "is unable to gain knowledge of God's ruling on this matter. What is the ruling concerning him?"

> Answer: He should do *wuḍūʾ* because fulfilling the law to its optimum (*ishtighāl al-dhimma*) is known to be dependent upon an act of purification. He does not know precisely what this act of purification is. Performing *wuḍūʾ* after *ghusl* would be an innovation only if he knew that *ghusl* rendered [*wuḍūʾ*] superfluous. The situation here is that he is ignorant of God's rule, and has doubts concerning it. The same is true of one who doubts whether he has urinated or ejaculated, though

[81] Astarābādī, *al-Fawāʾid*, pp. 347–348. This is one of a series of "questions concerning what we have presented and asserted concerning the statements of our Imams" (Astarābādī, *al-Fawāʾid*, p. 312).

[82] Other schools considered *ghusl al-jumʿa* recommended or merely permitted. See Nawawī, *al-Majmūʿ*, v. 2, p. 201 for an account of the variety of views.

> he knows he has done one of the two.[83] He should do both acts of purification, in accordance with what has already been said.[84]

Astarābādī's reasoning here is compressed, but can be expanded as follows. The person concerned is unsure about whether the special *ghusl* for Friday is effective in eliminating both minor and major ritual purity infractions (that is, whether *wuḍū'* is necessary as well as *ghusl* as part of the special Friday purifications). The most cautious course of action would be for him to perform both *wuḍū'* and *ghusl*. This is what the person who is ignorant of the law should do. He should act in such a way that he maximises his chances of fulfilling the law. In fact, performing the *wuḍū'* after the *ghusl* (when there is no independent need to perform the *wuḍū'*) is a recommended action, but not required.[85] If he had known this, then he would be failing to obey the law if he then performed it, considering it to be obligatory. What is required by the law, and what is required in order to maximise one's chances of fulfilling the law are distinct entities. The former depends on the law itself; the latter on the individual's knowledge of the law. Such a distinction is broadly in line with Astarābādī's differentiated epistemology outlined above.

Another case mentioned in *al-Fawā'id* concerns water which has come into contact with an impure substance (such as blood or semen). Say one is not sure that this water was of a sufficient quantity (*al-kurr*) to render it suitable for ritual purification (that is, the person is unsure whether the impure substance is sufficiently diluted to mitigate any impurifying effects). If one then comes into contact with this water/impure substance mixture, is one then obligated to perform both *wuḍū'* and *ghusl* serially (because of the risk that one has suffered either or both of the major and minor purity infractions through coming into contact with it)? Alternatively, presuming there is no other available water, should one perform *tayammum* (ritual purification with sand which substitutes for water purification of either type when water is not available), or is it enough simply to perform *wuḍū'* with the liquid? Astarābādī's answer is as follows:

---

[83] Each of the two emissions requires different purification procedures—emission of urine requires *wuḍū'*, whilst emission of semen requires *ghusl*.

[84] Astarābādī, *al-Fawā'id*, p. 348.

[85] That Astarābādī thought so, is clear from his answer to al-Shaykh al-Ẓahīrī concerning all types of *ghusl* other than *ghusl al-janāba* (see Astarābādī, *al-Fawā'id*, p. 575). This opinion is attributed to al-Sayyid al-Murtaḍā.

> The reports which relate to the *kurr* imply that one can declare such water impure when one knows that it has not reached a *kurr*. [They also say] that one can declare the water pure when one knows it has reached the *kurr*. So the implication of these two reports, and also of the other reports which say one should suspend judgement on all occasions when one does not know the actual ruling, is that it is obligatory to suspend judgment (*tawaqquf*) concerning the two rulings. It is known that *tawaqquf* necessitates avoiding [the water] and hence it is specified that *tayammum* is sufficient here.[86]

Once again, Astarābādī opts for the safest option—treat the water as if it is impure, and perform sand purification. However, there is an important difference between this second question and the first on *ghusl al-jumʿa*. The first concerns the law itself; the second issue concerns a fact. This means that the person's ignorance has a different subject in each case. In the first case, the person is ignorant of the law. In the second, the person knows the law but is unsure whether it applies in this case or not. Astarābādī subtly changes the terms of the second case from ignorance of the facts of the case to ignorance of the law itself. He argues that the *akhbār* are not entirely clear whether a body of water of uncertain quantity should be considered pure or impure after it has come into contact with an impure substance. Hence the question concerns the requirements of the law in such cases, rather than being a matter of the amount of water involved. By doing this, Astarābādī enables the case to fall more easily into his methodological schema. His theory is primarily concerned with cases where the law is uncertain or not known, rather than cases in which the law is known, but the facts of the case are uncertain.[87]

In Astarābādī's answers to the questions set by al-Shaykh Ḥasan al-Ẓahīrī,[88] there are yet more individual rulings on specific issues. Most answers are mere statements of Astarābādī's opinion with little legal reasoning. Whilst this makes the answers less than ideal sources for a comparison of Astarābādī's legal theory and positive law, they do illustrate Astarābādī's position on a number of important legal questions. On the much-debated issue of Friday Prayer during the

[86] Astarābādī, *al-Fawāʾid*, p. 348.

[87] Later Akhbārīs were more willing to deal with the necessity of *iḥtiyāṭ* in cases of factual uncertainty and not merely uncertainty concerning the ruling. With respect to al-Baḥrānī, see Gleave, *Inevitable Doubt*, pp. 105–112.

[88] An edition of these questions and answers is found in Gleave, "Questions and Asnwers" and in Astarābādī, *Jawāb*.

ghayba,[89] Astarābādī states simply that he is in agreement with Shahīd II in considering it individually obligatory. Concerning the opinion that Friday Prayer becomes obligatory only when there is a suitably qualified jurist present (a position known as *al-wājib al-takhyīrī*), Astarābādī tersely notes that "this opinion and those like it are opinionated fancy that should not be relied upon."[90] Whilst Astarābādī delineates a distinct role for the scholars during the absence of the Imam, this role does not extend to quasi-sacerdotal roles, such as the legitimisation of Friday prayer.[91] Interestingly, Astarābādī notes here that if it is not possible to hold Friday Prayer, then the individual must migrate (*muhājara*) to a land where it is possible. In the current situation, this cannot happen anywhere other than Iran (*ghayr bilād al-ʿajam*). Even inside Iran it is not possible, he adds, because "*taqiyya* is obligatory [there] because of the ferocity of those jurists who have not read the *aḥādīth* deeply."[92]

Another controversial question relates to the qualifications for the recipients of *zakāt* and whether or not they must have moral probity (*ʿadāla*) or not. It is clear that Astarābādī does not consider the *zakāt* tax to be lapsed (*saqiṭ*) during the *ghayba*. Astarābādī's answer is that the recipients need not have moral probity, but they must observe prayer and not be sinful (*fisq*—minors are exempt here).[93] Astarābādī also expresses opinions on a number of other issues including bequests, a wife's inheritance, the purity status of baked mud from Karbala and purifications other than the usual *ghusl al-janāba*. Unfortunately, Astarābādī's reasoning is not laid out in even a minimal form here.

---

[89] This, it will be recalled, was one of Newman's criteria for determining a scholar as Akhbārī or Uṣūlī (see his "Development and Political Significance", p. 26, amplified in the rest of his introduction, pp. 26–56).

[90] Astarābādī, *Jawāb*, p. 573: "opinionated fancy" (*al-khayālāt al-ẓanniyya*) here is a reference to Astarābādī's perception that the *mujtahid*'s category of *ẓann* is merely a product of their wish to justify their own opinion (see above, p. 88).

[91] Unfortunately, Astarābādī does not outline crucial questions such as how the *khaṭīb* is chosen, what qualifications are necessary in him etc. These determine the nature of the *ʿulamā*'s leadership role, and (to an extent) the relationship of the Shīʿa with political power. See below, p. 174.

[92] Astarābādī, *Jawāb*, p. 573, a reference to the dominance of Uṣūlī jurisprudence in the first half of the Safavid period. It is interesting that Astarābādī considers *taqiyya* from both Uṣūlī and Sunnis necessary, whilst Shīʿī thought normally restricts *taqiyya* to the latter group only.

[93] Astarābādī, *Jawāb*, p. 575. Belief in the invalidity of *zakāt* was also one of Newman's criteria for Akhbarism (see above, p. 16).

Hence one is unable to come to any significant conclusions concerning the relationship between Astarābādī's legal theory and his *fiqh*. If there is evidence of his Akhbarism in these answers, it is to be found in his constant appeal to the *akhbār*, though the relevant source is rarely mentioned, and one finds such references in *mujtahid* works of *fiqh* also.

Astarābādī's interpretive method with respect to the *akhbār* can be obtained most easily from his various commentaries on the "Four Books".[94] These works are inevitably a rather haphazard collection of comments, and were clearly originally marginalia on Astarābādī's personal copies. They became separate works through the collating efforts of later Akhbārī scholars.[95] A number of points can be made when comparing Astarābādī's hermeneutic theory and the interpretive practice found in these commentaries. Firstly, notwithstanding Astarābādī's theoretical commitment to the canonicity of the *akhbār* collections, he is not adverse to occasionally correcting the collators (particularly, Ibn Bābūya). An example of this is found in Astarābādī's comment on Ibn Bābūya's version of the prayer to be said when performing *ghusl al-jumʿa*. Ibn Bābūya's version runs:

> Oh God, purify me, and purify my heart. Accept my purification, and make [expressions of] love for you proceed from my tongue.[96]

Astarābādī corrects this version, stating:

> In some *ḥadīth* on this topic [it is recorded]: "make adoration of you and praise for you proceed from my tongue." This is better.[97]

The reliability of Ibn Bābūya's extraction of the prayer from the *akhbār* is here questioned. Astarābādī also corrects Ibn Bābūya's recording of names in *isnāds* (for example, Yaḥyā b. Saʿīd is corrected to

---

[94] Two of these have been edited and published by ʿAlī Fāḍilī: Astarābādī, *al-Ḥāshiya ʿalā Uṣūl al-Kāfī* and *Ḥāshiya Man lā Yaḥḍuruhu al-Faqīh*. The former was collected by the Akhbārī thinker Khalīl al-Qazwīnī (d. 1089 and on whom see below, p. 000). Whether these marginalia are distinct from Astarābādī's *shurūḥ* (commentaries) on the *akhbār* collections (see above, p. 165) is not clear.

[95] Apart from Mullā Khalīl al-Qazwīnī's efforts mentioned in the previous note, Astarābādī's marginalia on Ṭūsī's *al-Istibṣār* and *al-Tahdhīb* were collected by Muhammad b. Jābir al-Najafī (see Astarābādī, *al-Ḥāshiya alā Uṣūl al-Kāfī*, pp. 241–242 (editors introduction).

[96] Ibn Bābūya, *Man lā*, v. 1, p. 111.

[97] Astarābādī, *Ḥāshiya Man lā Yaḥḍuruhu al-Faqīh*, p. 460.

al-Ḥusayn b. Saʿīd).[98] More serious is Astarābādī's criticisms of *ḥadīth*. He considers a *ḥadīth* from Imam Jaʿfar:

> The first to place the sermon before the Friday Prayer was ʿUthmān.[99]

Astarābādī remarks that this "is a unique opinion [of Ibn Bābūya—*gharīb*] which we do not know of from anyone other than this author". According to the other collections of *ḥadīth*, ʿUthmān was the first to place the sermon before the ʿĪd prayers (but not the Friday prayers). On the subject of the ʿĪd prayers, Astarābādī criticises Ibn Bābūya's understanding of the word *sunna* in the *ḥadīth* from al-Imām al-Bāqir, "The prayer for the two ʿĪds is *sunna*". Ibn Bābūya states that *sunna* here "means they are the least of the obligatory duties (*sughār al-farāʾiḍ*)". Astarābādī, however, asserts:

> Interpreting the word '*sunna*' as being proven by something other than the text of the Book is better than this interpretation.[100]

Examples such as these are, in the main, aimed at Ibn Bābūya's own comments (rather than the *akhbār* themselves).[101] However, there are occasional doubts raised concerning the content of the *akhbār*, as well as corrections to Ibn Bābūya's recording of *isnāds*. Despite the correction, Astarābādī (in line with his Akhbārī principles) does not comment on the authenticity of reports. He merely notes occasionally that a report is unusual.

Whilst Astarābādī argues that the *akhbār* give knowledge "in the language of the people" (*lughat al-raʿiyya*) in *al-Fawāʾid*, he regularly feels the need to explain the meaning of reports. Phrases such as *yaʿnī* ("this means..."), *murāduhu* ("his intended meaning..."), *qaṣduhu* ("his intention..."), *kaʾanna* ("it is as if [to say]...") or *yufham min dhālika* ("from this it is understood that...") appear at the outset of nearly every comment. Whilst the *akhbār* are clear, there is a need for interpretation in order to make them so. Some of these comments are mere glosses, identifying the meaning of obscure words in the

[98] Astarābādī, *Ḥāshiya Man lā Yaḥḍuruhu al-Faqīh*, p. 461. Other examples of minor corrections in a the wording of *akhbār* can be found on p. 673 (*bi-qaṭʿ sayl/bi-baṭn musīl*, though he states that "God knows best" which is correct).

[99] Ibn Bābūya, *Man lā*, pp. 432–433.

[100] Astarābādī, *al-Ḥāshiya Man lā Yaḥḍuruhu al-Faqīh*, p. 478. Another example is Astarābādī's bold statement that Ibn Bābūya's interpretation of inheritance law is simply not correct (*laysa bi-ṣaḥīḥ*). See Astarābādī, *Ḥāshiya Man lā Yaḥḍuruhu al-Faqīh*, p. 510.

[101] See Gleave, "Between *ḥadīth* and *fiqh*".

reports. For example, al-Imām al-Ṣādiq was asked about a man who prays with a vial of *naḍūḥ* (*ṭawr fīhi naḍūḥ*) in his hand.[102] Astarābādī considers the term *naḍūḥ* here to be obscure, and he explains it as being a type of liquid perfume made from water infused with dates, sugar, cloves, apple or saffron and left to ferment. It is popular amongst the women of Mecca and Madīna and is called *naqūʿ* today (*fī hādhā al-zamān*).[103] On other occasions, Astarābādī feels the need to clarify a perceived ambiguity in a report. For example, the statement by al-Imām al-Ṣādiq that "It is permitted when travelling to rise [to prayer] without [making or hearing] the call to prayer"[104] is taken to refer to the second of the two prayers (*fī'l-ṣalāt al-thāniyya*, that is not the *ʿaṣr* but the *ʿishā'* prayer) which can be conjoined when travelling. The straightforward understanding of the report is that the call to prayer is not necessary at all when travelling.[105] Here, Astarābādī is adjusting the apparent meaning of a report to ensure it does not conflict with a previously decided legal position. This is, therefore, an example of the failure of his hermeneutic principles to lead to an inevitable legal opinion.

Astarābādī also rebuts accusations that he has misunderstood a report, as in his identification of the recipients of *zakāt* termed *al-mu'allafa qulūbuhum* in the Qur'ān (4.69). This term was subject to a number of interpretations in classical Imāmī jurisprudence,[106] and Astarābādī argues that it refers to those who have heard the call of the Prophet, accepted the unity of God, but ignored the remainder of the Prophet's message. They may receive gifts from the *zakāt* in order to secure the message "firmly in their hearts". This view is then challenged by an unknown interlocutor, who asks whether this contradicts another *ḥadīth* which talks of God "splitting the hearts" of those who do not obey his message. Astarābādī replies:

> No—this is because it is possible to interpret [the second *ḥadīth*] as referring to those who have discernment after having considered the call [of the Prophet] and what it demonstrates. They [that is, *al-mu'allafa qulūbuhum*] do not possess such discernment.[107]

---

102 Ibn Bābūya, *Man lā*, v. 1, pp. 254–255.
103 Astarābādī, *Ḥāshiya Man lā Yaḥḍuruhu al-Faqīh*, p. 467.
104 Ibn Bābūya, *Man lā*, v. 1, p. 291.
105 Astarābādī, *Ḥāshiya Man lā Yaḥḍuruhu al-Faqīh*, p. 468.
106 See Gleave, "Intra-Madhhab *Ikhtilāf*", pp. 127–134.
107 Astarābādī, *al-Ḥāshiya ʿalā Uṣūl al-Kāfī*, p. 389.

A potential clash in the *akhbār* is avoided by Astarābādī by him arguing that it is possible to interpret (*iḥtimāl al-ḥaml*) the reports as referring to different groups of people (the crucial difference here between being those who have heard and considered the Prophet's message and those who have simply heard it). Although Astarābādī has a procedure for dealing with conflicting *akhbār*, he clearly wishes to rely on legal nuance in order to avoid a conflict. His argument here is not particularly convincing and he himself admits that there is only a "possibility of interpreting" the *ḥadīths* in this manner. The fact that a mere possibility of interpreting the *ḥadīth* in a particular manner solves a potential conflict between reports does not (surely?) lead inevitably to the conclusion that this is what the reports actually mean.[108] In *al-Fawā'id*, Astarābādī criticised those *mujtahids* who rely on far-fetched explanations to join (*jamʿ*) together contradictory reports. There he argued that the contradictions should stand (one is *taqiyya* and the other not). When faced with *akhbār* contradictions in his commentaries on the *akhbār*, however, he seems more willing to re-interpret the reports to produce a harmony within the revelatory sources.

At times, then, Astarābādī's actual interpretive practice conflicts with his hermeneutic theory. He does, though, conceive of these commentaries as thoroughly "Akhbārī" works. Not only do the collections of *akhbār* (and commentaries upon them) supplant the need for works of *fiqh*,[109] Astarābādī also uses his commentaries to expound his Akhbārī principles. At the opening of his commentary on Ibn Bābūya's *Man La Yaḥḍuruhu al-Faqīh*, he embarks on a long exposition of Ibn Bābūya's use of the term *ṣiḥḥa*, explaining (as he did in *al-Fawā'id*) how this flatly contradicts al-ʿAllāma's use of the term. More explicit still is his commentary on the opening section of al-Kulaynī's *al-Kāfī*. Astarābādī outlines the four positions of the "*ʿulamā' al-uṣūl*" on those areas for which no revelatory indicator is

[108] For al-Baḥrānī's criticism of the joining of *akhbār*, see Gleave, *Inevitable Doubt* p. 114. At Astarābādī, *Ḥāshiya Man lā Yaḥḍuruhu al-Faqīh*, p. 455, Astarābādī states that a certain *ḥadīth* concerning the ritual wash has one of two intended meaning (*al-murād aḥad al-amrayn*). Other reports, however, confirm that the second of the two is correct. The details need not concern us here, what is clear is that the perceived plain meaning of other *ḥadīth* (*al-taṣrīḥ*) can effect the interpretation of the report under consideration.

[109] See above, pp. 39–40 and Astarābādī, *Jawāb*, p. 578. "I have chosen to comment on *al-Kāfī*, and I mention in [my commentary] what the author himself did not mention. This is sufficient."

found. These are (1) that God has no ruling here at all; (2) that God has revealed a general principle which can be applied here; (3) that God has revealed an indicator which is less than certain (*ẓannī*); and (4) that God has a ruling here, and that he has revealed a decisive indicator of that ruling. The *akhbār* cited by al-Kulaynī "demonstrate that the first three opinions are invalid, and the fourth is correct," according to Astarābādī. The fourth position is described as "that of the *Ahl al-Bayt* and our earliest Akhbārī colleagues."[110]

A final example of Astarābādī's interpretive practice that is available is his *fatwā* and subsequent treatise on the purity of wine.[111] This is Astarābādī's only *risāla* on a specific area of *fiqh*, and in it he mentions that he is merely applying here the method laid out in *al-Fawā'id al-Madaniyya*. The *risāla* was written in reply to criticisms of a *fatwā* he gave to Shāh ʿAbbās, in which he had declared that wine is not an impure substance, though, of course, its consumption is prohibited.[112] His evidence for this controversial opinion comprises twelve reports in which the Imams permit believers to pray in garments splattered by wine (he also makes mention of reports which declare wine impure, but says that these are fewer in number, and does not list them). If, so the argument goes, the Imams allowed prayer in wine-stained garments, then wine cannot in itself be impure (*najis*).[113] Even by the *mujtahid* tests of dissemination

---

[110] Astarābādī, *al-Ḥāshiya ʿalā Uṣūl al-Kāfī*, pp. 291–292. A full examination of Astarābādī's hermeneutic methodology in these *ḥadīth* commentaries awaits the publication of his other commentaries on *akhbār* collections.

[111] An edition of this *risāla* is found in Gleave, "Purity of Wine" and an edition of the *fatwā* can be found in Appendix 3.

[112] Details of the decree are to be found in the editor's introduction to Astarābādī, *al-Ḥāshiya ʿalā Uṣūl al-Kāfī*, p. 241 and pp. 247–248. In the *Risāla* itself he refers to Shāh ʿAbbās having become a wine drinker at an early age and worried that he was therefore unable to pray. He asked Astarābādī for a decree on this matter. Astarābādī stated that wine drinking was, indeed, forbidden, but that it was more serious for him to neglect prayer. He can, therefore, perform prayer even if his clothes have been splattered with wine. This view appears to contradict Astarābādī's own position (expressed in *al-Fawā'id*) and on which his theory of *iḥtiyāṭ* is based (namely, that performing a forbidden act is more serious than neglecting an obligatory one). See above pp. 80–82.

[113] Of course, Astarābādī's argument here relies upon an interpretive leap which he is not willing to highlight, since his hermeneutic explicitly avows interpretation. The leap from permission of the Imam being granted to pray in a wine-stained garment to the declaration that it is not impure is based on the assumption that the Imam would, on occasions, not have allowed prayer to take place in a garment which has come into contact with an impure substance. The argument is, of course, based on a demand for coherence in the law which is not dissimilar to that demanded by the partisans of *qiyās* and the Shīʿī *mujtahids*. Underlying Astarābādī's argument is the

(*shuhra*), the *ḥadīths* which declare wine impure (collectively) provide a stronger indicator of God's law on this matter. That there are slight discrepancies between the relevant reports gives Astarābādī the opportunity to embark on a detailed exposition of the process whereby contradictions within the *akhbār* are solved. A test as to which *ḥadīth* to accept is "agreement [or rather disagreement] with the Sunnis" (*muwāfaqat al-ʿāmma*). All the Sunnis say that wine is impure, and the *akhbār* from the Imams are contradictory. If this test is applied, then wine must be considered pure.[114] Furthermore, the Sunnis argue that the prohibition on the consumption of wine inevitably leads to the conclusion that wine is impure. This, he argues, is *qiyās* between purity (*ṭahāra*) and consumption (*akl*). He reiterates the standard Shīʿī position (that *qiyās* is a forbidden hermeneutic practice). Furthermore, since the Sunnis are unable to understand the argumentation which leads to the declaration that wine is pure (that is, they are unable to understand why *qiyās* is forbidden), disputation with them is pointless and to be avoided. Of all Astarābādī's writings on *fiqh*, this is perhaps the clearest expression of how the Akhbārī hermeneutic might be put into practice in the interpretation of *ḥadīth*. The match is not perfect,[115] but the *risāla* is a conscious effort to demonstrate the application of Akhbārī jurisprudence to a particular area of *fiqh* (rather than *uṣūl*). Sections of the *risāla* are, inevitably, discussions of the merits or otherwise of *uṣūl* issues (and not matters of *fiqh per se*). Since Astarābādī presents his position as based squarely on a particular approach to the sources of the law, anyone who disagrees with his conclusions, in fact, disagrees with his hermeneutic. Therefore, Astarābādī perceives the need not only to justify his *fatwā*, but also to justify the method by which he reached it. This technique was not always present in the treatment of similar issues amongst the *mujtahids*. Their method allows for more extensive differences of opinion arising from adherence to a single methodology.

---

belief (unexpressed, but nonetheless present, in the *Risāla*) that the above assumption is more likely to be the case than the assumption that anything which it is forbidden (to consume) is also impure.

[114] Note that Astarābādī is here carrying out a test to determine which of the reports are *taqiyya*, and which are not. In most cases, the tests (listed by al-Kulaynī and referred to by Astarābādī) cannot be carried out for practical reasons. However, in this case, all the Sunni schools of law agree that wine is impure, hence one can, here, compare the Imams' decrees with those of the Sunnis.

[115] See above, n. 114.

The *mujtahid*'s personal opinion is ultimately based on his personal assessment of the evidence (from sources, from reason and from the application of rational interpretive procedures to the sources). For this reason, *mujtahids*, when writing detailed treatises on individual legal issues, rarely feel the need to argue for their interpretive methodology and merely argue for their own opinions. Hermeneutic issues usually emerge in the *mujtahids' fiqh* works when their opponents' views are based upon an entirely different conception of the derivation of the law from the sources.

Conclusions

From the above analysis, a number of common characterisations of Astarābādī's theory can be corrected or modified. Astarābādī's Akhbarism was not anti-clerical. His legal theory appears (at first) to extend access to religious knowledge to the population at large in a quasi-Protestant manner. However, a detailed analysis of his theory demonstrates that Astarābādī wishes to retain the *muftī* and *qāḍī* as the only religiously authoritative figures. Where he differs from his Uṣūlī counterparts is in his assertion that knowledge of *ḥadīth* was their principal qualification.[116] He was critical of certain elements of the *ʿulamāʾ* class, but this was due to their failure to follow the *akhbār*, not because of their scholarly authority *per se*. Indeed, the *ʿulamāʾ* (aka *ḥadīth* scholars) were given the important roles of (not only) preserving the *akhbār*, but also of preserving their correct interpretation. They were not simply memorisers and collators of *ḥadīth*. In order to apply the legal regulations found in the *akhbār* to specific circumstances, some juristic (that is, interpretive) skills were necessary. Whilst Astarābādī does not call the scholars "*fuqahā*'" (that is, jurists), it is clear that individual *muftīs* and judges are personally responsible for deducing rulings when the sources are unclear or seem to offer no answer. The mechanisms of *iḥtiyāṭ/takhyīr*

[116] Whether this makes Astarābādī's theory one of "juristic" or "traditionist" authority is interesting (Stewart, "The Genesis"). As I present it here, there is still work for the scholar to perform in order to ensure the law is obeyed, and this is not merely policing the implementation of the law. Determining the law requires exegetical effort, and therefore I would maintain, Akhbārīs can still be considered jurists, at least in some senses of the word.

(caution/choice), the application of the procedures for solving *ḥadīth* contradiction and the advocacy of *tawaqquf* did not necessarily imply scholarly indolence.[117]

Furthermore, difference of opinion is quite possible in Astarābādī's theory, and (to a degree) is to be tolerated. Agreeably this tolerance is restricted to areas where the indicators conflict, are ambiguous or are not perceived by the *muftī*. In practice, however, one can imagine these categories comprising a large number of cases. Those who are yet to perceive a *dalīl* are not to condemn those who perceive one. Different opinions (one based on *iḥtiyāṭ* and one on a perceived *dalīl*) are, then, acceptable. Unfortunately, Astarābādī does not lay out criteria for identifying conflicting and ambiguous texts. Furthermore, it is not entirely clear whether general statements in the *akhbār* (such as "all things are pure until you know them to be impure") are applicable without exception, and if not, then how the exception process works.[118] Astarābādī's rather haphazard application of Akhbārī principles in his extant *fiqh* works is, perhaps, evidence of the difficulty inherent in consistently applying a juristic methodology which rejects, rather than embraces, interpretation. These areas of Akhbārī theory were, to an extent, expanded upon by later scholars.[119]

Finally, the characterisation of Astarābādī (and Akhbarism more generally) as promoting a rigid epistemology based around certain knowledge is not entirely accurate.[120] Astarābādī introduced a flexible concept of *ʿilm*, in which knowledge was divided into various categories based on both the object of knowledge (the ruling itself, the indicator in the text, the most cautious course of action) and the quality of this knowledge (unqualified *ʿilm* as against *al-ʿilm al-ʿādī*). It is perhaps the elasticity of Astarābādī's concept of knowledge which

[117] Stewart's criticisms of Moussavi's analysis mentioned above (n. 116) are based, of course, on a difference between jurists and *ḥadīth* specialists in which the former use reasoning and the latter do not. Astarābādī is, however, concerned primarily with the difference between Akhbārī jurists and *mujtahid* jurists. The former base their rulings solely on *ḥadīth* (though as we have seen, this requires exegetical effort on the jurist's part), and the latter on *ḥadīth* and a number of interpretive principles which aim to extend the meaning of the *akhbār*, filling those gaps in the law which have no (or uncertain) evidence in the *ḥadīth* corpus. See above, pp. 83–84.

[118] These issues are normally covered in works of *uṣūl al-fiqh* under the rubrics of *ʿāmm* and *khāṣṣ*, *muṭlaq* and *muqayyad*. See Gleave, *Inevitable Doubt*, pp. 175–178.

[119] See below, pp. 268–296.

[120] This point has already been noted by Kohlberg ("Akhbārī", p. 135).

prompted some later scholars to argue that the Akhbārī-Uṣūlī dispute was terminological (*lafẓī*): the Akhbārīs merely replaced the variable *ẓann* of the *mujtahids* with variable *ʿilm*. Finally, the admission that interpretive processes were necessary to comprehend the meaning of the *akhbār*, and the assertion that only certain types of scholars (i.e. *ḥadīth* experts) are qualified to perform this interpretation (thereby maintaining the *muftī*/layperson distinction) inevitably led later scholars to argue that while Astarābādī may have formally denied *taqlīd*, he actually recommended its use, and merely employed different terms to describe it.

CHAPTER FOUR

ASTARĀBĀDĪ'S THEOLOGICAL AND PHILOSOPHICAL THOUGHT

Astarābādī was best known for his writings on legal theory and, in particular, the denunciation of *ijtihād* and the *mujtahids* in *al-Fawā'id al-Madaniyya*. In terms of volume, however, most of his surviving work is in the area of theology and philosophy. Most of this output remains in manuscript form and has not, to my knowledge, been extensively used in previous descriptions of Astarābādī's thought. All his surviving "pre-Akhbārī" works concern theological and philosophical matters,[1] as does a significant proportion of work dating from after his conversion to Akhbarism. Indeed, one section of *al-Fawā'id al-Madaniyya* (a work usually cited for its juristic content) concerns matters of strictly theological and philosophical import.

The works consulted for the following summary of Astarābādī's theological and philosophical views comprise *risālas, ḥadīth* commentaries and "*fawā'id*" works. "*Fawā'id*" works are divided into separate sections, each named *fā'ida* ("useful comment"). Each *fā'ida* is best considered as a discrete study, and Astarabādī only occasionally shows concern for the relationship between the ideas of different *fā'idas* in a single *fawā'id* work. Neither comprehensive coverage of theological and philosophical issues, nor the construction of a coherent system of ideas appear to be his priority. *Al-Mabāḥith al-Thalātha, al-Fawā'id al-I'tiqādiyya, al-Fawā'id al-Makkiyya*[2] and *Dānishnāmah-yi Shāhī* all fall into the category of "*fawā'id*" works. Astarabādī's *risālas* (treatises) on particular theological issues include a *risāla* on the doctrine of *badā'*, a distinctively Shī'ī doctrine concerning God's ability to change his mind.[3] Finally, Astarābādī wrote a number of

[1] The only possible exception to this is his undated commentary on Ḥillī's *Ma'ārij al-Aḥkām* (see above, p. 38).

[2] By which I mean the work on theology and not Astarābādī's commentary on al-Ṭūsī's *al-Istibṣār*. See above, p. 38.

[3] The extant copy of this *risāla* is, in fact, a collection of Astarabādī's comments on *akhbār* relating to *badā'* alongside the commentary of other scholars. It is possible

commentaries on *ḥadīth* collections. The most significant of these in theological terms is his commentary on the *Uṣūl al-Kāfī* (the first section of al-Kulaynī's *al-Kāfī* in which theological reports from the Imams are recorded). In addition to these sources, both the theological section of *al-Fawāʾid al-Madaniyya* (entitled, "Errors of the Philosophers and Theologians of Islam")[4] and Astarābādī's answers to the questions posed by al-Shaykh al-Ẓahīrī[5] contain rudimentary arguments for his theological and philosophical doctrines.

My principal concern in the following analysis is to examine the extent to which Astarābādī's conversion to Akhbarism might be said to have influenced his theological and philosophical ideas (as opposed to his jurisprudence). As already established, his conversion to Akhbarism produced a new (or perhaps, revived an old) juristic method. Whether or not there are signs that Astarābādī considered the method equally applicable to both doctrinal and legal issues is a question I answer in the course of the following analysis. In short, was the Akhbārī method, as conceived by the founder of Akhbarism, considered simply a matter for law or was it to be applied to theological and philosophical speculation also?[6] To answer this question, one needs to trace Astarābādī's theological development, and this, unfortunately, is hampered by a number of factors. Firstly, not all of his literary output has survived, and when it has survived, it is not easily accessible. His refutation of Jalāl al-Dīn al-Dawwānī and Ṣadr al-Dīn al-Shīrāzī al-Dashtakī, for example, appears to be lost,[7] as does his linguistic work *Fawāʾid Daqāʾiq al-ʿUlūm*. Until such

he wrote another work with the same title which has not yet come to light. On this *risāla*, see below, pp. 107–109.

4 Astarābādī, *al-Fawāʾid*, pp. 471–515.

5 These are edited in Gleave, "Questions and Answers".

6 I am not so concerned here with an author's formal description of his theology as being identical with that of the Imams. This was a common (perhaps inevitable) motif for Shīʿī writers, and can usually be seen as decorative. There is nothing distinctly "Akhbārī" in such a claim, as all Shīʿī authors, explicitly or implicitly, claim to propose "the Imam's theology". More significant for my argument here is whether the reasoning employed in demonstrating the validity of theological and philosophical doctrines is reliant on revelatory sources and their interpretation, or doctrines are primarily justified by rational argumentation (with revelatory sources used as embellishments). On this, with respect to Astarābādī, see below, pp. 104–107.

7 A large number of criticisms of al-Dawwānī and al-Shīrāzī can be found, however, in the extant works of Astarābādī. Al-Dawwānī, in particular, is the subject of vehement criticism found in works written both before and after Astarābādī's conversion to Akhbarism. See below, pp. 126–127.

works are found, a full account of Astarābādī's theological views is not possible. Secondly, a number of Astarābādī's works cannot be dated accurately, and a chronology of his compositions (such as that given in the Chapter 2)[8] is inevitably provisional. This tempers any assessment of the development of both his compositional skills and his religious views, though I make some provisional remarks on the basis of the extant data in the conclusion of this chapter.[9] Thirdly, Astarābādī wrote works of different literary genres (*uṣūl al-fiqh, kalām*, commentaries, answers to questions and short *risālas* on specific issues). Whilst at times topics from earlier works re-appear, only a handful of theological issues are subject to sustained and continuous examination throughout Astarābādī's academic career. These recurrent discussions provide useful, though inevitably limited, evidence in a description of Astarābādī's intellectual method and its development. One such issue is examined below in detail, and forms the evidence base for my conclusions.[10]

Astarābādī's Approach to Theology and Philosophy

It is, perhaps, surprising that Astarābādī wrote any works of *kalām* or *falsafa* after his conversion to Akhbarism. When asked about the validity of these academic disciplines by al-Shaykh al-Ẓahīrī, he wrote:

> The *akhbār* from the pure Imams are transmitted in a widespread manner saying that it is forbidden to rely upon the thoughts of the intellect; it is forbidden to study the science of *kalām* and to teach it unless one is referring to the *kalām* derived from Their [the Imams'] words.[11]

However, a survey of Astarābādī's theological output demonstrates that whilst he may have maintained a rhetoric critical of theology and philosophy, he was quite willing, on occasions, to enter into theological and philosophical argumentation without any reference to revelatory material. His mastery of philosophical and theological

---

[8] See above, pp. 31–60.

[9] See below, pp. 137–139.

[10] See below, pp. 117–137, on the question of *ʿilm al-Wājib biʾl-mumkināt*.

[11] Astarābādī, *al-Fawāʾid*, p. 571 (see also Gleave, "Questions and Answers"). See also, Astarābādī's rejection of logic as a useful tool in juristic deduction found in Astarābādī, *al-Fawāʾid*, p. 471.

terminology is demonstrated in these discussions, and it is clear that he had studied both *falsafa* and *kalām* in some depth. The standard structure of his approach to philosophical and theological issues is common to much Muslim theological argumentation:

1. Outline the different scholarly opinions on the issue at hand.
2. Select one of these scholarly opinions (for Astarabābādī, Jalāl al-Dīn al-Dawwānī's opinion is usually selected), and cite the relevant passage from this scholar's works.
3. Demonstrate the incoherence of this scholar's opinion.
4. Propose a coherent alternative.[12]

In his criticisms of past scholarly discussions (section 3 above), Astarābādī is usually concerned with the appropriate manner in which certain theological doctrines might be expressed. Hence when discussing God's attributes, the concern is not to determine whether God "knows", but rather to find the correct way in which one might say that he has the attribute of knowledge.[13] Similarly, the discussion over free-will and predestination concerns what it might mean to describe an entity (God or human) as "free" or having "will".[14] Of course, these linguistic concerns are uppermost in Astarābādī's mind because he considers language to be the main tool provided for us by God with which we might describe both him him and his laws.[15] We must then understand how language works, and use it correctly in our attempt to formulate theological doctrine.

Furthermore, Astarābādī's discussions cannot generally be described as distinctively Shīʿī; central Shīʿī doctrines (such as, the Imamate) are not discussed at any great length. Instead, Astarābādī's focus is upon the broader issues that a monotheistic theology poses for philosophical analysis: proofs for God's existence, exploration of God's nature, depictions of his attributes and the problem of predestination and free-will. His discussion of these issues is almost always derivative (that is, unlike his juristic thought, his theological work does not appear to be particularly ground breaking), and couched in

[12] One finds this pattern throughout *al-Mabāḥith, al-Fawāʾid al-Iʿtiqādiyya* and *al-Fawāʾid al-Makkiyya*. The *Dānishnāmah-yi Shāhī* (on which see below, pp. 134–137), represents a shift from a reactive presentational model to an independent discussion of the issues at hand. However, this is patchy, and this model of presentation still represents Astarābādī's dominant mode of presentation.

[13] See below, pp. 136–137.

[14] See below, pp. 114–115.

[15] See above, pp. 72–74.

philosophical or theological terms which do not demand the reader to have an explicit commitment to Islam, let alone Shiʿism. References to the Imams and reports of their theological doctrines are infrequent, and when they do occur can be described as decorative or (*ex post facto*) justificatory. He does not allow *ḥadīths* related from the Imams to control his discussion. Rather, it is the tradition of theological and philosophical enquiry which dictates the structure of his discussion, and references to the *akhbār* are introduced only after he has derived his position and demonstrated his conclusions.[16]

This does not, however, mean that there is no development in Astarābādī's method of presentation. One finds more regular reference to the *aḥādīth-i ahl-i bayt* in *Dānishnāmah-yi Shāhī* than one does in his earlier works. However, as will be demonstrated below, the introduction of this revelatory material did not affect either Astarābādī's conclusions, or his argumentation. Rather, an analysis of his treatment of a single doctrinal problem (*ʿilm al-Wājib*) at different points in his life, shows that Astarābādī reached his final conclusions through argumentation independent of any consideration of revelation. These conclusions, presented in *al-Fawāʾid al-Makkiyya*, are reproduced in his only post-conversion theological work, the *Dānishnāmah-yi Shāhī*. There is, indeed, some development in Astarabādī's presentation. In the *Dānishnāmah*, for example, he writes in (relatively) straightforward Persian rather than the frequently abstruse and technical Arabic of his earlier works. Furthermore, he is more willing to relate his conclusions to revelatory material in a formal manner, though few *ḥadīth* are cited even in *Dānishnāmah-yi Shāhī*. The connection to revelatory material is expressed through the insertion of phrases such as "the *akhbār* of the Imams clearly support" a conclusion (the principal proof for which consists of rational argumentation). There is, however, little evidence of doctrinal development between *al-Fawāʾid al-Makkiyya*

[16] This is the case even when, structurally speaking, a reference to an Imam precedes the discussion. See, for example, the reference to Imam Jaʿfar at the outset of the discussion of *ʿilm al-Wājib* in *al-Mabāḥith*, f.3a.6–7. This feature does not seem to have been influenced by Astarābādī's conversion to Akhbarism; see, for example, the reference to the "*aḥādīth-i ahl al-bayt*" at the outset of the different modes of existence in *Dānishnāmah-yi Shāhī*, f.30a.11–12. As we shall see, there is little evidence of a change in theological doctrine brought about by an increased reliance on revelatory material as a substitute for argumentation. The relevant statements of the Imams are rarely cited in full, and only occasionally analysed directly for theological or philosophical content. Only in works designed as commentaries on *ḥadīth* collections is this pattern broken.

(composed 1018) and *Dānishnāmah-yi Shāhī* (composed sometime after 1032). Astarabādī's juristic Akhbarism calls for a greater reliance on the *akhbār* as legal proofs. His conversion to Akhbarism may have brought about a greater prevalence of decorative *ḥadīth* references within his theological works, but it does not appear to have altered either his theological conclusions or his method of enquiry. His later theological compositions, then, demonstrate a common feature of Islamic theological writings: namely, the inability of revelation (be it Qur'ān, *sunna* or the *akhbār* of the Imams) to provide unequivocal dogmatic guidance such that revelation can be used to justify a variety of conflicting theological views.

There are, however, examples where revelatory material controls Astarabābādī's presentation (though not necessarily his conclusions). The most obvious examples of this are his commentaries on collections of *akhbār*, mentioned previously.[17] As is indicated below, however, Astarābādī's commentaries on theological *akhbār* show little sign of a distinctive "Akhbārī" exegetical method. In fact, his choice of genre in these cases (that is, *akhbār* commentary) does not lead to an exegesis which might be termed Akhbārī in style or conclusions. There is, for example, little emphasis on a "plain" or literal reading of the texts and the *akhbār* are most often springboards for complex theological discussions, characterised by sophisticated philosophical argumentation.[18] One work which can be included in the category of *akhbār* commentary—even though it presents itself as an independent treatise, is his *al-Risāla fī'l-Badā'* and some comment on this work is pertinent here. It is not clear that the work given the title *al-Risāla fī'l-Badā'* by manuscript cataloguists is, in fact, a separate treatise authored by Astarabādī. The only surviving copy is found in Mashhad, and has, as its colophon:

> Here ends the marginalia (*ḥawāshī*) from the comments of our Mawlā Muḥammad Amīn al-Astarābādī and others concerning *badā'*. I[19] copied them from the copy of one who copied them from the copy of...Muḥammad b. al-Ḥasan al-Ḥurr al-ʿĀmilī on 8th Jumāda al-Thānī 1129.[20]

---

[17] See above, pp. 37–38.
[18] See below, pp. 118–119.
[19] The actual copyist is not named here or elsewhere in the *Risāla*.
[20] Astarābādī, *Badā'*, f.16b.17–20.

At one point in the text of the treatise, the author/collator concludes a comment on a *ḥadīth*, saying: "This is one of the useful comments made by al-Marḥūm... Mawlānā Muḥammad Amīn, may God have mercy on him."[21] The indication from the colophon is, then, that this work is, in fact, a selection of Astarabādī's comments on *badā'*, collected by al-Ḥurr al-ʿĀmilī, the surviving manuscript being a second generation copy of this collection. The work, as we have it, is not a treatise by Astarabādī then, but a collection of his comments (along with those of other scholars) on the *akhbār* relating to *badā'*.

Astarābādī's comments on *akhbār* are followed in the treatise by the comments of other scholars (al-Karājikī, al-ʿAllāma al-Ḥillī, al-Shaykh al-Ṭūsī and others), and it is unclear whether these comments were included in the work by Astarābādī or al-Ḥurr. Furthermore, the work has no introduction. The *basmala* is followed by a barrage of *ḥadīth* citations with no *ammā baʿd* or identification of the author. Al-Ḥurr al-ʿĀmilī mentions, in his *Amal al-Āmil*, that he has seen a treatise on *badā'*, written by Astarābādī,[22] and whilst the surviving manuscript may contain part (or perhaps all) of this treatise, it is certainly not identical to it. The comments attributed to Astarābādī in the so-called *al-Risāla fī'l-Badā'* are also not identical to those found in his commentary on *Uṣūl al-Kāfī* on the same *akhbār*[23] (though they do not contradict each other).

There are, it seems, two possibilities here: either the surviving manuscript is a copy of Astarābādī's marginal comments (*ḥawāshī*) on the *ḥadīths* related to the theological problem of *badā'*, collected with others and included in single treatise by al-Ḥurr al-ʿĀmilī (or someone previous to him), or there was indeed a *Risāla fī'l-Badā'* by Astarābādī which is lost, and al-Ḥurr (or the collator) quotes from this work in his collection. Whilst the surviving work displays a quite different method to Astarābādī's other theological works (with revelatory material controlling the discussion), it is not clear that it represents an independent treatise as such. It may be better classified as a collection of Astarābādī's comments on *ḥadīth*, and so whilst it

---

[21] Astarābādī, *Badā'*, f.4b.5–6. The quote indicates both that Astarābādī is dead at the time of the treatise's collation and that the comments are taken from a variety of different writers, and not Astarābādī alone.

[22] Ḥurr, *Amal*, v. 2, p. 35. Indeed, the only evidence we have for this treatise on *badā'* is al-Ḥurr al-ʿĀmilī's reference here. When reference is made to the treatise in later works, it is contained within a citation from al-Ḥurr.

[23] Astarābādī, *al-Ḥāshiya ʿalā Uṣūl al-Kāfī*, pp. 319–320.

may appear to be an exception to the above characterisation, it is not.

In conclusion, then, Astarābādī's method of presentation in his works of theology and philosophy is dominated by rational argumentation. *Al-Risāla fi'l-Badā'* and Astarābādī's *ḥadīth* commentaries provide exceptions to this general characterisation, as revelatory material forms the controlling element of the presentation. However, even here (as we shall see) rational argumentation predominates, as Astarābādī clearly considers the theological *akhbār* to be in need of rational exegesis.

Astarābādī's Theological Views

As mentioned above, Astarābādī did not write a systematic work of *kalām* or *falsafa*, nor anything approximating to a creed. His philosophical theology, therefore, cannot be constructed with perfect coherence. Of course, Astarābādī may never have intended to produce a coherent theology, and for us to attempt to do so may be disingenuous. Furthermore, his opinions on certain issues may have changed, and it is possible that we only possess earlier (or later) views, giving an impression of constancy when there may have been significant (but unrecorded) change. It may, then, be more appropriate to talk of different theologies, held at different points in his life, though my analysis below reflects the patchy nature of the sources. Harmonising one doctrine with another seems to have been abandoned in his later works, in favour of a concentrated theological and philosophical interest in the elucidation of individual doctrines. The disjointed structure of his theological writings, jumping from topic to topic, indicates a writer who is interested in discovering and analysing past arguments, but not in synthesising his own doctrinal positions. He concentrates on theological precision (investigating the most appropriate manner in which a doctrine might be expressed, for example) rather than a system of theological beliefs. One issue (namely, God's knowledge) was examined on at least four occasions, and these treatments are analysed in the next section. Firstly, however, I provide a brief overview of Astarābādī's theological views, notwithstanding the incompleteness of the material available to us.[24]

---

[24] The rather patchy coverage of his writings means that for many issues, if

In general terms, Astarābādī's philosophy is Avicennan (or more particularly, a version of Naṣīr al-Dīn al-Ṭūsī's interpretation of Ibn Sīnā); his theology owes much to the incorporation of Muʿtazilī ideas into Imāmī Shiʿism.[25] These elements combine to produce interesting, and sometimes novel, hybrid doctrines.[26] In terms of proofs for God's existence, Astarābādī describes God as the first cause—not in temporal terms, but in terms of the "need" of all contingent (*mumkināt*) entities for an active cause (*ʿilla fāʿiliyya*). Reason dictates that the contingent entity's need (*iḥtiyāj*) for a cause is not only generative but constant, and that the existence of contingent entities is maintained by God willing them to be so (*aʿdām al-mumkināt muntahiya ilā ʿadam irādat al-Wājib li-wujādātihā*).[27] The reason why God might no longer will the existence of a contingent entity is because it no longer conforms to his intended good (*al-maṣlaḥa al-maqṣūda*). One can see here a certain amount of philosophical expertise, in the form of a rough and ready reproduction of Ibn Sīnā's famous proof. However, it is blended with Muʿtazili doctrines concerning God continually willing the good for his creation. Furthermore, it accords with Astarābādī's conception of *badāʾ*, in that God's "decrees and desires are renewed to him each day, on account of the benefits (*al-maṣāliḥ*) which he perceives."[28] His perception of the benefits may change, and therefore his "decrees and desires" may also change.

Astarābādī changed his position during his life, there is rather scant evidence on which to draw. In what follows, I concentrate on arguments and positions present at least once in Astarābādī's writings, noting any discernable development at the appropriate points.

[25] Of course, such a characterisation of Astarābādī's theology demonstrates the stark difference between his own position and that of the so-called early *akhbāriyya* (see above, pp. 14–16). The earlier *akhbāriyya* are recorded as having critical of *kalām* as a discipline, being the opponents of the *kalāmiyya*. Notwithstanding Astarābādī's own rhetoric concerning *kalām*, his writings demonstrate learning and interest in theological and philosophical subjects.

[26] See below, concerning his ideas on God's knowledge, pp. 117–137.

[27] Astarābādī, *al-Fawāʾid al-Makkiyya*, f.2a.5–6. For a comprehensive discussion, see f.1b.6–2a.8. In *al-Fawāʾid al-Madaniyya* (p. 508), Astarābādī also argues against the theologians who say that the existence of the world does not prove that God is its maintaining cause (*mūjib*).

[28] Astarābādī, *al-Ḥāshiya ʿalā Uṣūl al-Kāfī*, p. 309. That such a position might imply change in the Sharīʿa (either by *naskh* or some other means) is not explored, further indicating the separation of jurisprudence and philosophy in Astarābādī's writings. Concerning *badāʾ* more generally, see Ayyoub, "Divine Preordination and Human Hope", though Ayyoub's analysis is entirely based on revelatory material and does not enter into the theological problem of how an omniscient being can change its mind.

Concerning God's attributes (that is, the debate concerning their subsistence in or identity with God's essence), Astarābādī claims in the *Dānishnāmah-yi Shāhī* that both rational and revelatory proofs demonstrate that God's attributes are identical with his essence:

> The doctrine of the philosophers and the Muʿtazila is that it is absurd to propose that there are additional attributes which subsist in the essence of God, the Holy One. They mention rational proofs of this position. The statements of the Ahl al-Bayt also explicitly confirm the validity of the philosophers' position.[29]

These attributes, which are not entities subsisting in God's essence but his essence itself, are divided into two types: essential attributes (*ṣifāt-i dhātī*) and attributes of his actions (*ṣifāt-i afʿāl, ṣifāt al-fiʿl*). This distinction is, of course, known from Muʿtazilī *kalām*, and Astarābādī's formulation of the distinction, found in his commentary on *al-Kāfī*, owes much to the Muʿtazilī ideas which seeped into Imāmī *kalām*:

> Any attribute for which both the attribute and its opposite exist in God's truth, is one of the *ṣifāt al-fiʿl*. Any attribute which is not like this is one of the *ṣifāt al-dhāt*.[30]

The distinction can, then, be summarised:

1. *ṣifāt al-dhāt*: Some of God's qualities cannot be negated and then applied to him (that is, they are his essential attributes). These include attributes such as power and knowledge. God cannot be both powerful and weak; neither can he be both knowledgeable and ignorant.
2. *ṣifāt al-fiʿl*: Other attributes are susceptible to negation and yet can still be applied to him, such as being the "approving one" and being the "disapproving one" (that is, both the attribute and its negative can be attributed to God).[31]

The reason why the attributes of God are divided in this way is due to their different origins (*manshaʾ*):

1. The attributes which can be ascribed to God on account of his essence (*ṣifāt al-dhāt*) have their origin in God's essence alone.

---

[29] *Dānishnāmah*, f.32b.3–5. Here revelatory material appears as a tie-breaker, ending the deadlock between theological camps, and supporting one over the other.

[30] Astarābādī, *al-Ḥāshiya ʿalā Uṣūl al-Kāfī*, p. 309.

[31] Astarābādī, *al-Ḥāshiya ʿalā Uṣūl al-Kāfī*, p. 309.

2. The attributes which can be ascribed to God on account of his actions (*ṣifāt al-fiʿl*) have their origin in God's essence in relation to the things he creates.

God cannot be other than his essence, though he can, if he desires, do other than his actions.

Both categories of attributes are merely verbal descriptions of God's essence, and do not represent "qualities" which subsist in him. However, the mode of definition of the attributes differs between the *ṣifāt al-dhāt* and *ṣifāt al-fiʿl*:

> The meaning of the "powerful one" (*al-qādir*) is not "he in whom power subsists" (*man qāma bihi al-qudra*) and the meaning of the "knower" is not "he in whom knowledge subsists". Rather all they mean is "he who is not weak" and "he who is not ignorant".[32]

The essential attributes, then, are best described apophatically. To say God is powerful is merely to say he is not weak and so on.[33] It does not mean that he has a quality or attribute of "powerfulness". In the case of the *ṣifāt al-fiʿl*, Astarābādī defines these as follows:

> It is possible to refer all the *ṣifāt al-fiʿl* to an existential meaning. For example, the meaning of "wanting" or "will" or "decree" [when applied to God] is merely that he creates writing on the heavenly tablet which are given these names.[34]

As in Muʿtazilism, the simple unity of God precludes the existence of any attributes subsisting in his essence. Rather, attributes describe God's essence (*ṣifāt al-dhāt*) as being of a particular character or him doing (*ṣifāt al-fiʿl*) particular things. Whilst Astarābādī claims that this theological position is derived directly from the *aḥādīth* of the Imams,[35] it is clear that it is a reproduction of the Basran Muʿtazilī doctrine attributed to Abū ʿAlī al-Jubbāʾī and his followers, and portrayed as having an Imāmī pedigree later.[36] Apart from this claim,

---

[32] Astarābādī, *al-Ḥāshiya ʿalā Uṣūl al-Kāfī*, p. 309.

[33] A full discussion of God's attribute of "powerfulness" is found in *al-Fawāʾid al-Makkiyya*, f.16b.8–26a.5.

[34] Astarābādī, *al-Ḥāshiya ʿalā Uṣūl al-Kāfī*, pp. 309–310. See also below, n. 42, on the question of *badāʾ*.

[35] Astarābādī, *Dānishnāmah*, f.33a.3–4. Indeed his discussion of the *ḥadīths* in *al-Kāfī* concerning God's essential and other attributes (Kulaynī, *al-Kāfī*, v. 1, pp. 107–108) is intended to support such a derivation (Astarābādī, *al-Ḥāshiya ʿalā Uṣūl al-Kāfī*, pp. 309–311).

[36] On which, see Van Ess, *Theologie und Gesellschaft*, v. 4, pp. 436–438.

there is little distinctively Shīʿī about Astarābādī's claim. Similarly, the Muʿtazilī (and later, Shīʿī) doctrine of *badāʾ* is linked with the division of the attributes of God. The attributes which are involved in the process of him performing an act are listed (*mashīʾa, irāda, qadar* and *qaḍāʾ*). These, Astarābādī explains, are *ṣifāt al-fiʿl* because it is possible for their nature to change. It is possible for God to desire and wish different (and contradictory) things at different times.[37]

Similarly, Astarābādī argues for an essentially Muʿtazilī position with regard to *al-taḥsīn waʾl-taqbīḥ* (that is, the dispute over whether moral qualities are externally extant or "placed" upon an action by a decree of God). In the *Dānishnāmah-yi Shāhī*, Astarābādī outlines the basic parties to the dispute. The Muʿtazilīs, philosophers and Imāmīs argue that good and evil are essential to action (*dhātī*). Reason can perceive them to be beneficial or distasteful.[38] The Ashʿarīs disagree saying that God determines the moral qualities of these actions (and the one who performs them). Astarābādī's position is that of the Muʿtazila, though he refuses to go so far as to identify these essential characteristics as necessarily linked to the five legal categories (namely: obligatory, recommended, permitted, disapproved and forbidden). Some Muʿtazilīs, he claims, have done this, and some misguided Imāmīs have followed them in this. For Astarābādī, reason can recognise the good and bad qualities in actions, but the five categories are God's own deontic classification. An action, then, has two sets of moral qualifications, the moral (*ḥusn/qubḥ*) and the legal (*aḥkām-i khamsah*). The latter are not essential to the action, but are "brought about by the action of God" (*aḥkām-i khamsah...dhāt-i afʿāl nīst balkih bi-jaʿl-i shāriʿ ḥāṣil shud*). Once again, this position is attributed to a plain reading of the *aḥādīth* of the Imams.[39] In *al-Mabāḥith*, there is also

[37] Astarābādī, *Badāʾ*, f.4b.9–10.

[38] Astarābādī seems unconcerned about which terminology one chooses to describe this essential nature: "Some actions are of such a condition that reason and wisdom (*ʿaql va ḥikmat*) judge their *māhiyyah-yi nawʿiyyah, jawhar-i shakhṣiyyah* or *ṣifāt-i intizāʿiyyah* [all of which can be translated here as "essential nature"] and the one who performs them to be distasteful (*nifrat kunad*)." (Astarābādī, *Dānishnāmah*, f.43a.12–14).

[39] Astarābādī, *Dānishnāmah*, f.43b.12–13. There is, perhaps, a link here with Astarābādī's juristic method, though it is not made explicit. If the five categories are not inherent to an act, then they can be adjusted in accordance with circumstances much more easily. The principal example of this is *taqiyya*. In his *al-Fawāʾid al-Madaniyya* (see above, p. 84), Astarābādī argues that believers are justified in following *taqiyya ḥadīth* even though the rules contained within them contradict the true Sharīʿa. In

a discussion of the problem of how God "punishes and rewards the actions of his servants" (that is, human beings). In this, the external qualities of good and evil (and their potential recognition by the *ʿaql*), are proposed as the means whereby God can punish a person who has not come into contact with the law.[40]

The issue of punishment and reward, central themes of Muʿtazilī theology and adopted by Imāmīs after the *ghayba* of the Twelfth Imam, naturally impinge upon the question of human free-will. Astarābādī's position, once again, accords with the dominant Muʿtazilī and Imāmī opinion concerning a human being's ability to create their own acts. An agent, however, does not simply choose to perform an action in any random fashion. Rather, there must be within him a "will" to perform that action. This will (*irāda*) to perform the action comes about because of a motive (acting as a sort of prompt—*dāʿī*) within him. Any free action is dependent upon the agent's will, and that will is dependent upon the presence of a prompt. The prompt is a knowledge of the final cause (*al-ʿilla al-ghāʾiyya*) of the action (that is, the reason for its performance and what it is intended to achieve). There is a causal chain here, for when all the prerequisites of an action being performed are present, and there is nothing to bar it coming into existence, the agent performs the action. Any free action, then, is a combination of the agent, his knowledge of the ultimate effect of the action, and the will to perform it. To say that it is possible for an agent to perform one of two courses of action is to say that it is possible for these component parts of an action to be differently arranged. Now God knows how these components will be arranged on any particular occasion, but the agent does not act in a particular manner because God knows (in an eternal sense) this arrangement:

> God knows that the agent will act in a particular way. [But] the agent does not act in a particular way because God knows.[41]

In sum, then, Astarābādī's position is that whilst God has foreknowledge of an agent's actions, the agent can still be said to have power

---

this way, the believer may perform a morally reprehensible act which is, nonetheless, legally permitted (and perhaps obligatory).

[40] Astarābādī, *Mabāḥith*, f.39b.3–44b.10. In the course of the discussion al-Dawwānī's position is discredited and that of Naṣīr al-Dīn al-Ṭūsī receives qualified support. See generally on this question, Reinhart, *Before Revelation*.

[41] Astarābādī, *al-Fawāʾid*, p. 503.

over his actions in any meaningful sense of the phrase.[42] We are, it seems, constrained by language to use words such as "power" in a particular manner, and once this constraint is recognised, describing agents as having power over their actions is unproblematic.[43] Our actions, then, can truthfully be ascribed to us, hence we are judged to be the cause of them and hence the responsibility for their performance lies with us.

Finally, Astarābādī's doctrine of prophecy accords with commonplace Muʿtazilī and Imāmī theology. Whilst the Prophet is recognised by his miracles (and this constitutes the only means of determining a Prophet's identity), these are not in themselves the reason for God sending prophets.[44] Rather the Prophet brings knowledge of the five legal categories which overlay the moral categories of good and bad recognised by human reason. Before the Prophet's arrival, agents can be punished or rewarded for performing actions which are bad or good according to reason (*al-taḥsīn wa'l-taqbīḥ*). After his coming, they are to be punished or rewarded in accordance with the manner in which they respond to his message and the new, legal classification of acts.[45] With regard to the necessity for the Imamate, Astarābādī also utilises primarily rational proofs, and these are laid out in the *Dānishnāmah-yi Shāhī*.[46] God does not wish there to be discord within his community, hence he must provide an individual whose decrees are not personal opinion (*ra'y*) but are the truth. For this reason, the person selected must be free of error (*maʿṣūm az ghalaṭ*), and

[42] In terms of God's knowledge of his own future actions, this is subject to *badā'*. According to Astarābādī, God writes upon the preserved tablet events which will certainly happen and events which will happen "if I will" it (*in shi'tu*). The latter are potentially subject to *badā'*. See Astarābādī, *al-Ḥāshiya ʿalā Uṣūl al-Kāfī*, p. 320.

[43] That this position is ultimately determinist is argued by al-ʿĀmilī in his comments on *al-Fawā'id al-Madaniyya* (p. 503). The prompt, it seems, does not merely make an action more likely to take place. It necessitates its occurrence.

[44] In his answers to al-Shaykh al-Ẓahīrī, Astarābādī mentions that the Prophet's miracles compel an individual who is true to himself to accept his prophecy (See Astarābādī, *Jawāb*, f.2b.5, Astarābādī, *al-Fawā'id*, p. 569 and Gleave, "Questions and Answers").

[45] Astarābādī's summary of these positions is to be found in Astarābādī, *al-Mabāḥith*, f.39b.3–44b.10 (third *baḥth*) and the relevant section in the *Dānishnāmah* (f.43a.11–43b.13). There is a lack of general argumentation for them, he merely states his position, and refutes his opponents.

[46] The relevant sections are the 35th and 36th *fā'idas* (Astarābādī, *Dānishnāmah*, f.46a.7–17 and f.46a.17–52a.14 respectively).

his word must be that of God's revelation and not the opinion of men (*kalām-i ū az rū-yi vaḥy-yi ilāhī nah ra'y-yi bashar*).[47] Once this is established, the debate over the identity of this individual is primarily historical, and that the Prophet designated ʿAlī, and ʿAlī designated Ḥasan (and so on down the list of twelve Imams) is, for Astarābādī, the best attested historical record. Furthermore, God taught (*taʿlīm kard*) ʿAlī and his descendents all the injunctions of the law, through his intermediary, the Prophet (*bi-vasāṭ-i sayyid al-mursalīn*).[48] The presence of such an Imam is, then, a rational necessity, since God's people must be able to access the law. This is the case even if the Imam is not able to operate freely. In the event of him not being able to operate (an extreme example of this is the *ghayba*), it is also a rational necessity that the Imam provide his people with tools by which sufficient knowledge might be gained. These tools are, of course, the books of the Imams' sayings and actions, which by rational necessity therefore, must be accurate reflections of the Imams' words and deeds. One sees here how Astarābādī's doctrine of the Imamate begins to blend with his Akhbārī jurisprudence. In his jurisprudence, there is an historical justification for the probative force of the *akhbār* of the Four Books. Here a rational justification for their accuracy is also established. The two doctrines cohere, although the coherence is not explicitly stated. On the question of so-called extremist (*ghālī*) beliefs, Astarābādī, in his commentary on *al-Kāfī*, naturally does not challenge the authenticity of the *ḥadīths* which talk of the Imams being addressed directly by angels,[49] being created from spirits in a quite different manner to the rest of humanity and as having a different bodily character to other human beings. However, his primary emphasis in discussion of *imāma* is the Imam's knowledge (particularly, of the law), which the Imam gains through a number of mechanisms (from God directly, from angels, from the Prophet—as teacher, whether physically present or not—or through the previous Imam). These views are broadly in line with the Twelver Shīʿī theological orthodoxy of the time. Astarābādī's Akhbarism is, therefore, quite distinct from the *akhbariyān* mentioned in earlier texts (such as al-Qazwīnī's *Kitāb al-Naqḍ*). There, a link was made between

[47] Astarābādī, *Dānishnāmah*, f.46b.4.
[48] Astarābādī, *Dānishnāmah*, f.50a.3.
[49] On this doctrine, see Kohlberg, "Muḥaddath".

the *akhbāriyya* and the *ghulāt*;[50] here Astarābādī's views squarely accord with non-*ghālī* Twelver conceptions of the Imamate.

Astarābādī's Philosophical and Theological Development

An issue which occupied Astarābādī's thoughts on more than one occasion concerns God's knowledge. As is well known, this topic was an area of dispute between theologians (*mutakallimūn*) and philosophers (*falāsifa*). In general terms, the philosophers were accused of constructing the conception of a god who was not intimately involved in the workings of creation, being a remote "primary cause" or "necessary being". This distance from creation implied not so much disinterest, but an inability to know and intervene in the operations of his creation. God's knowledge, then, was to be restricted to "generalities", and the specifics (particulars) of the workings of the world were unavailable to him. This was, of course, one of the points on which al-Ghazālī declared the philosophers unbelievers, and was the subject of a virulent response by Ibn Rushd.[51] Astarābādī, writing firmly in the tradition of Ibn Sīnā's Aristotelianism, focuses his concern on God's knowledge of possibly existent things (*mumkināt*, hereon "possibles"). The term could also be translated as "contingent entities", however, Astarabādī is clearly interested in God's knowledge of things which may not actually exist (in the external sense), but could exist (if God were to will them so).

On four separate occasions in his theological works, Astarābādī discusses the nature of God's knowledge, concentrating on God's knowledge of possibly existent entities (*ʿilm al-Wājib biʾl-mumkināt*) on each occasion. His earliest examination is in *al-Mabāḥith al-Thalāth* completed in 1014. He tackles the issue in *al-Fawāʾid al-Iʿtiqādiyya* (completed in 1017) and again in *al-Fawāʾid al-Makkiyya* (written in 1018). His final treatment of the subject is found in his Persian work *Dānishnāmah-yi Shāhī*—probably completed in the early 1030s. In the last of these he accurately records that he "has studied this question for many years, and consulted [both] the *ḥadīth* of the Ahl al-Bayt and the thought of the philosophers and

---

[50] See above, pp. 18–19.
[51] See Davidson, *Medieval Islamic Philosophy*, pp. 108–120.

theologians."[52] He credits his final solution to the problem to "God placing an understanding of the issue in his heart", rather than any rational breakthrough. Astarābādī's preoccupation with the question of God's knowledge probably began with his time in Shiraz, studying with Muḥammad al-Nassāba.[53] Whilst Muhammad al-Nassāba is not credited with any specific literary output in the biographical sources, Āghā Buzurg records that he did write a work entitled *al-Risāla fī ʿilm Allāh taʿālā biʾl-maʿdūmāt waʾl-ḥawādith*,[54] and Muḥammad al-Nassāba is in all likelihood the "teacher" (*ustādh*) whom Astarābādī mentions on numerous occasions in *al-Mabāḥith al-Thalāth*.[55]

In structural terms, Astarabādī's examinations of this issue show marked development:

(1) In *al-Mabāḥith*, Astarābādī is content to give a brief introduction, followed by a long quotation from Jalāl al-Dīn al-Dawwānī's commentary on *al-ʿAqāʾid al-ʿAḍudiyya* (of ʿAḍud al-Dīn al-Ījī).[56] This quotation itself contains citations from Ibn Sīnā and Naṣīr al-Dīn al-Ṭūsī, and hence serves the double purpose of outlining Astarābādī's opponent (al-Dawwānī) and the tradition of thought with which Astarābādī identifies (Ibn Sīnā and Naṣīr al-Dīn al-Ṭūsī). This is followed by a dissection and refutation of al-Dawwānī's passage and a defence of Naṣīr al-Dīn al-Ṭūsī's views.

(2) In his next analysis found in *al-Fawāʾid al-Iʿtiqādiyya*, direct citations from Ibn Sīnā and Naṣīr al-Dīn al-Ṭūsī are used, and only after an exposition of Astarābādī's interpretation (and defence) of these thinkers are al-Dawwānī's ideas introduced and refuted. Here, though, al-Dawwānī's passage is not cited in full, but divided and refuted point by point.

(3) In the third exposition, found in *al-Fawāʾid al-Makkiyya*, Astarābādī's direct citations are extended further to include Ibn al-ʿArabī and Shams al-Dīn al-Khafrī. The refutation of al-Dawwānī appears as a conclusion (*khātima*) to the discussion.

[52] Astarābādī, *Dānishnāmah*, f.19A.11–19A.12.

[53] See above, p. 32.

[54] Ṭihrānī, *al-Dharīʿa*, v. 15, p. 323 #2074. The *risāla* was actually written by a student (Muḥammad Sharīf al-Kashmīrī) from al-Nassāba's notes.

[55] See, for example, Astarābādī, *Mabāḥith*, f.3b.10 (*al-ustād, madda ẓilluhu*—indicating this person is alive at the time of writing), f.27b.9, f.30b.4, f.34a.6, f.42a.11 (*ustādunā al-muḥaqqiq*).

[56] On al-Ījī's epistemology generally, see van Ess, *Die Erkenntnislehre des ʿAḍudaddīn al-Īcī*.

(4) Finally, in *Dānishnāmah-yi Shāhī*, citation is kept to a minimum and the issue is discussed in a comprehensive manner with less dependence upon the actual text of any previous author (though regular reference is made to the ideas of past scholars). The references which are made concern a yet broader range of scholars, including al-Rāzī, al-Taftāzānī, al-Jurjānī and al-Qūshjī.

One can, then, recognise in these the progression from a scholar who uses past texts as the basis for discussion, to one who is willing, with greater confidence, to develop his own ideas referring to established sources but not (structurally speaking) being bound by them.

In terms of content, Astarābādī's concern when discussing these matters is to demonstrate that God's knowledge of things (*ashyā'*) is not limited simply to those that happen to exist. Writing in the tradition of Ibn Sīnā (whom he regularly defends from attack), he considers all that exists to be merely "contingent". Things need not have been as they are, and there is nothing necessary about the state of existence as it stands. If things could have been different, then there are states of affairs which could have existed, but do not. These possible states of affairs ("possibles") do not exist in the external sense (*khārij*), but they do have some sort of existential status, being the mental ideas of possibly existent states of affairs. The number of these possible states of affairs is infinite. The question which occupied Astarābādī's attention on these occasions is whether God can be said to know all the possibles, and if so, then how does he know them? Is his knowledge of these possibles the same as his knowledge of actually existent things? If these types of knowledge differ, how do they differ? This question, like most theological and philosophical questions, cannot be dealt with in isolation. Other issues are involved in its investigation, including the attribute of God's knowledge (and its relationship to God's essence), the manner in which God "knows" anything (existent or non-existent) and whether his knowledge of events predetermine their occurrence (and the implications of this for any theory of moral responsibility). It is, therefore, necessary in analysing Astarābādī's thought on the question of God's knowledge of possibly existent things to (at times) explain first his thinking on these attendant issues. The purpose of the following analysis is not merely to explore an interesting element of philosophical theology. It also illuminates the development, such as it is, in Astarābādī's philosophical theology. From this we can gauge whether Astarābādī's conversion to Akhbarism can be discerned in his theological writings, as his *oeuvre* in this genre of religious literature (unlike his extant *ḥadīth* commentaries and writings on *uṣūl al-fiqh*) covers the periods before and after his conversion to Akhbarism.

Astarābādī's engagement with previous scholarship on this issue involves detailed examinations of the positions of an array of past theologians and philosophers, beginning with Ibn Sīnā.[57] Jalāl al-Dīn al-Dawwānī is a particular *bête noir* of Astarābādī. The former is regularly criticised, not only with regard to his doctrine concerning God's knowledge of possibles, but also on other theological and philosophical issues.[58] In Astarābādī's various examinations of the question of *ʿilm al-Wājib*, al-Dawwānī's refutation of Naṣīr al-Dīn al-Ṭūsī is cited and (in turn) refuted on each occasion. We have, then, Astarābādī's refutation of al-Dawwānī's refutation of Naṣīr al-Dīn al-Ṭūsī. The work of Naṣīr al-Dīn al-Ṭūsī used is his commentary on *al-ʿAqāʾid al-ʿAḍudiyya* (of ʿAḍud al-Dīn al-Ījī). However, in the section which concerns al-Dawwānī (and hence Astarābādī), Naṣīr al-Dīn al-Ṭūsī is presenting an interpretation of Ibn Sīnā's ideas, as found in *al-Ishārāt* and *al-Shifāʾ*. An understanding of Astarābādī's ideas on *ʿilm al-Wājib*, then, draws us into the complex network of refutations, counter-refutations and commentaries which is characteristic of classical Muslim theological and philosophical writings generally.

The most convenient place to begin this analysis is where Astarābādī himself initiates discussion: with the general account of knowledge which these scholars are said to share. For an individual (be it God or another) to know a thing, the form of the thing must be present within him (*al-ṣūra al-ḥāḍira ʿinda al-mudrik*).[59] Form (*ṣūra*) is that element of a thing which distinguishes it from others.[60] The forms of

---

[57] As already mentioned, in the course of his consideration of this topic, Astarābādī cites Naṣīr al-Dīn al-Ṭūsī (d. 672/1274), Fakhr al-Dīn al-Rāzī (d. 606/1210), Saʿd al-Dīn Masʿūd al-Taftāzānī (d. 793/1390), al-Sayyid al-Jurjānī (d. 816/1413), ʿAlī b. Muḥammad al-Qūshjī (879/1474), ʿAḍud al-Dīn al-Ījī (d. 756/1355), Ibn al-ʿArabī (d. 638/1240), Shams al-Dīn al-Khafrī (d. 956/1550) and (critically) Jalāl al-Dīn al-Dawwānī (d. 908/1502).

[58] As mentioned above (p. 118), Astarābādī wrote a refutation of al-Dawwānī's logical method, and a refutation of the commentary on the creed of Naṣīr al-Dīn al-Ṭūsī by al-Dashtakī and al-Dawwānī.

[59] *al-Mabāḥith*, f.4.5.

[60] *al-Mabāḥith*, f.4.10–11: "the meaning of the *ṣūra* is the thing, from the perspective of it being a representation whereby one thing comes into existence and not another." This I take to be the same definition, though in a different form of words, to that used

particular things, then, distinguish them from one another, whether they are both mental ideas and external realities, or merely the former. How the individual acquires this form is discussed below, but the basic understanding of knowledge is to be found in the presence of forms in the prospective knower. The question which had troubled philosophers before Astarābādī is whether or not this attribute of knowledge can be predicated to God.

(i) *Ibn Sīnā's position*

Ibn Sīnā supposedly argued that God cannot have this type of knowledge, for if he was to have this type of knowledge, there would be multiple forms in his essence and his simple unity would be compromised. This was one element (amongst many) of Ibn Sīnā's position that al-Ghazālī found objectionable.[61] Ibn Sīnā's position is, in fact, more complex. According to al-Dawwānī (in a passage quoted by Astarābādī), Ibn Sīnā seems to have held two positions. In *al-Ishārāt*, Ibn Sīnā seems to assert that the forms "subsist" in God's essence (*qāmat bi-dhātihi taʿālā*). In *al-Shifāʾ*, however, a different process of God's knowledge is presented:

> He, the Most High, intellects all things in one instant, without there being a multiplicity of forms in his essence, or the true nature of his essence being affected by those forms. Rather an intellected form emanates from [his essence], and therefore [this emanated form] is an intellect which comprehends those forms which have emanated from him, thereby comprehending his essence. He considers his essence, and that he is the origin of all things, so he therefore comprehends all things through [a consideration of] his essence.[62]

This abstruse passage appears to indicate that rather than the forms subsisting in God's essence, they in fact proceed from him, being separate but dependent entities. The orthodox dogma that he knows all things is preserved by God being a "comprehending" being. When he comprehends his essence, he sees that his essence is the origin of

---

by al-Shahrastānī in his summary of the philosophers' position: "knowledge is the form of the object of knowledge with the knower" (*al-ʿilm ṣūrat al-maʿlūm ʿinda al ʿālim*) (Sharastānī, *Kitāb al-Musāraʿa*, pp. 67–8 (Arabic), p. 60 (English)).

[61] See Ghazālī, *Tahāfut al-Falāsifa*, pp. 109–218.

[62] Dawwānī, *Sharḥ al-ʿAqāʾid*, p. 119.

all things (*mabda' li-kull shay'*). The forms of these things proceed from God contemplating himself as the origin of all things.

(ii) *Naṣīr al-Dīn al-Ṭūsī's position*

The object of al-Dawwānī's criticism though is not Ibn Sīnā, but Naṣīr al-Dīn al-Ṭūsī who elaborates on Ibn Sīnā's position in his *Sharḥ al-Ishārāt*. In the passage cited by al-Dawwānī (and found embedded in Astarābādī's citation of al-Dawwānī's refutation), Naṣīr al-Dīn al-Ṭūsī outlines his own position.[63] There are, he argues, two ways of comprehending a thing (that is, of acquiring its form). One is to be the originator of that thing—the creator of a thing has knowledge of its form since he fashioned it (that is, gave it the form). In this instance, the form is said to have "presence" (*ḥulūl*) within the creator, and the thing emerges from him. The other manner in which one can come to know a thing is to be an observer of it (*qābiluhu*). One acquires the form of the thing not through it being pre-existent within oneself, but through some other means. It is important to recognise that Naṣīr al-Dīn al-Ṭūsī considers both of these ways of coming to know a thing as being on an "equal" level of knowledge. Being the origin of a thing does not necessarily lead to a superior kind of knowledge of the thing than being its observer (or, indeed, vice versa). In both cases, the form is acquired and knowledge is achieved. The difference between them is not in terms of form, but purely in terms of perspective (*iʿtibāriyy[an] maḥḍ[an]*). Furthermore, in order to gain the form of a thing, one does not need to gain the form of its creator, that is, one can acquire knowledge of a thing (through acquiring its form) without necessarily knowing its origin. Now, in Naṣīr al-Dīn al-Ṭūsī's emanationist schema, the Intellectual Essences (*al-jawāhir al-ʿaqliyya*) which emanate from God are caused (*maʿlūl*) by him, and it is they which eventually lead to externally existent things. God's comprehending of his own essence is performed by his intellect (*ʿaqluhu li-dhātihi*), and this intellect understands God's essence through *ḥulūl* (that is, through the form of God's essence being present within it). However, the intellect comprehends the Intellectual Essences through some other

[63] In al-Dawwānī's words, Naṣīr al-Dīn al-Ṭūsī "hovers around the clear meaning of the words in *al-Shifā'*" (Dawwānī, *Sharḥ al-ʿAqā'id*, p. 120).

means (that is, as an observer), and at the same time, it comprehends them as being created by God. Since the two means of coming to a position of knowledge are equal, God's knowledge of the Intellectual Essences (acquired through his intellect) is of the same status as his knowledge of his own essence. This is the case even though the latter came about through self-reflection, and the former came about through observation of the created things and only then recognising them as being created by the self. In this way, God knows all things; "nothing escapes him, even the weight of an atom" (*lā yaʿzib ʿanhu mithqāl dharra*).[64] Does this lead to a multiplicity in God's essence? In other words, does the emergence of his intellect through which he comprehends his essence lead to two things (God's essence and his intellect which comprehends his essence)? Naṣīr al-Dīn al-Ṭūsī argues that it does not, because the essence is the cause (*ʿilla*) of the intellect, and in existential terms (*fī'l-wujūd*), an effect is contained within its cause, therefore to know the cause is to know the effect. Similarly, the intellect is now the cause of the multiplicity of forms which make up the Intellectual Essences (the first caused thing and subsequent caused things). Since the intellect is the cause of these caused things, and there is an existential unity between cause and effect, there is no multiplicity in God's essence, since existentially they are a unity. The multiplicity which we perceive is, once again, purely perspectival (*iʿtibārī*).

(iii) *Al-Dawwānī's refutation of Naṣīr al-Dīn al-Ṭūsī*

Al-Dawwānī's objections to this position, whereby God knows all things as the origin of all things, and yet there is no multiplicity in his essence, are numerous. They comprise of eight discrete points, in each of which there is a citation from Naṣīr al-Dīn al-Ṭūsī's passage, followed by a refutation. Al-Dawwānī's refutations are a philosophical defence of elements of Sunni dogma against Naṣīr al-Dīn al-Ṭūsī's perceived heresy, and can be distilled into three principal points.

Firstly, Naṣīr al-Dīn al-Ṭūsī mentions a means of coming to know the form of something which is not through its innate presence within the knower (*ḥulūl*). How this second means of gaining a form operates is not made clear, and it is not obvious how this might come

[64] Dawwānī, *Sharḥ al-ʿAqā'id*, p. 123.

about. How does one gain knowledge of a thing (that is, acquire its form, and know that the form is instantiated on a particular occasion) when there is no form present within oneself prior to comprehension? This, al-Dawwānī states, is "not clear" (*ghayr ẓāhir*).[65]

Secondly, if God is to comprehend all things—both the general and the particular—through the forms of the Intellectual Essences, then there is an inevitable delay (or indirectness—*ta'akhkhur*) in God's knowledge of these particulars and generalities. God must first comprehend himself. Through that comprehension of himself, he comes to know the forms of the things of which he is the originator. Al-Dawwānī is not arguing that the "delay" here is temporal—rather God's knowledge of himself is logically prior to his knowledge of the forms of the Intellectual Essences he originates. Similarly, there is a delay as the forms of the bodily particulars become "imprinted" (*irtisām*) upon the immaterial Intellectual Essences. God's knowledge of particulars is, then, demoted to a logically lower position than his knowledge of himself, and this, al-Dawwānī implies, compromises the integrity of this knowledge.[66]

Finally, rational forms—that is, the Intellectual Essences—which subsist in God's essence without resulting in multiplicity, must be attributes of that essence (*ṣifāt dhātihi*). Attributes of an essence are necessarily present when the essence is present. However, the presence of the first caused thing (and hence subsequent caused things) is not necessitated by the presence of God's essence.[67] God's causal activity is a matter of his choice. Under Naṣīr al-Dīn al-Ṭūsī's proposal, God is robbed of the power of choice in the creation of his intellect and all that flows from its reflection on God's essence.

The issue of God's knowledge of possibly existent states of affairs is an extension of this last point concerning God's power of choice. Naṣīr al-Dīn al-Ṭūsī's position does not, al-Dawwānī argues, permit possible existence, since God's knowledge of the forms in the Intellectual Essences (and through them the externally existent particulars) is a knowledge of things which already exist. It is, if you like, a

[65] Dawwānī, *Sharḥ al-ʿAqā'id*, p. 127.

[66] Dawwānī, *Sharḥ al-ʿAqā'id*, pp. 130–131.

[67] "The first caused thing is not one of the attributes of God such that the presence of one is necessitated by the presence of the other, and perceiving one necessitates perceiving the other" (*laysa al-maʿlūl al-awwal min ṣifāt al-Wājib ḥattā yakūna ḥuḍūruhu mustalzam*[an] *li-ḥuḍūrihi wa-idrākuhu mustalzam*[an] *li-idrākihi*, Dawwānī, *Sharḥ al-ʿAqā'id*, p. 124).

secondary knowledge which comes about through self-reflection. Dogmatically, one must (surely) be committed to the view that God knows all states of affairs, both those that exist (externally) and those that do not exist. This is the doctrine of *ʿilm al-Wājib biʾl-mumkināt*. Naṣīr al-Dīn al-Ṭūsī's doctrine prevents God from knowing possibles. Al-Dawwānī considers a possible modification of Naṣīr al-Dīn al-Ṭūsī's theory which might solve the problem. One could postulate two forms of existence—one as an idea in the mind of God (*al-wujūd al-ʿilmī*) and the other as an externally existent thing (*al-wujūd al-khārijī*). The former includes all possible and actual states of affairs, and God is compelled (*biʾl-ījāb*) to bring these into existence and to know them all. The latter comprises merely of those states of affairs which achieve existence and God knows these by choice, since it was he who brought them into existence through choice (*biʾl-ikhtiyār*). An externally existent thing is known in both ways by God, but a possibly existent thing is known only in the first. Al-Dawwānī does not consider this modification to solve the problem. The reason for its failure lies in the fact that God does not know an existent thing in two ways (once as an idea and then again as an existent thing). Rather he knows these things (both possible and actual) in one way: he has their forms in an identical manner. The difference between a possible and an actual thing is that, although God knows them both, he knows the latter to have the additional attribute of existence.[68] This modification, then, leads to further incoherence in Naṣīr al-Dīn al-Ṭūsī's scheme.

(iv) *Astarābādī's al-Mabāḥith al-Thalātha*

As already mentioned, Astarābādī's writings on *ʿilm al-Wājib biʾl-mumkināt* (God's knowledge of possibles) can be found in four separate works, three of which were written before his conversion to Akhbarism. Despite the variation in structure noted earlier,[69] the content of these early refutations is similar, and the wording is often identical, making them appear to be re-workings of a single treatise. Astarābādī's refutation is a defence of Naṣīr al-Dīn al-Ṭūsī, and can

[68] Al-Dawwānī's own position is, he says, to be laid out in a separate treatise (Dawwānī, *Sharḥ al-ʿAqāʾid*, p. 146), and is not directly relevant to our purpose here, which is to analyse Astarābādī's reaction to al-Dawwānī's criticism of Naṣīr al-Dīn al-Ṭūsī.

[69] See above, pp. 118–119.

be summarised as follows. To begin with, al-Dawwānī has made the error of thinking that Naṣīr al-Dīn al-Ṭūsī is arguing that God actually comes to know particulars, both possible and actual, through the mechanisms he mentions (that is, God's essence being contemplated by God's *ʿaql*, which then recognises *al-jawāhir al-ʿaqliyya* as caused by God). Rather (Astarābādī argues), Naṣīr al-Dīn al-Ṭūsī is attempting to establish the *rational possibility* that God knows particulars through these mechanisms, and therefore refute the suggestion that Ibn Sīnā holds the heretical view that it is impossible for God to know particulars.[70]

Concerning the first of the specific points made by al-Dawwānī *contra* Naṣīr al-Dīn al-Ṭūsī, Astarābādī argues as follows. Al-Dawwānī thinks that it is not clear how an intellect might come to know a form without that form being already innately present in the individual. Astarābādī argues that Naṣīr al-Dīn al-Ṭūsī is not arguing that there is such a means. He is arguing that it is *rationally possible* that there is such a means. That is, it is not necessary for a form to be innately present in oneself (*ḥulūl*) in order to comprehend a thing, though it is necessary to have knowledge of that form. Astarābādī argues that Naṣīr al-Dīn al-Ṭūsī is suggesting that it is rationally possible that there are means of gaining knowledge of a form other than it being innately present within oneself. Astarābādī is presumably referring to a process initiated by an entity external to the individual, through which knowledge of the forms is implanted in the person.[71] It is merely a rational possibility (*jawāz ʿinda al-ʿaql*) that these alternative means exist, and that one can come to know the forms by means other than those restricted to innate presence of the forms.

Astarābādī also refutes al-Dawwānī's other two main criticisms of Naṣīr al-Dīn al-Ṭūsī (that is, that there is a "delay" in God's knowledge of particulars, and the schema negates the doctrine of God's choice in creative acts). However, to understand his refutation of these two criticisms, an understanding of Astarābādī's general position is necessary. In his earliest exposition, Astarābādī argues that there are

[70] That such a suggestion does not withstand a detailed analysis of Naṣīr al-Dīn al-Ṭūsī's comments here is obvious. Naṣīr al-Dīn al-Ṭūsī is clearly proposing a theory which he considers to be logically sound (and, more importantly, true), and which he is attempting to demonstrate to his reader as the preferred manner in which God's knowledge can be conceived.

[71] The intellect's ability to comprehend these forms, both in the uncomplicated manner (*ḥulūl*) and in these other ways, is termed *khuṣūṣiyyāt*.

two ways in which God can know a thing, and that God knowing a thing necessarily brings it into existence. Firstly, God knows in a comprehensive manner (*ʿilm ijmālī*). This knowledge is identical to God's essence. It is eternal, and is not dependent upon what exists and what does not exist (*ghayr maqṣūr ʿalā al-mawjūdāt*). It is not knowledge of a particular thing; it is simply a comprehensive knowledge. Secondly, God knows those things which have come (or will come) into existence (*ʿilm tafṣīlī*).[72] In this first treatise, Astarābādī argues that the accusation of a delay in God's knowledge can be refuted because God already has a comprehensive knowledge. This comprehensive knowledge exists before the Intellectual Essences come into existence, and perhaps before he even contemplates himself (Astarābādī is not clear on this point). What is clear is that this comprehensive knowledge enables one to say that there is no "delay" in God's knowledge. Al-Dawwānī only thinks there is a delay here because he has a monodimensional view of knowledge (that is, it can only refer to things). Once one understands that God has both *ʿilm ijmālī* and *ʿilm tafṣīlī*, the idea of a delay disappears. On the second issue raised by al-Dawwānī (God's creative acts being compelled in Naṣīr al-Dīn al-Ṭūsī's schema), Astarābādī turns the argument on its head. Al-Dawwānī had argued that choice is dependent upon God already having knowledge, power and will. If, as is stated, God knows particulars after having brought them into being (through the Intellectual Essences), then God cannot have had knowledge of them before the creative act, and therefore cannot have created them willingly. Astarābādī argues that God did have knowledge before the creative act, but it was comprehensive knowledge, and not specific knowledge of existent things.

Turning to the problem of God's knowledge of possibles, Astarābādī naturally states that God's knowledge of them is comprehensive, and therefore the problem of God not knowing possibly existent states of affairs melts away. There is no need for supporters of Naṣīr al-Dīn al-Ṭūsī's argument to turn to a distinction between two sorts of knowledge (one which God is compelled to have and one which he chooses to have).[73] Rather the distinction is between God's eternal

[72] In his *al-Fawāʾid al-Makkiyya*, f.3a.9–3b.4 Astarābādī explores types of knowledge between *ʿilm tafṣīlī* and *ʿilm ijmālī* which can be predicated of God. The discussion here is complex and requires separate treatment.

[73] This distinction, the reader will recall, was refuted by al-Dawwānī (above, p. 125).

knowledge which is an attribute which exists before his self contemplation, and God's acquired, specific knowledge which is obtained at the point of him causing all things to exist. That is, God already knows that he is to be the originator of all things (both possible and actual) by his comprehensive, eternal knowledge. This attribute he has even before he contemplates himself, giving rise to the intellect which contemplates his essence. He learns again of all things once his intellect contemplates his essence and finds, again, that his essence is the origin of all things. The distinction between comprehensive and specific knowledge is put to task in explaining how God knows both possible and actual states of affairs. This, then, is the position Astarābādī holds in his first discussion of God's knowledge of possibles, as found in *al-Mabāḥith*.

(v) *Astarābādī's al-Fawā'id al-I'tiqādiyya*

Astarābādī's second discussion of this issue is found in *al-Fawā'id al-I'tiqādiyya*, much of which is a re-ordering of the material in *al-Mabāḥith* with little additional information. The principal additional piece of argumentation is the insertion of *ḥadīths* from al-Kulaynī's *al-Kāfī*, which are said to establish the doctrine that God's knowledge does not change.[74] This, in turn, establishes God's comprehensive knowledge, whilst not impinging on Astarābādī's division of God's knowledge into comprehensive and specific. The introduction of *ḥadīths*, even in this supporting role to rational argumentation, could indicate that Astarābādī's Akhbārī proclivities were beginning to influence his theological writings. *Al-Fawā'id al-I'tiqādiyya* was written soon after his arrival in Mecca when presumably he first came into contact with Ṣāḥib al-Rijāl (who is said to have commanded Astarābādī to revive the Akhbārī school).[75] Whilst Astarābādī's theological position has not changed from *al-Mabāḥith*, his mode of argumentation now includes more extensive *ḥadīth* citation. How-

[74] The reports are from al-Imām al-Bāqir and al-Imām al-Riḍā. In the former, the Imam states, "God was, and there was nothing other than him, and he did not cease to be a knower of all that is. His knowledge of [all that is] before it came into being is the same as his knowledge of it after it came into being." In the latter, the Imam is asked whether God has knowledge of things before they were created or not. The Imam answers, "God's knowledge of things never ceases to be, whether before they are created or after." Kulaynī, *al-Kāfī*, v. 1, p. 107 and Astarābādī, *al-Fawā'id al-I'tiqādiyya*, f.4b.13–5a.8.

[75] See above, pp. 33–36.

ever, these are used to support positions already established through rational argumentation.[76]

(vi) *Astarābādī's al-Fawā'id al-Makkiyya*

In his third examination of the issue, *al-Fawā'id al-Makkiyya*, there is a significant change in Astarābādī's position. The criticism of al-Dawwānī is not omitted in this work, but it is relegated to the end of the discussion (where it is reproduced almost verbatim from *al-Mabāḥith*). The new material takes the form of citations and references to additional past scholars, particularly Ibn al-ʿArabī and Shams al-Din al-Khafrī. The discussion here is not so much a commentary on previous works, but an independent examination of the problem of God's knowledge of possibles, using the works of previous scholars as prompts in the discussion. Consequently, the section opens with a general discussion of whether or not the attribute of knowledge can be applied to God at all. The answer (unsurprisingly) is that he must be able to be described as knowing, for he is greater than his creation, and created beings are described as knowing. There is a reference to the Qur'ān to support this argument.[77] The objection concerning the delay in God's knowledge (found in al-Dawwānī's refutation of Naṣīr al-Dīn al-Ṭūsī and mentioned above) is discussed here, however the objection is not attributed to a particular scholar, and is simply stated in an *in qulta…fa-qultu* (if you say…then I say…) passage. Astarābādī's refutation of the objection is also new. If a man has foreknowledge of a thing he is about to create and God is knowledgeable about what he is to create only after its creation,

[76] This is not to say that reference to the Imams is absent from *al-Mabāḥith*. Astarābādī states at the outset of his discussion of God's knowledge in *al-Mabāḥith* that "what is reported from the Imam…Abū ʿAbdallāh [al-Imām al-Ṣādiq], and the established position amongst a group of theologians, according to al-Muḥaqqiq al-Dawwānī in his commentary on *al-ʿAqā'id* [*al-ʿAḍudiyya* of al-Ījī], and what is heard from the Teacher [a probable reference to Muḥammad al-Nassāba] is that God's essence is knowledge itself in relation to all possibles." (*al-Mabāḥith*, f.3b.5–11). However, no *ḥadīths* are cited, and specifically Shīʿī doctrines (such as appeal to the Imam's status or reports from him) are not mentioned again. So-called "traditional proofs" (such as *hadīth* citation) clearly play a greater, though still not predominant, role in *al-Fawā'id al-Iʿtiqādiyya*.

[77] The reference is to Q 16 (al-Naḥl) 68 where bees are said to know how to build houses. A creator cannot be deficient in relation to his creation, so if the bees know this, the creator must also know this, therefore, the attribute of knowledge must be applicable to God (Astarābādī, *al-Fawā'id al-Makkiyya*, f.2b.16).

then this would make man superior to God in this respect. This, as has already been proven, is logically incoherent, and therefore God must have foreknowledge of the things he is to create.[78] This argumentation did not appear in Astarābādī's previous discussions.

Another novel element in *al-Fawā'id al-Makkiyya* is a more complex typology of God's knowledge. In earlier discussions, God's knowledge had been divided between comprehensive and specific. In this later work, Astarābādī outlines five different "categories" (*marātib*) of God's knowledge. This classification is not his own, he says, but taken from others. The first and the fifth categories are already known to us:

1. comprehensive knowledge (*ʿilm ijmālī*)

5. specific knowledge (*ʿilm tafṣīlī*)

However, between these two there are three other types of knowledge which can be predicated of God. Each is general in relation to what is below it, and specific in relation to what is above it:

2. knowledge of things which are existent in the intellect—which refers to a knowledge gained by the intellect of those things which exist in the mind of God.
3. knowledge of the simple celestial souls (*al-nufūs al-mujarrada al-falakiyya*) which emanate from the intellect.
4. knowledge of the "drawing out" (*intiqāsh*, the beginnings of giving them bodily substance) of these souls before they form discrete things which are extant in the external world.

The scheme, of course, owes much to emanationism, and Astarābādī does not object to it, though he does says that concerning the issue of God's knowledge of possibles, only the first and the fifth categories (that is, those previously covered in *al-Mabāḥith* and *al-Fawā'id al-Iʿtiqādiyya*) are relevant. The significant development from Astarābādī's earlier discussions is that whilst previously God's knowledge of possibles was described as a comprehensive (and not specific) knowledge, here Astarābādī attempts to demonstrate that

---

[78] Astarābādī, *al-Fawā'id al-Makkiyya*, f.2b.17–3.7. As an aside, Astarābādī also mentions that this argument demonstrates that God is a free agent, since he creates free agents. If he was not a free agent, then his creation would be greater than him and "it is not permitted for a caused thing to be greater that its creator, and a free agent is greater than one who is not free." (Astarābādī, *al-Fawā'id al-Makkiyya*, f.3a.8–9).

this knowledge of possibles is a type of specific knowledge also. So, in technical terms, God's knowledge of possibles is not only *ijmālī*, but also *tafṣīlī*. Crucially, though, it is *tafṣīlī* in a manner different to the *tafṣīlī* knowledge of actual existents, and this new type of *tafṣīlī* knowledge springs from a sophisticated epistemological typology.

Astarābādī's argument runs as follows: Knowledge is knowledge of something. Even God's comprehensive knowledge is God's knowledge of himself as having the attribute of knowledge. So knowledge must be directed at something, and this something must, at the point of knowledge, be distinguishable (or separate—*intizāʾī*) from the knower. If God is to be described as knowing possibles, then these possibles must be, in some sense, separate from him. If this is the case, then him knowing them as separate is a type of detailed knowledge rather than simply comprehensive knowledge. The question that remains is "How are these possibles separate from God's essence when they exist merely as ideas in God's mind and never attain external existence?" It is this question which occupies an elaborate and complex discussion in *al-Fawāʾid al-Makkiyya*.

Being "knowledgeable" or "a knower" is an attribute, and there are different types of attribute.

(1) There are attributes which are absolute, and which require the existence of no other attribute (or a thing with that attribute): the example used by Astarābādī is black. They are termed *inḍimāmī* ("combined").

(2) There are also attributes which require the existence of another attribute (or thing with that attribute, the example used by Astarābādī is "high"). They require another distinct (*intizāʾī*) thing.

The difference between the attributes "black" and "high" is that to describe a thing as black is to describe an attribute of the thing itself. To describe it as high is to say it is high *in relation to something else* (and therefore imply the existence of this something else). Within the latter category, there are subcategories:

(2a) There are attributes which require the existence of another thing, but that thing is self subsistent (the example given is "just").

(2b) There are also attributes which require the existence of another thing, but that thing is not itself self subsistent, but requires, in turn, other things (the example given is being a father).

Attributing the qualities of justice or fatherhood to a person is similar in that they both require the existence of another thing. In the case of justice, it can only be attributed to someone who performs just actions. In the case of fatherhood, it can only be applied to someone who has a son. A thing is black in itself, but a person is only just or a father if other things exist. However, there is a difference between being just and being a father. Being just requires the existence of just actions, but these just actions are just in themselves: that is, they have an external quality of justice which holds a similar status to that of black—an external, self subsistent quality.[79] Being a father requires the existence of a son, but a son, in turn, requires the existence of another thing (that is, the father or the mother). The attribute of fatherhood depends on another thing, which in turn depends on yet another thing. (That this third thing is actually identical with the first thing is not relevant here; Astarābādī is only interested in the truthful application of "son" being dependent upon something other than itself.)

When we come to the attribution of "being a knower" or "being knowledgeable" (*ʿālim*), we need to ascertain into which of these categories an attribute falls. Firstly, it could fall into either category (1) or category (2) above. When God is said to have the attribute *ʿilm ijmālī* (comprehensive knowledge), it is an attribute of type (1) above. When his knowledge is *ʿilm tafṣīlī* (specific knowledge), it is of type (2). However, this second type of knowledge can be further subdivided, in that knowledge of actual existents is of type (2a), and of possible existents is type (2b). God's knowledge of actual existents is not dependent upon yet another thing to exist to be truthfully applied. God's knowledge that possible existents do not have external existence does, however, require the existence of other things. Possible existence implies existence in the way that "high" implies "low"; but actual existence does not (logically speaking) imply the possible existence of anything. Actual existence is like the attribute "black"—an externally, self-contained attribute which does not imply possible existence.

To complicate matters further, Astarābādī wishes to further modify this typology. For him, actual existence should be classified as type (2b) because it implies non-existence (*maʿdūm*—which is to be distinguished from possible but not external existence), in the way "high"

[79] This, of course, was a position with which the Ashʿarīs would disagree, but Astarābādī, writing within the Muʿtazilī-influenced tradition of Imāmī theology, treats this as uncontroversial.

implies "low". Possible existence is, in fact, a further subcategory of (2b) in that knowledge of it does not only imply knowledge of external existence, but (furthermore) knowledge of external existence implies knowledge of non-existence. Hence we have an additional subcategory:

(2c) There are attributes which require the existence of another thing, and that thing itself requires the existence of yet another thing.

It is in this way that God can be said to know possibles. Possibly existent things may not have an external existence, but they do have existence as objects of God's knowledge which can be truly attributed to him in the manner labelled above as (2c).

If this is how the attribute of knowing possibles can be accurately ascribed, how can possibles be said to exist? Astarābādī argues that actual existents have an essence (*dhāt*) which is known by God after they have come into existence (that this essence is ultimately dependent upon God's causal power is not important here). Possibly existent things, by virtue of them being possibly existent, must have the potential to be existent (this distinguishes them from non-existent things). The attribute of existence must be (potentially) ascribable to them, even though it is, in fact, not. There must, then, be a cause which could, potentially, bring about them attaining external existence (if there was no such potential cause, they could not be possibly existent). They do not, in themselves, exist eternally. What exists eternally is the potential cause of them becoming externally existent things. God's knowledge of himself as the potential, but not actual, cause of possibly existent things is the only manner in which these possibly existent things exist eternally. It is only in this form that they can be objects of God's knowledge, and it is only in the manner described in (2c) above that God can be said to know them.

In the course of arguing for this position, Astarābādī cites Ibn al-ʿArabī, Shams al-Din al-Khafrī, Ibn Sīnā and ultimately al-Imām al-Ṣādiq. It is clear that this represents a major theological advance on his earlier position that God can only know possibles through his *ʿilm ijmālī*. In *al-Fawāʾid al-Makkiyya*, Astarābādī argues that God knows possibles not only in this general, comprehensive manner, but also in a specific manner (*ʿilm tafṣīlī*). The difference between this type of *ʿilm tafṣīlī* and the type by which he knows externally existent particulars is supported by different modes of eternal existence for actuals and possibles.

(vii) *Astarābādī's* Dānishnāmah-yi Shāhī

The position established in *al-Fawā'id al-Makkiyya* is maintained (in slightly different terms) in Astarābādī's final theological treatise, the *Dānishnāmah-yi Shāhī*. The position he had argued for in *al-Mabāḥith* was that God's knowledge of possibles is purely comprehensive (*ijmālī*) and not specific (*tafṣīlī*). Astarābādī rejects this position in the *Dānishnāmah-yi Shāhī* (though without reference to his earlier works):

> It is mentioned in the thought of the *ḥukamā* and some of the scholars of Islam that the Lord, the Most High, has two sorts of knowledge—one is a comprehensive knowledge which is identical with his essence and comes before the creation of things, and the other is a specific (*tafṣīlī*) knowledge of existence, which is the things themselves. This [position] is now known to be contrary to the truth.[80]

His point here is that God's knowledge is unchanging, and this is proven by reference to the *ḥadīth* from the Imams concerning God's knowledge remaining the same before and after creation. This constitutes a revelatory proof, though it is also supported by a rational proof. In an extended passage (the twelfth *fā'ida* of the work), the views of various past scholars are referenced (Ibn Sīnā, al-Rāzī, al-Taftāzānī, al-Qūshjī and al-Jurjānī) and rejected in favour of Astarābādī's own formulation. To begin with Ibn Sīnā's position, this seems to involve a denial of the logical possibility of God knowing possibly existent things. The reasoning used by Ibn Sīnā involves the argument that possibly (but not actually) existent things do not, by definition, exist. If God is to know them, then, he has attached his knowledge to a non-existent thing, and this is invalid. Or as Astarābādī puts it:

> If the limitless ideas [that is, limitless in numerical terms] which are known in eternity to God do not come into existence, then this means that [God's] knowledge is attached to total non-existence, and this is invalid.[81]

Now Astarābādī says he has considered the *ḥadīths* of the Ahl al-Bayt, and the arguments of the theologians and philosophers "for many years, until God, the creator and knower, placed in the heart

[80] Astarābādī, *Dānishnāmah*, f.22a.14–17.

[81] Astarābādī, *Dānishnāmah*, f.18b.15–17. *lāzim miyāyad ta'alluq-i 'ilm bi-lā shay' maḥḍ, va īn bāṭil-ast.*

of this lowly being [that is, Astarābādī himself]" a solution. The solution is that:

> total non-existence is simply that [the attributes] of being (*hastī*) do not subsist in [the thing]. God's consideration (*mulāḥaẓah*) of it does not come into existence. However, why is it not possible that all the limitless ideas, both simple and having extension, are present in the consideration of God in this sense: his consideration is attached to them without solidity, existence and being (*thubūt va ḥuṣūl va hastī*) being subsistent in them, and that this consideration is God's knowledge itself, and his essence?[82]

God knows ideas, even though they be infinite in number, and this knowledge consists of him considering (*mulāḥaẓah*) them. For things that exist, he considers them to be existent, and for things that do not, he considers them to be non-existent. The point here is that God considering them to be non-existent in the external sense (*khārijī*), does not preclude them from being existent in the mental (*dhihnī*) sense. The difference between these two types of existence is that whilst with the former the thing has existence "in itself" (*fī nafsihi*), the latter does not (acquiring existence purely in the consideration of God).[83] Though expressed in different terms, this position is consonant with that reached in *al-Fawā'id al-Makkiyya*, where what is known to God of possibles is God's potential to bring them into existence, and he knows this of each individual possible state of affairs. In the *Dānishnāmah*, God's consideration of possibles as non-existent, and his knowledge of himself as the being with the facility to bring things into existence (*imtiyāz*—that is, to give them separateness) are the elements which constitute the mental existence of the possibles. This position is backed up with a citation from al-Sayyid al-Jurjānī's commentary on al-Urmawī's (Maḥmūd b. Abī Bakr, d. 681/1283) *Maṭāliʿ al-anwār*.

Once again, the principal opponent in the remainder of Astarābādī's discussion is al-Dawwānī. He is cited as accusing Ibn Sīnā of muddled and contradictory statements on the issue, and, in this, he is following similar accusations made against Ibn Sīnā by Fakhr al-Dīn al-Rāzī, al-Taftāzānī and al-Qūshjī. Furthermore, whilst the passage cited in the earlier works is not found in the *Dānishnāmah-yi Shāhī*,

---

[82] Astarābādī, *Dānishnāmah*, f.19a.11–17.

[83] *mujarrad-i ḥuṣūl dar mulāḥaẓah dāshtah bāshad va ḥuṣūl fī nafsihi nadāshtah bāshad*. Astarābādī, *Dānishnāmah*, f.19a.3.

al-Dawwānī is recorded as criticising Naṣīr al-Dīn al-Ṭūsī's thoughts on Ibn Sīnā's position. Astarābādī replies that his opponents have misunderstood Naṣīr al-Dīn al-Ṭūsī here:

> We distinguish between a thing existing in itself (*fī nafsihi*) and a thing existing in the mind. We say that a thing which exists in itself either subsists in something else or in itself. However, the existence of a thing which exists in the mind does not subsist in itself. Rather consideration is attached to it. We say that the eternal consideration is the actual essence of God Almighty, and ideas do not exist in themselves in eternity.[84]

The point here is that al-Dawwānī "erroneously thinks that existence in the mind is one of the types of existence which a thing has in itself."[85] Externally existent things have existence "in themselves", mentally existent things have a different sort of existence which is not "in itself", but is determined by them being perceived by a mind. The example given is of a "fifth wife". She cannot exist in reality because a fifth wife is a contradiction in terms (the fifth marriage would be *bāṭil*, and the woman would not be the man's wife). There is no bar, however, to a fifth wife being a possibly existent item which exists purely in the sense that it is perceived by God as a possibility that could have come about and did not.[86]

God, then, does know all things—both what has attained existence, and what has not. It could be said that he knows both through his actions (*bi'l-fi'l*), for just as he knows himself as the origin of all that has attained existence, so he knows himself as the one who has prevented possibles from attaining external existence. Hence God knows all particulars, and to deny this is to believe something (as some philosophers do) "contrary to rational proofs, and contrary to the necessary element of all religions."[87] Then again, to deny that mental existence is a type of existence through fear of compromising God's unity (as some theologians do) is also invalid. Astarābādī proposes a third way (*ṭarīq-i thālith*) in which the mental existence

---

[84] Astarābādī, *Dānishnāmah*, f.20b.14–18.

[85] *Fāḍil-i Davānī khayālī kurdah-ast kih vujūd dar mulāḥazah az ānhā-yi vujūd-i ashyā-ast dar nafs al-amr*. Astarābādī, *Dānishnāmah*, f.21a.11–12.

[86] A fifth wife would not be possible even in the mind of God if legal categories (from prohibited to obligatory, and invalid when referring to contracts) were external features of the item (as the Mu'tazilīs are made to claim elsewhere in the *Dānishnāmah*). See above, p. 115.

[87] *khilāf-i adillah-yi 'aqliyyah va khilāf-i ḍurūrat-i adyān-ast*.

of possibles is a type of existence, but not one in which the thing exists "in itself". Rather the thing exists as a perceived thing only. These things, therefore, do not have properties in themselves, eternally (and God's unicity is not compromised), but they can be described as having the properties they would have had, had they attained existence.[88]

Conclusions

What, then, can we deduce from this extended analysis of Astarābādī's theological writings, and in particular his reflections on the problem of *ʿilm al-Wājib biʾl-mumkināt* (God's knowledge of possibly existent things)? A number of comments are appropriate here. Firstly, Astarābādī's argumentation is complex and sophisticated, and demonstrates extensive study of *falsafa* and *kalām*. This complexity does not disappear after his conversion to Akhbarism, though there is a tendency to express complex ideas in a simpler manner. This coincides with a change in his language of choice (from Arabic to Persian), and these two events are probably related. The conception of Astarābādī as a simple traditionalist, unwilling to delve into the intricate questions of theology and prepared to accept all doctrine as stated in the *ḥadīth* has to be rejected.

Secondly, there is development in Astarābādī's theology. The crucial change, however, does not coincide with his conversion to Akhbarism. His position on *ʿilm al-Wājib biʾl-mumkināt*, for example, develops from that expressed in *al-Mabāḥith* and *al-Fawāʾid al-Iʿtiqādiyya* to a more nuanced version established in *al-Fawāʾid al-Makkiyya* (and much later in *Dānishnāmah-yi Shāhī*). *Al-Fawāʾid al-Makkiyya* was written in 1018, three years after his arrival in Mecca, when he was probably under the tutelage of *Ṣāḥib al-Rijāl*. This was before Astarābādī retired to Mādīna to develop his Akhbārī juristic

[88] As an aside, Astarābādī enters into the debate concerning whether or not these mental ideas exist if there is not a mind to perceive them. Al-Rāzī and his followers have said they do, whilst al-Dawwānī and Ṣadr al-Dīn al-Shīrāzī al-Dashtakī (not Mullā Ṣadrā as intimated by Sefatgol, *Sākhtār-i Nihād*, p. 524) say that they do not. Astarābādī argues that these mental ideas do, indeed, have an existence, and that it is impossible to conceive of them not being perceived by a mind, since the possibles considered by God are limitless (*ghayr intihāʾī*). See Astarābādī, *Dānishnāmah*, f.22a.17–23a.1 (*fāʾida* 15).

methodology and write *al-Fawā'id al-Madaniyya*. Furthermore, instead of a sudden change in theological views brought about by Astarābādī's conversion to Akhbarism, there is a gradual development of theological argumentation and increasingly careful expression of his theological views. Crucially though, these changes had mostly taken effect before his conversion to Akhbarism, and they do not seem to have been altered significantly by it.

Thirdly, Astarābādī's theological writings bear witness to a developing scholarly mind. He moves from being a mere commentator on previous writings (*al-Mabāḥith*), to a more confident scholar, re-fashioning past formulations of a problem (*al-Fawā'id al-I'tiqādiyya*), to an independent author, who nonetheless acknowledges his debt to previous scholars (*al-Fawā'id al-Makkiyya*). Astarābādī's *Dānishnāmah-yi Shāhī* represents the end point of this process, containing an uninhibited exploration of a theological issue. If anything, Astarābādī's conversion to Akhbarism enabled him to enter into independent theological speculation and compose a treatise in which his individual views are expressed. His juristic commitment to Akhbarism did not make him an unadventurous follower of the plain meaning of revelatory tradition, as the common image of Akhbarism might imply.

Finally, if differences are to be found between Astarābādī's early and later theological works, they lie primarily in a greater clarity of expression and an increased willingness to cite *ḥadīth* as supporting evidence for theological opinions. However, *ḥadīth* are never a substitute for detailed theological and philosophical argumentation. They are presented merely as further evidence of the validity of Astarābādī's position.

Astarābādī adopted a juristic methodology which was critical of rational argumentation and deduction (including *ijtihād*). However, this was not (it seems) accompanied by a concomitant stunting of his theological and philosophical endeavours. His jurisprudence was combined with a complex theology, developed through rational argumentation and using established philosophical concepts. This is the case despite his criticism of philosophers and theologians in *al-Fawā'id al-Madaniyya*, and his assertion that the rational sciences are useful only for determining the direction of the *qibla*. That subsequent Akhbārīs studied his jurisprudence, and not his theology, serious skewed the subsequent Akhbārī movement, making it a traditionalist juristic movement. Astarābādī's combination of philosophical theology and juristic traditionalism was not entirely ignored by later

Akhbārīs,[89] but "philosophical Akhbarism" certainly struggled to establish itself as a major trend during the growth of the movement following Astarābādī's death.

[89] See Gleave, "Scriptural Sufism", with particular reference to Muḥsin Fayḍ al-Kāshānī.

CHAPTER FIVE

THE SPREAD OF AKHBARISM AFTER ASTARĀBĀDĪ

The importance of Astarābādī's critique of *ijtihād* found in *al-Fawā'id al-Madaniyya* was rapidly recognised in the intellectual world of Eleventh/Seventeenth Century Shiʿism. I have already referred to the swift identification of Astarābādī with the foundation of a new (or revived) school of jurisprudence,[1] termed *al-akhbāriyya* and challenging the established Uṣūlī-*mujtahid* juristic methodology of the day.[2] In the subsequent two centuries and beyond, groups of Akhbārī scholars were evident throughout Safavid Iran, southern Iraq, eastern Arabia, Jabal ʿĀmil in southern Lebanon and India. This chapter outlines the probable routes whereby Astarābādī's critique came to be known throughout the Shīʿī world, the growth of the school's influence and the activities of the best known Akhbārīs who either identified themselves as such or were described as such by subsequent authors.[3]

---

[1] The discussion of the emergence of legal schools (*madhhab*, pl. *madhāhib*) in early Islam has been developed by Makdisi (in particular in *The Rise*), and then, more specifically, by Melchert (*The Formation*). When one can say that a *madhhab* emerged is, of course, dependent upon one's preferred criteria. Melchert (*The Formation*, p. xvi) argues that the term *madhhab* was used in a number of different (and often unconnected) ways in the biographical literature, but that the criteria for a school consists of the recognition of a chief scholar, the production of commentaries on standard legal works and the regular transmission of legal knowledge (delineated through an *ijāza* system), to which I might add the establishment of a relatively stable set of distinctive school doctrines. When these are present, a *madhhab* can be said to have come into existence. These criteria are used in this chapter in an oblique manner in order to propose a date of inception for the Akhbārī school. Stewart (*Islamic Legal Orthodoxy*, pp. 175–208), however, has argued that the late classical Akhbāriyya cannot be considered a *madhhab* because they rejected the notion of consensus (and the discipline of *uṣūl al-fiqh* more generally). I have argued that the term *madhhab* was not only used within the Shīʿī tradition to refer to the Akhbāriyya, but also that the term "school" can be legitimately ascribed to them on the basis of Melchert's criteria (see Gleave "Intra-*madhhab ikhtilāf*"). The biographical and *ijāza* evidence examined in this chapter enables us to date the formation of an Akhbārī school/*madhhab* according to these criteria.

[2] See above, p. 47.

[3] Of course, the identification of a scholar as an Akhbārī served, at times, a polemic purpose (to praise or denounce a scholar depending on the biographer's perspective). Whether or not a scholar is best described as an Akhbārī is, in the absence of personal

The manner in which Astarābādī's ideas were developed, adapted and criticised by subsequent Akhbārīs are described in the following chapters. Here I am concerned with establishing an (approximate) historical account of how Akhbārī ideas took root, beginning with Astarābādī's teaching in Mecca and then spreading to most of the Twelver Shīʿī world.

The most appropriate place to initiate an account of the spread of Astarābādī's ideas is with scholars who are recorded as having studied under him. The terms used for these scholarly relationships are numerous, and are recorded in biographical works (*ṭabaqāt*/*tarājim*) and "licences to teach" (*ijāzāt*, sometimes as separate documents, often as passages recorded on manuscript copies). As I have outlined elsewhere, these lines of transmission (*asānīd*), which still today form an important part of an individual Shīʿī scholar's pedigree, serve to link the individual scholar/pupil to past scholarly tradition, through the teacher (whose own authority is, in turn, validated by his own lines of transmission).[4] The multiple and multifarious Shīʿī *isnād* chains found in *ijāza* documents generally present a unified and coherent tradition. The *ijāza* recipient (*mujāz*) is initiated through the granting of the *ijāza* by his teacher (*mujīz*). There was an irenic character to the *ijāza isnāds*, and debates which had often led to mutual deprecation and declaration of unbelief in the past were submerged by the overall objective of establishing the scholarly class (*ʿulamāʾ*) as the preservers of the true message of the Shīʿī Imams. The *ʿulamāʾ*, through the *ijāza* system, demonstrate that it is they who hold primacy in the interpretation of religion and in the regulation of community life more generally. Sometimes the links may be tendentious, and relationships of both great intimacy and passing acquaintance are subsumed under an *ijāza* or in an *isnād* by the formula "so-and-so relates (*yarwī*) from his teacher, so-and-so". Astarābādī features extensively in late Shīʿī *ijāzāt* as a link in these *isnāds*. Other scholarly relationships are contained within entries in *ṭabaqāt* works which detail the achievements of Astarābādī and his pupils. It is these relationships which form the principal recorded avenues for the dissemination of

identification, admittedly problematic. In this chapter I take later identifications generally as sound. In subsequent chapters, I explore whether or not such identifications are justified. On the relatively late identification of some scholars as Akhbārīs, see Newman, "Anti-Akhbārī Sentiments", pp. 156–158.

[4] See Gleave, "The *Ijāza*". See also Schmidtke, "The *Ijāza*", pp. 67–69.

his Akhbārī ideas.[5] They were almost certainly not the only means whereby his ideas were spread. There may have been many other pupils and acquaintances who did not receive *ijāzas*, did not "relate" from him or were not remembered as his pupils. Their activities, though, are not recorded in currently available documentation. Furthermore, Astarābādī's ideas may have been inadvertently spread through those who publicly criticised him. However, even with these provisos in mind, it does seem most likely that those who studied under Astarābādī were responsible for the immediate dissemination of his ideas since, by all accounts, these pupils presented Astarābādī's Akhbarism in a highly sympathetic manner.

The Transmission of Knowledge in Imāmī Shiʿism

Phrases such as *lahu ijāza min*...("he has an *ijāza* from" so and so), *yarwī ʿanhu* ("he relates from him") and *min mashāyikhihi kāna* ("amongst his teachers was" so and so) in the biographical literature indicate, then, a scholarly relationship between two individuals which presents itself as the most likely conduit for the spread of the senior party's ideas. Before examining the network of such relationships springing from Astarābādī himself, one should take note of the different relationships indicated by the various terms found in the relevant literature. Astarābādī's relationships with his own teachers have already been touched upon,[6] and provide a good example of the variety of terms used in describing and establishing the relationship between two scholars. These are described with a number of stock terms, both by himself (in his own works) and by later biographers (in *ṭabaqāt* works) after his death. As has been mentioned already, Astarābādī describes Muḥammad al-Nassāba as someone with whom

---

[5] Of course, receiving an *ijāza* does not necessarily indicate that the *mujāz* agreed with the *mujīz* on all issues of jurisprudence. There are numerous cases of Akhbārīs receiving *ijāzas* from Uṣūlīs (and vice versa). A large number of the early Akhbārīs relate from the Uṣūlī, al-Shaykh al-Bahāʾī, for example. Baḥr al-ʿUlūm was an Uṣūlī who related from Yūsuf al-Baḥrānī (an Akhbārī). See Gleave, "The *Ijāza*". Evidence of transmission linkages need, then, to be triangulated with other evidence from an author's own writings and biographical literature generally. When this is possible, a route for the dissemination of the *mujīz's* ideas is established.

[6] See above, pp. 32–33.

"he read" (*qaraʿa ʿalā*)[7] and as "the greatest of the scholars".[8] Ṣāḥib al-Maʿālim Ḥasan b. al-Shahīd al-Thānī (Ḥasan b. Shahīd II) is referred to as "our shaykh" (*shaykhunā*),[9] Ṣāḥib al-Madārik al-ʿĀmilī as "the first of my shaykhs in two sciences of *ḥadīth* and *rijāl*"[10] and Ṣāḥib al-Rijāl al-Astarābādī as "the last of my shaykhs in sciences of *fiqh*, *ḥadīth* and *rijāl*".[11] He also records that he "relates from" Ṣāḥib al-Maʿālim's son Muḥammad b. al-Ḥasan (father of the author of *al-Durr al-Manthūr*) "by way of *ijāza*" (*bi-ṭarīq al-ijāza*).[12] Known to exist are documents recording Astarābādī's *ijāzas* from Ṣāḥib al-Madārik and Ṣāḥib al-Rijāl. In both *ijāzas*, Astarābādī is given permission to pass on "all that [the teacher] himself was permitted to transmit, by transmission lines which are [already] established".[13] The *ijāzas* are, then, permission to transmit works, and are not necessarily a sign of doctrinal agreement between the teacher and pupil.[14] Although names of works are mentioned, they are clearly meant as examples of the breadth of the scholarly tradition which is being bestowed upon Astarābādī, the *ijāza* recipient. Teachers, after decreeing the licence (by uttering the phrase *ajaztu lahu*), list their own *ṭuruq* or *asānīd* (that is, their transmission lines). These establish linkages between the teacher and the earliest generations of Shīʿī scholars, and ultimately the Imams themselves. The transmission lines now become the pupil's

[7] Astarābādī characterises the relationship thus himself (see above, p. 32), and it is repeated by al-Ḥurr al-ʿĀmilī (*Amal*, v. 2, p. 310). Al-Ḥurr's note concerning Astarābādī's relationship with al-Nassāba (itself a reference to *al-Fawāʾid*) is supposedly an addition to the original text, but is found in the oldest manuscript of *Amal al-Āmil* used in the edition, corrected by al-Ḥurr himself (see *Amal*, v. 2, p. 310, n. 2. See also the description of this manuscript by the editor, v. 1, pp. 61–62 of the editor's introduction).

[8] See Astarābādī, *al-Fawāʾid*, p. 265.

[9] Astarābādī, *al-Fawāʾid*, p. 283, p. 293 (Fāḍilī notes he also refers to him as *Shaykhunā* in his commentary on *al-Tahdhīb*, see Astarābādī, *al-Ḥāshiya ʿalā Uṣūl al-Kāfī*, p. 234 (editors introduction)). The use of the phrase may be merely formulaic, as *shaykhunā* and *mawlānā* is often used when a direct scholarly relationship is not chronologically feasible.

[10] See Astarābādī, *al-Fawāʾid*, pp. 58–59.

[11] See Astarābādī, *al-Fawāʾid*, p. 59.

[12] Astarābādī, *al-Ḥāshiya ʿalā Uṣūl al-Kāfī*, p. 235 (editors' introduction). This *ijāza* is mentioned by al-Afandī, *Riyāḍ*, v. 2, p. 193 and v. 4, p. 86.

[13] The phraseology here is clearly formulaic: *jāmiʿ mā yajūzu lahu* (or *lī*) *riwāyatahu bi-ṭuruqihi al-muqarrara fi amākinihā* (see Fāḍilī, *Ijāzāt*, p. 521 and p. 524).

[14] This distinguishes them from *ijāzāt al-ijtihād* (or similar formulations such as *ijāzāt al-iftāʾ*) in which a scholar is recognised as a *mujtahid* by his *mujtahid* teacher. This sort of *ijāza*, frequently referred to in the literature, is, by definition, not considered important for Akhbārīs. See Stewart, "The Doctorate", pp. 46–52.

own, to pass on to his pupils. As Ṣāḥib al-Madārik is recorded as saying, "I wish [Astarābādī] to be incorporated into the *silsila* (chain) of transmitters of the pure *ḥadīths*, recorded from the Ahl al-Bayt and the Illuminator of the Message [that is, the Prophet]."[15] The reference here to the transmission of *ḥadīth* is a reflection of one possible origin of the *ijāza* system. By granting an *ijāza*, the recipient is licensed to relate the *ḥadīths* (or books of *ḥadīths*) mentioned by the *ijāza* donor. In late Shīʿī *ijāzas*, the "Four Books" (*al-kutub al-arbaʿa*) are typically named as the material the recipient is permitted to relate. In the early period of Islam, of course, oral transmission was considered the prime means of *ḥadīth* transmission.[16] By Astarābādī's time, the emphasis on oral transmission was still in evidence. However, the theoretical elaboration (outlined below) of the different ways in which one scholar can be said to "relate" from another indicates that the *ijāza* system had become uncoupled from its original purpose concerned with the oral transmission of *ḥadīth* alone. Any book (indeed "all books", and not merely *ḥadīths*) can form the content of an *ijāza*, and the transmission need not be oral (though oral transmission still maintained a theoretical primacy). It is clear that what is being passed from teacher to pupil is more than merely a qualification in *ḥadīth* transmission. The later *ijāzas* provide the recipient with both scholarly linkages with the past (in the form of *asānīd*) and the literary residue of the scholarly tradition. The latter includes the collection and arrangement of *ḥadīths*, but other genres of literature also. Ṣāḥib al-Rijāl al-Astarābādī mentions some of the works he has studied with Astarābādī, and these include works of *ḥadīth* (the Four Books), but also works of scholarly biography (*rijāl*, including Ṣāḥib al-Rijāl's own famous work in this genre) and *fiqh* (the *Mukhtalaf al-Shīʿa* of al-ʿAllāma is, for example, mentioned).[17] Ṣāḥib al-Madārik refers to Astarābādī being able now to "relate" from him in the rational, traditional and jurisprudential sciences (*maʿqūl wa-manqūl wa-uṣūl*).[18] The *ijāzas*, then, are a measure of the esteem held by the teacher for the pupil, entrusting him with the transmission of the scholarly tradition. They form not only a permission to

[15] Fāḍilī, *Ijāzāt*, p. 520.
[16] See Schoeler, "schriftlichen oder mundlichen", pp. 201–205 for a summary of current opinions on this, and pp. 228–230 for Schoeler's own thesis.
[17] Fāḍilī, *Ijāzāt*, p. 524.
[18] Fāḍilī, *Ijāzāt*, p. 521.

transmit *ḥadīth* (or a particular work of *ḥadīth*), but also to redeploy them in the author's own works (of whatever genre), and ultimately to interpret them within the boundaries established by the Shīʿī tradition. The links, once established, become an essential element of any subsequent biography of a scholar. The *ṭabaqāt* works record Astarābādī as "relating from" Ṣāḥib al-Madārik[19] and Ṣāḥib al-Rijāl al-Astarābādī,[20] but do not describe the relationship as one of *ijāza*, *qirāʿa* or another of the modes of transmission (described below).

The vocabulary used in the description of Astarābādī's relationship with his teachers (that is, *ijāza, qirāʾa*) is well-established in the transmission of Muslim scholarship. The *ijāza* was, of course, the prime mode of knowledge transmission, which Makdisi associates with a *licentia docendi*.[21] Typologies of knowledge transmission amongst the Muslim scholarly elite generally, and the Shīʿī *ʿulamāʾ* in particular, are attempts to bring coherence to what was clearly an informal system of tutorship by modern standards. The earlier typologies of modes of transmission are well-known. Muhy al-Dīn Yaḥyā al-Nawawī (d. 676/1278), the Shafiʿī author, outlines eight modes of transmission which are formally linked to *ḥadīth* transmission, but were clearly used to preserve the scholarly pedigree of other works.[22] In al-Nawawī's schema, as with most other earlier schema, *ijāza* was only third in the hierarchy of modes of transmission. Typologies of knowledge transmission, couched in terms of *ḥadīth*, but clearly referring to religious knowledge more generally, are also to be found in Shīʿī works of *al-dirāya* (*ḥadīth* study). A typical exposition, roughly contemporary with Astarābādī, is found in Muḥammad Taqī al-Majlisī's (Majlisī I) Persian commentary on Ibn Babūya's *Man lā Yaḥḍuruhu al-Faqīh*. In the introduction to this work, Majlisī I outlines seven modes of transmission. These are obligatory for all who wish to transmit *ḥadīth*, and are all called "*ijāza*". These are:

---

[19] For example, Afandī, *Riyāḍ*, v. 5, p. 36.

[20] For example, Baḥrānī, *Luʾluʾa*, p. 119.

[21] Makdisi, *The Rise*, pp. 140–152.

[22] See Nawawī, *Taqrīb*, pp. 102–121, where there are eight types of *ḥadīth* transmission, the third of which is *ijāza* and which is, in turn, divided into seven different sub-types. See also Ibn Salāḥ, *Muqaddima*, pp. 62–87, of which al-Nawawī's *Taqrīb* is a summary (also described briefly by Robson in his article "The Study and Transmission of Ḥadīth", p. 24).

1. the teacher reads a work from beginning to end to his pupil.
2. the teacher reads part of a work to his pupil.

Both of these, Majlisī I claims, are termed "*qirāʾat-e shaykh*" (*qirāʾat al-shaykh*—"the Shaykh's reading").

3. the pupil reads the work to the teacher (termed *qirāʾat bar shaykh/ qirāʾa ʿalā al-shaykh*—"reading to the Shaykh").
4. the pupil is present when another pupil reads the work to the teacher.
5. the teacher gives a copy of the work to the pupil, saying "relate this work from me" (termed *munāwala*—"taking possession").
6. the teacher gives the pupil permission to relate a particular book from him, though (it seems) without the recitation of the work taking place (termed *ijāzah bi-maʿnā-yi akhaṣṣ*—the most specific meaning of the term *ijāza*).
7. an individual finds a work in the hand of a particular teacher, and then relates the work with this qualification (*wijādah*—"finding").[23]

It is clear that Majlisī I's order is one of preference, as he refers, for example, to 1. as the best (*bihtarīn*) and 4. as "not bad" (*bad nīst*) and "close to" (*qarīb bih*) 3. in value. There is, he mentions, some dispute concerning whether or not 3. should be ranked above 1. and 2. Majlisī I himself considers it most prudent (*aḥwaṭ*) to consider 1. and 2. as superior. He gives no reasons for his preference, though one can surmise what they might be. In these latter modes of transmission, the teacher himself speaks, whilst in 3. the pupil speaks (and hence there is a greater potential for erroneous transmission). The dogged primacy of oral transmission mentioned above is evident in Majlisī I's typology. However, there is also potential for ambiguity in any simple application of this typology to the descriptions of knowledge transmission found in *ijāza* documents and *ṭabaqāt* works. All the above seven modes are referred to as types of *ijāza*, though 6. is the "most specific meaning" of *ijāza*. As has been observed by others, *ijāza* as a specific term for the third ranked mode of transmission (after

[23] The typology is described in Majlisī I, *Lawāmiʿ*, v. 1, pp. 65–67, and is followed by Majlisī I's own transmission lines back to the Imams, acquired through his own *ijāza* (though of what type is unclear) from al-Shaykh al-Bahāʾī, his father Ḥusayn b. ʿAbd al-Ṣamad al-ʿĀmilī, ʿAbd Allāh al-Tustarī and others. The *ijāza* is not merely for *ḥadīth*, but also works of *fiqh, rijāl, tafsīr, qirāʾāt* and other.

qirā'a and *samā'a*, equivalent to 3. and 4. above) has given way to *ijāza* being used as a general term for all modes of transmission. Its original meaning is now reserved for the sixth ranked mode, where recitation by either pupil or teacher is not a prerequisite.

In *ṭabaqāt* works, the relationship between scholars is most regularly described in terms of a junior scholar "relating from" the senior. This appears to signify an *ijāza* in the general sense (and not type 6. above) referred to by Majlisī I. Which of the above seven modes of transmission is designated by phrases such as *yarwī 'an* is rarely explicitly signified. For example, one of Astarābādī's pupils, Zayn al-Dīn b. Muḥammad b. al-Ḥasan (whose father had previously given Astarābādī an *ijāza*), is recorded as "relating from" Astarābādī in al-Ḥurr al-'Āmilī's important *ṭabaqāt* work, *Amal al-Āmil*.[24] Elsewhere, al-Ḥurr records the relationship as one of *qara'a 'alā* ("reading to")—that is, Majlisī I's third type of knowledge transmission.[25] One might, then, suppose that the phrase "relating from" designates a "reading" relationship, as many of the relationships recorded in al-Ḥurr's *'Āmil* (and elsewhere) are referred to as "related from" in one place, and "read to" in another. However, there are exceptions. Al-Ḥurr's paternal uncle Muḥammad b. 'Alī al-Jub'ī is recorded as having "read *with*" (*qara'a 'indahu*) Zayn al-'Ābidīn b. Muḥammad al-Nabāṭī.[26] Al-Ḥurr feels it necessary to add that Muḥammad b. 'Alī relates "from him" as well as having "read with" him. What such locutions might signify is rarely explicitly stated. It could mean that they studied together under another shaykh, as al-Nabāṭī is recorded as having studied with Ṣāḥib al-Ma'ālim (though not as relating from him). Alternatively, their relationship may have been of type 1 or 2 above. Whichever is the case, the stock phrase *yarwī 'an* can clearly indicate a relationship other than *qara'a 'alā* ("reading to"). Further examples could be elaborated to demonstrate that the phrase *yarwī 'an* is general and indicates a scholarly relationship different in character from "studentship" (*tilmīdh*). Al-Nabāṭī, for example, was a student of Ṣāḥib al-Ma'ālim, but neither al-Ḥurr nor subsequent *ṭabaqāt*

[24] See Ḥurr, *'Āmil*, v. 2, p. 246. The linkage is recorded in an account of al-Ḥurr's own linkage with Astarābādī: "We relate from Zayn al-Dīn b. Muḥammad b. al-Ḥasan from him [i.e. Astarābādī]".

[25] Ḥurr, *Amal*, v. 1, pp. 92–93.

[26] Ḥurr, *Amal*, v. 1, p. 99.

writers designate this as a "relating from" relationship. Rather like the manner in which Majlisī I describes *ijāza* as having become a general term to describe any of the six modes of knowledge transmission, *yarwī ʿan* (at least in al-Ḥurr's *Amal*) designates a general scholarly connection. Though his relationship is more specific and is stronger than merely being a pupil, it could also be more precisely described ("reading with", "reading to" etc.). For ease of reference I refer to *ijāza* recipients as "pupils" even though this does not always describe a particular relationship between a *mujāz* and a *mujīz*. The typologies of knowledge transmission recorded by Majlisī I and his contemporaries appear rather rarefied and theoretical, bearing a tangential relationship to the multiplicity of scholarly relationships (and the terminology used to describe those relationships) found in other literature of the period. The difficulty in assessing the relationship signified by phrases such as *yarwī ʿan*, or indeed which type of *ijāza* was given (if any), should be borne in mind in the course of the following analysis.

Astarābādī's Scholarly Network

The variety of scholarly linkages exemplified by the description of Astarābādī's connections with his teachers can also be seen in the depiction of Astarābādī's relationships with his pupils. It is primarily through an analysis of these relationships, in which Astarābādī was the senior party, that the spread of his ideas concerning the illegitimacy of *ijtihād* can be initially mapped. Ten scholars are recorded as "relating from" (or more explicitly, as receiving an *ijāza*) from Astarābādī. These ten scholars certainly do not exhaust Astarābādī's pupils. They represent only those relationships which were deemed significant for documentation, and for which documentation has survived. There were, surely, many other pupils who have not been documented. However, contained within the ten names below are scholars of some importance who themselves are recorded as having carried out teaching activities. Notwithstanding the likelihood of unrecorded pupils, the list below probably exhausts the most important pupils of Astarābādī. Some of the listed scholars receive extensive coverage in the *ṭabaqāt* and *ijāzāt* literature in their own right. A possible network of routes through which Astarābādī's Akhbarism may have been disseminated in the early period is thereby established.[27] In (approximate) chronological order of their period of study with Astarābādī, the ten scholars are:

1. Muḥammad Maʿṣūm b. Aḥmad al-Dashtakī (d. 1015) is said to "relate from" (*rawā ʿan*) Astarābādī.[28]
2. Mājid al-Baḥrānī (d. 1028) received an *ijāza* in 1020 from Astarābādī.[29]
3. ʿAbd al-Hādī al-Ḥusaynī al-Tustarī (d. not known) who received an *ijāza* from Astarābādī in 1029.[30]
4. Ḥusayn b. al-Ḥasan b. Yūnis al-Ẓahīrī (alive in 1051) who is said to have "read with" Astarābādī in Mecca.[31]
5. Zayn al-ʿĀbidīn b. Nūr al-Dīn al-Kāshānī (d. after 1040), known as Muʾassis Bayt Allāh,[32] whose own *ijāza* to ʿAbd al-Razzāq al-Māzandarānī (d. not known) mentions being a pupil of Astarābādī and relating from him.[33]
6. Ibrāhīm b. ʿAbd Allāh al-Astarābādī known as al-Khaṭīb (d. after 1081) who is said to "relate from" Astarābādī in various later *ijāzas*.[34]

---

27 A similar type of analysis has been carried out by Schmitdke (see her, "The *ijāza*").

28 The linkage is found in the *ijāza* of Aḥmad b. Muḥammad Maʿṣūm al-Dashtakī (d. 1085), father of al-Madanī (author of *Sulāfat al-ʿAṣr*) to Muḥammad b. ʿAbd al-Ḥusayn al-Dashtakī (d. not known). See Majlisī II, *Biḥār*, v. 107, p. 30. Fāḍilī, in his list of Astarābādī's pupils (Astarābādī, *al-Ḥāshiya ʿalā Uṣūl al-Kāfī*, pp. 237–241, editor's introduction), fails to mention this linkage, probably because the reference in the *ijāza* is to Muḥammad Amīn al-Jurjānī (rather than al-Astarābādī). See below, pp. 150–151.

29 A copy of the *ijāza* is found in the Malik Library in Tehran: *Malik Fihrist*, v. 5, p. 236, MS#1118/8.

30 Al-Ṭihrānī has seen a copy of the *ijāza*, written in Mecca and found on a manuscript "amongst the books of Sayyid Muḥammad al-Yazdī in Najaf". He refers to the relationship as *qaraʾa ʿalā*, with regard to the legal sections of *al-Kāfī* only. Ṭihrānī, *Ṭabaqāt*, v. 5, p. 264.

31 Afandī, *Riyāḍ*, v. 2, p. 44. A note records that in the manuscript, al-Ẓahīrī's name is deleted, and a fuller entry is given a few pages later. Ḥasan al-Ṣadr includes al-Afandī's entry in his *Takmila Amal al-Āmil* (pp. 178–179). Al-Ẓahīrī's questions to Astarābādī and Astarābādī's answers are edited in Gleave, "Questions and Answers" (and in different form in Astarābādī, *al-Fawāʾid*, pp. 547–567).

32 "The restorer of God's house", a reference to him placing the Black Stone in place during the rebuilding of the Kaʿba following a flood in 1039. See Afandī, *Riyāḍ*, v. 4, pp. 399–400. See also below, n. 59.

33 Majlisī II, *Biḥār*, v. 10, p. 14; Kashmīrī, *Nujūm al-Samāʾ*, p. 97.

34 The link is well-established, for example, in the *ijāzas* of Muḥammad Bāqir al Majlisī to Aḥmad al-Baḥrānī (d. 1100 or 1102) and Muḥammad Fāḍil al-Mashhadī (d. after 1092). Majlisī II, *Ijāzāt al-Ḥadīth*, p. 24 and p. 243 respectively.

7. Zayn al-Dīn b. Muḥammad b. al-Ḥasan al-ʿĀmilī (d. 1064) relates from Astarābādī according to al-Ḥurr.[35]

8. Fakhr al-Dīn Ḥaydar al-Lankarī (d. after 1031) who studied with Astarābādī in 1031.[36]

9. ʿĪsā al-Duzmārī (or al-Dizamārī, d. not known) received an *ijāza* from Astarābādī in 1032.[37]

10. Ṣafī al-Dīn Muḥammad al-Shīrāzī (d. after 1033) received an *ijāza* in 1033 from Astarābādī.[38]

Information on a number of these scholars is minimal. Ḥaydar al-Lankarī, ʿAbd al-Hādī al-Tustarī, Ṣafī al-Dīn al-Shīrāzī and ʿIsā al-Duzmārī were clearly minor scholars, whose names survive merely as *ijāza* recipients or students of Astarābādī. Little else is known of their scholarly careers. As more copies of manuscripts with which Astarābādī had a connection (usually through him having corrected a copy) are found, it seems likely that additional names will come to light. Furthermore, the reliability of some of the linkages might be questioned. For example, the first mentioned linkage above (Muḥammad Maʿṣūm) refers to the grandfather of the author of *Sulāfat al-ʿAṣr*, who is said to have died in 1015. He relates from one Muḥammad Amīn al-Jurjānī, who relates from Mīrza Muḥammad al-Astarābādī (Ṣāḥib al-Rijāl). Muḥammad Amīn al-Jurjānī is the name by which Astarābādī is known in the *Sulāfat al-ʿAṣr*, and there is no record of any other Muḥammad Amīn studying with Ṣāḥib al-Rijāl. It seems highly likely, then, that the reference here is to Astarābādī. However, the chronology does not fit. Muḥammad Maʿṣūm's death date (1015) forces one to the conclusion that he must have studied with Astarābādī early in the latter's career when one would have expected Astarābādī to have been junior to Muḥammad Maʿṣūm. Furthermore, this al-Jurjānī is recorded in the *ijāza* as relating from Ṣāḥib al-Rijāl, whom we know he did not meet until relocating to Mecca in 1015, the year of Muḥammad Maʿṣūm's death. These points might cast

[35] See n. 24 above.

[36] Kashmīrī, *Nujūm al-Samāʾ*, p. 97.

[37] The *ijāza* is found on a copy of al-Kulaynī's *al-Kāfī*, corrected by Astarābādī himself, found in the Gowharshad Library in Mashhad (*Gowharshad Fihrist*, v. 1, p. 240, MS#282).

[38] The *ijāza* is found on the cover of a copy of *al-Fawāʾid al-Madaniyya* in the Marʿashī Library in Qum, MS#423 (*Marʿashī Fihrist*, v. 2, p. 28). Fāḍilī transcribes the *ijāza* in his introduction to Astarābādī, *al-Ḥāshiya ʿalā Uṣūl al-Kāfī*, p. 240.

doubt on the reliability of the identification of Muḥammad Amīn al-Jurjānī with Astarābādī, though an explanation is available. Firstly, Muḥammad Maʿṣūm may have died young, and studied with Astarābādī during his time in the ʿAtabāt or in Shiraz before Astarābādī's move to Mecca in 1015. Secondly, seniority in age does not necessarily preclude a scholar from being an *ijāza* recipient. Mājid al-Baḥrānī, for example, received an *ijāza* from Astarābādī (of which we have a copy), though he was almost certainly older than Astarābādī at the time. Thirdly, the fact that Aḥmad, Muḥammad Maʿṣūm's son traces his *isnād* through Jurjānī/Astarābādī, and then Ṣāḥib al-Rijāl, implies only that the final (that is, death bed) *isnād* of a teacher can be used by the pupil, even if their period of study together predates the teacher's later *ijāza* relationship.[39] Astarābādī received his *ijāza* from Ṣāḥib al-Rijāl in 1017, two years after Muḥammad Maʿṣūm's death. This, however, did not prevent Aḥmad, Muḥammad Maʿṣūm's son, using the Astarābādī-Ṣāḥib al-Rijāl linkage in his own *ijāza* to Muḥammad b. ʿAbd al-Ḥusayn al-Dashtakī.[40] If this is the case, then this demonstrates well that what is conferred in an *ijāza* is not a technical transmission of material, but a mark of approval in which the scholar is awarded a place in the scholarly hierarchy. Even if this link is established, it is not directly relevant to our study here, since Muḥammad Maʿṣūm would have studied with Astarābādī well before the latter's conversion to Akhbarism, and therefore could not have been responsible for disseminating Akhbarism.[41]

It is, then, with the remaining six of Astarābādī's pupils that his immediate influence can be traced. These pupils formed the principal conduits for the initial transmission of Astarābādī's ideas to a major

[39] The link is further confirmed by a reference in al-Madanī's *Riyāḍ al-Sālikīn*: "my father Aḥmad Niẓām al-Dīn, related to me from his father Muḥammad Maʿsūm, who related from his Shaykh Muḥammad Amin *al-Astarābādī*, who related from his Shaykh Mīrzā Muḥammad al-Astarābādī" (Madanī, *Riyāḍ al-Sālikīn*, v. 1, p. 31, emphasis added).

[40] See above, n. 28.

[41] It is of passing interest, though, that Muḥammad Maʿṣūm was a descendent of Ṣadr al-Dīn al-Dashtakī, whom Astarābādī heavily criticises in his theological works, see above, p. 39. The link may have been established during Astarābādī's time in Shiraz. Furthermore, Aḥmad b. Muḥammad Maʿṣūm is said to have relocated to Haydarabad whilst it was under Quṭbshāhī control and had close relations with the royal household there (see Madanī, *Sulāfat*, pp. 10–22 and Ḥurr, *Amal*, v. 2, p. 27). This linkage may have been the reason for Astarābādī dedicating the *Dānishnāmah-yi Shāhī* to Muḥammad Quṭbshāh (see above, p. 36). Muḥammad Maʿṣūm himself had close relations with Shāh ʿAbbās II (r.1052/1642–1077/1666).

part of the Shīʿī world. It is reasonable to assume that the centres of Akhbārī thought which developed in the century following Astarābādī's death were initiated by these scholars (or by the pupils of these scholars). However, none of Astarābādī's pupils are explicitly described as Akhbārīs in subsequent *ṭabaqāt* literature.[42] Considering my earlier comments concerning the general functions of *ṭabaqāt* literature this is not particularly surprising. Even the staunchest Akhbārīs are rarely designated as such in Safavid *ṭabaqāt* literature. The genre did not allow such explicit and controversial designation until the late Twelfth/Seventeenth Century.[43] Astarābādī's own pupils, then, are not remembered in *ṭabaqāt* works as Akhbārīs, and their works are not yet available in order that their school allegiance might be determined. However, many of their own pupils are known to have been famous Akhbārīs. Some identified themselves as such; others were designated as Akhbārīs by others; and others still can be described as Akhbārīs through an examination of both their extant works and summaries of the contents of works which have not yet been discovered. My point here is that the initial spread of Astarābādī's Akhbarism was brought about by the work of these scholars. However, the major intellectual elaboration of Akhbarism was carried out by their pupils (and their pupils' pupils) who did produce significant literary material and whose Akhbārī allegiance can be securely verified. These later scholars, two generations removed from Astarābādī, became the leading figures of the Akhbārī school in Safavid Iran, the ʿAtabāt and elsewhere.

If Muḥammad Maʿṣūm al-Dashtakī's studentship with Astarābādī is discounted as too early or of unreliable authenticity, the first pupil to receive an *ijāza* from Astarābādī (as an Akhbārī) was Sayyid Mājid al-Baḥrānī. This *ijāza* is dated 1020, three years after Astarābādī himself received his final *ijāza* from Ṣāḥib al-Rijāl. It was therefore given either after Astarābādī's conversion to Akhbarism or at least after Ṣāḥib al-Rijāl's call to him to revive the Akhbārī path. The recipient, Mājid al-Baḥrānī, was born in Bahrayn, where he held the position of judge (*qāḍī*) and prayer leader until he moved (at an unknown date) to Shiraz. From there, he made trips to the ʿAtabāt and the Hijaz (during one of which he received the *ijāza* from

[42] See above, p. 150.
[43] See above, pp. 48–54.

Astarābādī). He died in Shiraz in 1028.[44] It was he, it appears, who founded an Akhbārī teaching establishment in Shiraz, training a number of noteworthy Akhbārī scholars. The earliest biographies record his excellence in poetry (*shāʾir*^{an} *adīb*^{an}),[45] though works on *uṣūl al-fiqh, rijāl* and *ḥadīth* are also attributed to him. He is said to have been the "first to disseminate *ḥadīth* in Shiraz", which al-Ṭihrānī takes to be a reference to his Akhbārī leanings.[46] Whether or not he is best classified as an Akhbārī must await the discovery and publication of his religious works. One of his students calling him "the seal of the *mujtahids*" might seem to mitigate such an identification,[47] as might his close relationship with the *mujtahid* al-Shaykh al-Bahāʾī (d. 1031), from whom he also received an *ijāza* (though this was a linkage of a number of early Akhbārīs). He is perhaps best known as the first teacher of Muḥsin Fayḍ al-Kāshānī (d. 1091), a major Safavid religious thinker, who, for at least for part of his life, was a staunch Akhbārī (described as *akhbāriyy*^{an} *ṣalb*^{an} in later biographical compendia).[48] Fayḍ studied with Mājid al-Baḥrānī, receiving an *ijāza* from him.[49] When Mājid al-Baḥrānī died, Fayḍ moved to Isfahan, studying with al-Shaykh al-Bahāʾī for a year, before making his way to Mecca where he met Astarābādī,[50] though he does not seem to have studied with him or received an *ijāza*. It was also, probably, during

[44] Biographical references for Mājid (in chronological order of composition) include: Madanī, *Sulāfat*, pp. 500–504; Ḥurr, *Amal*, v. 2, p. 225 and p. 226; Afandī, *Riyāḍ*, v. 5, pp. 5–6 (identical with al-Ḥurr's entry); Baḥrānī, *Luʾluʾa*, p. 137 (the reference to 1022 as the date of death is a misprint); Khʷānsārī, *Rawḍāt*, v. 6, pp. 67–72; Baḥrānī, *Anwār al-Badrayn*, pp. 78–82; Tabrīzī, *Rayḥānat*, v. 1, p. 232; Ṭihrānī, *Ṭabaqāt*, v. 5, p. 482.

[45] Madanī, *Sulāfat*, pp. 492–500; Ḥurr, *Amal*, v. 2, p. 225 and p. 226 (there are two entries, though al-Ḥurr reckons them to be the same person despite them having different names).

[46] Fayḍ mentions him in this role in his introduction to his *ḥadīth* collection/commentary, *al-Wāfī* (v. 1, pp. 28–29); see also Baḥrānī, *Luʾluʾa*, p. 137. Presumably the science of *ḥadīth* was already present in Shiraz, and hence Ṭihrānī takes this to be a reference to Mājid al-Baḥrānī's Akhbārī proclivities. Ṭihrānī, *Ṭabaqāt*, v. 5, p. 483.

[47] The reference is quoted by Ṭihrānī, *Ṭabaqāt*, v. 5, p. 483.

[48] Baḥrānī, *Luʾluʾa*, p. 121. An account of his Akhbārī ideas can be found in Kohlberg, "Akhbārī", pp. 136–145. See also, Gleave, "Two Classical Shīʿite Theories of *qaḍāʾ*".

[49] The story of how Fayḍ came to study with Mājid al-Baḥrānī is found in Jazāʾirī, *Zahrāʾ al-Rabīʿ*, pp. 284–285.

[50] Astarābādī is, presumably, who is referred to when Fayḍ says he was "guided by one of our companions from the people of Astarābād, who lives in Mecca", *al-Ḥaqq al-Mubīn*, p. 12.

his period with Mājid al-Baḥrānī that Fayḍ wrote his first work of *uṣūl al-fiqh*, entitled *Naqḍ al-Uṣūl al-Fiqhiyya*. This brief work is, as its title suggests, a denunciation of the science of *uṣūl*, and whilst Fayḍ does not describe himself as an Akhbārī within it, his refutation of *ijtihād* covers all the principal points of Astarābādī's own critique in *al-Fawā'id al-Madaniyya*.[51] Fayḍ's *al-Uṣūl al-Aṣīla*, written later in 1044, is also clearly a work of Akhbārī jurisprudence,[52] as is his *Safīnat al-Najāt*, analysed by Kohlberg.[53] Finally, Fayḍ's identification of Mājid as his most important teacher in *ḥadīth*, given at the beginning of his *ḥadīth* collection/commentary *al-Wāfī* (in which he explicitly states his Akhbarism), is further evidence that he was initially introduced to Akhbārī ideas through his two years of study with Mājid al-Baḥrānī in Shiraz.[54] Fayḍ was surely not the only scholar in Shiraz to gain a grounding in Akhbarism under Mājid's tutelage, and Shiraz produced a number of important later Akhbārī scholars who probably studied in the establishment founded by Mājid. Amongst Mājid's other pupils was ʿAlī Naqī al-Shīrāzī (d. 1060) who "read to" Mājid al-Baḥrānī.[55] ʿAlī Naqī's allegiance to Akhbarism is not mentioned in the earliest biographical entries, though he did compose a treatise criticising the practice of *taqlīd*, and held (like many Akhbārīs) to the doctrine that smoking tobacco was prohibited.[56] He was a judge in Shiraz, under Qulī Khān Hākim, and then Shaykh al-Islām in Isfahan under Shāh ʿAbbās II. One of his pupils pioneered the "Akhbārī method" of *tafsīr* referred to by Lawson, namely ʿAbd ʿĀlī

---

[51] See below, p. 297.

[52] I refer to this work further below, p. 234.

[53] Kohlberg, "Akhbārī", p. 136. The earliest dated manuscript of this work was copied in 1069 (*Melli Fihrist*, v. 9, p. 394, MS#1386 (1)). This is the *terminus ad quem* for the composition, and indicates that Fayḍ was propounding Akhbārī ideas into his sixties. Abisaab erroneously attributes an Akhbārī work by this title to Astarābādī (Abisaab, *Converting Persia*, p. 106).

[54] Fayḍ, *al-Wāfī*, v. 1, pp. 28–29. Mājid al-Baḥrānī is described by Fayd as "my teacher, and upon whom I rely and depend in matters of legal knowledge."

[55] See Afandī, *Riyāḍ*, v. 4, p. 271.

[56] The reasons for Akhbārī preference for treating smoking tobacco as forbidden are discussed by Jaʿfariyān in his introduction to Ḥurr, *Risāla fī bayān ḥukm shurub al-Tutun*, pp. 81–87. ʿAlī Naqī's treatise on this topic is to be found cited at some length in Afandī, *Riyāḍ*, v. 4, pp. 273–276. One of his pupils refers to him as *Khātam al-mujtahidīn* (Muḥammad b. Maḥmūd al-Ḥabasī, *Nadadh al-Ta'rīkh* referred to in Ṭihrānī, *Ṭabaqāt*, v. 5, p. 419), though once again this may be of formulaic rather than sectarian significance. He is listed by Qummī with the "Akhbārīs of the time" who prohibited smoking tobacco (Qummī, *Favāyid*, p. 339).

al-Huwayzī (alive 1073), the Shīrāzī *muḥaddith*/Akhbārī and author of *Nūr al-Thaqlayn*.[57]

Whilst Akhbarism was taking root in Shiraz, other pupils of Astarābādī were establishing and maintaining Akhbarism elsewhere. An Akhbārī prescence in Mecca after Astarābādī's death was clearly maintained. Zayn al-ʿĀbidīn al-Kāshānī (d. after 1040), known as "the restorer of the House of God"[58] was an important figure amongst the Shīʿa of Mecca. His prestige was recognised in him being the one chosen to re-lay the Black Stone of the Kaʿba after the flood in 1039.[59] He was "martyred for his Shīʿī belief" in Mecca, though the date is not recorded,[60] and was buried in a plot he had reserved in the prestigious graveyard of ʿAbd al-Muṭallab and Abū Ṭālib. His respect for Astarābādī, perhaps indicating his commitment to maintaining Astarābādī's intellectual legacy, is indicated by his decision to be interred next to Ṣāḥib al-Rijāl and Astarābādī. His major work, however, did not concern theology or law, but instead was the result of his research into the construction (and reconstruction) of the Kaʿba, entitled *Mufarraḥat al-anām fī taʾsīs bayt Allāh al-Ḥarām*.[61] His most famous pupil was Muḥammad Muʾmin Dawsat (or Dūst) al-Astarābādī (d. 1087) who was Astarābādī's son-in-law and to whom Zayn al-ʿĀbidin gave an *ijāza*. Muḥammad Muʾmin is, in turn, best known as one of the teachers of Muḥammad Bāqir al-Majlisī (d. 1111) whose eminence is such that he is claimed by both Akhbārīs and Uṣūlīs as one of their own.[62] Muḥammad Muʾmin also gave an *ijāza* to Aḥmad b. Muḥammad b. Yūsuf al-Māqābī al-

---

[57] Lawson, "Akhbārī *Tafsīr*", pp. 178–180. His biography can be found in Ḥurr, *Amal*, v. 2, p. 153; Khʷānsārī, *Rawḍāt*, v. 4, p. 209. In the latter, al-Huwayzī's Akhbarism is discussed at length, citing al-Jazāʾirī's *al-Maqamāt*, in which al-Huwayzī is identified as a strict Akhbārī (p. 210).

[58] See above, n. 32.

[59] On this flood, see Sibāʾī, *Taʾrīkh Makka*, v. 1, pp. 22–23.

[60] This may have been as late as the suppression of Shīʿītes in Mecca in 1087 or 1088. For a full discussion of his life and work, see Amīn, *Aʿyān al-Shīʿa*, v. 7, pp. 168–169.

[61] This work has been edited by Rasūl Jaʿfariyān (see Zayn al-ʿĀbidīn al-Kāshānī, *Mufarrahat al-Anām*, pp. 368–393). These biographical details can be found in Afandī, *Riyāḍ*, v. 4, pp. 399–400. His admirer, Fatḥ Allāh, describes him as "the *mujtahid* of the age", though, as already mentioned (n. 56), this need not indicate him being critical of Astarābādī's position. See Fatḥ Allāh's *Abniyat al-Kaʿba*, including an Arabic translation of this work (Ṭihrānī, *al-Dharīʿa*, v. 1, p. 73 #360).

[62] Muḥammad Bāqir al-Majlisī's position in the Akhbārī-Uṣūlī dispute is discussed below, pp. 241–244 and pp. 264–266. On him as an anti-Akhbārī, see Naṣīrī, "ʿAllāmah-yi Majlisī", p. 46 and Malikī, "ʿAllāma Majlisī".

Baḥrānī (d. 1102) who held the distinctively Akhbārī doctrine that Friday prayer was individually obligatory.[63] In his own scholarship, Muḥammad Mu'min was clearly orientated to *ḥadīth* collection. His *Kitāb al-Rajʿa*—which has survived—is a selection of *ḥadīth* relating to the return of the Mahdī, with minimal personal comment.[64]

The Akhbārī school took roots in Jabal ʿĀmil, initially through the scholarly activities of Astarābādī's pupils, Zayn al-Dīn b. Muḥammad al-ʿĀmilī and Ḥusayn al-Ẓahīrī. Zayn al-Dīn al-ʿĀmilī was the grandson of the great *mujtahid* scholar, Ṣāḥib al-Maʿālim, Ḥasan b. Shahīd II (d. 1011/1602) who, according to some reports, had given an *ijāza* to Astarābādī. His family, then, had a strong Uṣūlī tradition, though he does not seem to have have followed Usulism. Born in Jabal ʿĀmil in 1009, he studied first with the pupils of his father and grandfather. He then spent time in Iraq before travelling to Iran. In Iran, he studied with al-Shaykh al-Bahā'ī for some time, "reading" and "hearing" the latter's works[65] and eventually travelling to Mecca with al-Shaykh al-Bahā'ī in 1030. It was probably during this trip that he studied with Astarābādī, deciding to stay in Mecca for sometime after Astarābādī's death, and dying there in 1064.[66] Despite his learning, he never wrote an organised work in any discipline. Al-Ḥurr al-ʿĀmilī, one of his pupils, attributes this to him "being cautious and fearful of fame" (*li-shidda iḥtiyāṭihi wa-li-khawf al-shuhra*). Whether this caution arose from piety or fear of Sunni opposition is unclear. However, al-Ḥurr does describe Zayn al-Dīn's intellectual approach in more detail:

> He used to say, "The modern scholars have written much, and in their works there are many errors—may God forgive them and us. This has even caused some of them to be killed." He was amazed that his grandfather, Shahīd II, and Shahīd I and al-ʿAllāma used to read to Sunni scholars (*qirā'atuhum ʿalā ʿulamā' al-ʿāmma*), following their [that is, the Sunnis] works of *fiqh, ḥadīth*, the two *uṣūls* [cf. *uṣūl al-fiqh* and *uṣūl al-dīn*] and studying with them. He used to criticise them for

---

[63] See Baḥrānī, *Lu'lu'a*, pp. 36–39.

[64] See Muḥammad Mu'min, *al-Rajʿa*, pp. 3–12 (introduction).

[65] The terms are obviously technical in the sense that they refer to types of *ijāza* donation discussed above, pp. 145–166.

[66] An account of Zayn al-Dīn's life is found in the famous work *al-Durr al-Manthūr* (v. 2, pp. 222–238), with extensive citations from his poetry. This work, composed by Zayn al-Dīn's brother, ʿAlī b. Muḥammad al-ʿĀmilī (d. 1103/1691), is probably the most reliable source for Zayn al-Dīn's life. Ḥihrānī notes he died in Mecca (Ṭihrānī, *Ṭabaqāt*, v. 5, p. 236).

this, saying, 'the results of this are as they are' [that is, the results of this are plain to see]. May God forgive them."[67]

The rejection of Sunni influence in the religious sciences was, as we have already seen, a major element of Astarābādī's Akhbārī methodology.[68] This, of course, does not incontestably identify Zayn al-Dīn al-ʿĀmilī as an Akhbārī. However, it does indicate that he adopted at least one of the major elements of the Akhbārī polemic—that is, the criticism of Sunni influence on Shīʿī thought, and it is likely that he developed this approach during his time with Astarābādī in Mecca.[69] It was this approach which he introduced back to Jabal ʿĀmil either directly or through intermediaries, and which he communicated to his pupil, al-Ḥurr al-ʿĀmilī (discussed below).

Ḥusayn b. al-Ḥasan al-Ẓahīrī was, perhaps, more influential in establishing an Akhbārī school in Jabal ʿĀmil, though little is known of his scholarly career. He spent much of his life in his home village of Jubʿa. He travelled to Mecca and studied with Astarābādī sometime between 1031 and Astarābādī's death in 1033 (or 1036), receiving an *ijāza* and studying *al-Fawāʾid al-Madaniyya*. His personal copy of *al-Fawāʾid* has survived.[70] He returned to Jabal ʿĀmil, dying there at an unspecified date (but certainly later than 1051). It seems likely that the questions he sent to Astarābādī (which have survived) were composed in Jabal ʿĀmil (that is, before 1036), on his return from Mecca. They may have been composed at the request of "the people" who were puzzled or eager to know more of this new doctrine of Akhbarism.[71] In the text of the questions, he praises Astarābādī profusely, and clearly considers Astarābādī's Akhbārī

[67] Ḥurr, *Amal*, v. 1, p. 93. Al-Ṭihrānī considers his criticism to be related to these scholars' "lack of *taqiyya*" (Ṭihrānī, *Ṭabaqāt*, v. 5, p. 236). It seems clear, though, that here is a principled objection to the incorporation of Sunni ideas into Shīʿī scholarship.

[68] See above, p. 98. Stewart identifies this as the major element of the Akhbārī message (see his *Islamic Legal Orthodoxy*, pp. 207–208).

[69] In this, he seems to have differed from his brother, ʿAlī b. Muḥammad al-ʿĀmilī (d. 1103), author of *al-Durr al-Manthūr* and the anti-Sufi and anti-Akhbārī work *al-Sihām al-Māriqa*. See, Abisaab, *Converting Persia*, pp. 110–112.

[70] Al-Ṭihrānī makes reference to it in *Ṭabaqāt*, v. 5, p. 174. It is dated 1047.

[71] Al-Afandī (*Riyāḍ*, v. 2, p. 49) refers to questions asked by the people (the Arabic in the printed edition is garbled—*saʾalahā ʿan al-nās*—and is corrected by Ṭihrānī, *Ṭabaqāt*, v. 5, p. 174—*saʾalahā ʿanhu al-nās*—who works from the manuscript of the *Riyāḍ*). Al-Afandī also contains two entries on al-Ẓahīrī (v. 2, p. 44 and v. 2, pp. 48–49), the latter being based on al-Ḥurr's entry (*Amal*, v. 1, p. 70).

approach to be the most appropriate methodology for the Shīʿa. He did not, however, fully understand the radical nature of Astarābādī's approach. For example, he asks Astarābādī to write a work of *fiqh*, whilst it would be immediately clear from a reading of *al-Fawāʾid al-Madaniyya* that Astarābādī has little time for the genres of either *fiqh* or *uṣūl al-fiqh*. Instead, Astarābādī answers that commentary on the *ḥadīth* is the most appropriate discipline for a true scholar.[72]

The introduction of Astarābādī's Akhbarism to Jabal ʿĀmil, then, most likely began with Zayn al-Dīn al-ʿĀmilī and Ḥusayn al-Ẓahīrī. Their importance to the history of Akhbarism lies primarily in their joint tutorship of Muḥammad b. al-Ḥasan al-Ḥurr al-ʿĀmilī (d. 1104), one of the most prolific Akhbārī authors of the Safavid period. Al-Ḥurr was born in the village of Mashghar in Jabal ʿĀmil, studied first with family members, and then later with Zayn al-Dīn al-ʿĀmilī and Ḥusayn al-Ẓahīrī amongst others. He gained *ijāzas* from them both.[73] He left Jabal ʿĀmil for Iraq in around 1072, and travelled on to Iran to complete his pilgrimage to the shrines of the Imams. He settled in Mashhad, founding an Akhbārī school there and dying in 1104.[74] His voluminous literary output and his distinctive brand of Akhbarism are analysed in later chapters.[75] Here it is sufficient to note that his Akhbarism dates from his time in Jabal ʿĀmil, before his relocation to Iran. This confirms the presence of Akhbārī scholarship in Jabal ʿĀmil, probably initiated and developed by Astarābādī's pupils, Zayn al-Dīn al-ʿĀmilī and Ḥusayn al-Ẓahīrī.

Bahrayn also became (and remains to this day) a centre of Akhbārī scholarship, and here also one finds scholarly links through which Astarābādī's ideas may have been transmitted. In particular, Astarābādī's pupil, Ibrāhīm b. ʿAbd Allāh al-Astarābādī, known as al-Khaṭīb, gave an *ijāza* to Aḥmad b. Muḥammad al-Maqābī in 1081.[76] As has already been mentioned, this scholar held the view that Friday Prayer was individually obligatory. This position was

---

[72] See Gleave, "Question and Answers" and Astarābādī, *al-Fawāʾid*, pp. 545–575.

[73] The *ijāza* from Zayn al-Dīn al-ʿĀmilī was given in 1051. See Ḥurr, *al-Wasāʾil*, v. 3, p. 170.

[74] His autobiography is included in his *Amal*, v. 1, pp. 141–154. His praise of and agreement with Astarābādī can be found at various points in his voluminous output (for a particular point of agreement, see Ḥurr, *Ithbāt al-Hudā*, v. 1, pp. 101–103).

[75] See below, pp. 246–249.

[76] Al-Ṭihrānī refers to the *ijāza*, and that the *isnād* is traced through Astarābādī (*Ṭabaqāt*, v. 5, p. 3).

popular amongst Akhbārīs, though not exclusively so (it was, as has already been noted, Astarābādī's own position).[77] Al-Khaṭīb also held an *ijāza* from Muḥammad Bāqir al-Majlisī.[78] These factors make it likely (though not certain) that he had Akhbārī leanings. Aḥmad al-Baḥrānī travelled outside of Bahrayn, though he completed most of his initial studies in Maqāba. He travelled to Mecca, received an *ijāza* from Ibrāhīm al-Khaṭīb, and returned to Bahrayn. He later travelled to Isfahan (where he met Majlisī II) and then on to Iraq. He died in Baghdad during the plague of 1102 and was buried in the Kāẓimayn shrine. Whilst teaching in Bahrayn before leaving for Isfahan and Iraq, he granted *ijāzas* to a number of Bahrayni scholars, thereby founding an Akhbārī teaching tradition there. Amongst his pupils was Sulaymān b. ʿAbd Allāh al-Baḥrānī al-Māḥūzī (d. 1121), counted an Uṣūlī by some, and an Akhbārī by others.[79] In turn, Sulaymān was the teacher of a famous later Akhbārī, ʿAbd Allāh b. Ṣāliḥ al-Samāhījī whose *Munyat al-Mumārisīn* is considered an important expression of later Akhbārī ideas and part of which has been edited and analysed by Newman.[80] Whilst geographical proximity, and the regularity of traffic between the Bahrayn region and Mecca, may have enabled Astarābādī's ideas to spread easily to this area, the earliest recorded scholarly link between Astarābādī and a Bahrayni scholar is through Aḥmad al-Maqābī.[81] The Akhbārī school in eastern Arabia developed from these initial activities, to dominate the cultural life of the Shīʿī community in the Gulf generally.

[77] See above, pp. 91–92 for Astarābādī's position.

[78] Majlisī II, *Ijāzāt al-ḥadīth*, pp. 19–25.

[79] Yūsuf al-Baḥrānī describes one of Sulaymān's works thus: "His work *al-ʿAshara al-Kāmila* contains ten treatises on *uṣūl al-fiqh*, which indicate that he was a vehement supporter of *ijtihād*. However, it can be understood from his later works that he returned to a position close to that of the Akhbārīs." (Baḥrānī, *Luʾluʾa*, p. 10). The fact that he "returned" to something approaching Akhbarism, possibly indicates that this was his doctrine early in life (ie during his time studying with Aḥmad al-Baḥrānī), before becoming a supporter of *ijtihād*, and then in later years returning to a modified form of Akhbarism. Sulaymān's pupil, al-Samāhījī, remembers having to hide his Akhbarism from his master (see Baḥrānī, *Luʾluʾa*, p. 37, n. 2).

[80] See Newman, "The Akhbārī-Uṣūlī Dispute, pt. 1" and "The Akhbārī-Uṣūlī Dispute, pt. 2".

[81] I am discounting here any influence of Mājid al-Baḥrānī in the establishment of a Bahrayni school of Akhbarism. Whilst his family certainly hailed from Bahrayn, extant records place him in Shiraz from the time of his *ijāza* from Astarābādī to his death in 1028. See above, pp. 152–153.

Of course, it is quite possible, indeed likely, that Astarābādī's ideas were disseminated through pupils who were insufficiently distinguished either to deserve biographical notice, or for their *ijāzas* to be preserved in the later collections of *ijāzāt*. The above account merely presents a summary of the avenues of distribution known through the currently available literature. An important scholar, such as Zayn al-Dīn al-ʿĀmilī, returning to his home town or sending his pupils back from Mecca, is likely to have established a more significant tradition than a minor scholar whose impact was inevitably less impressive. The prestige held by Astarābādī's pupils in Mecca, Lebanon, Shiraz and Bahrayn probably led to the spread of his ideas in these places first (between Astarābādī's death in 1036 and approximately 1050). After a short period of time, scholars trained in these nascent centres of Akhbarism travelled and spread Astarābādī's influence further a field (al-Ḥurr al-ʿĀmilī and Aḥmad al-Maqābī are examples of this second phase). By 1080, it seems, Astarābādī's ideas had been distributed further, giving rise to institutions of Akhbārī learning in Isfahan, the ʿAtabāt and ultimately in as distant a place as Haydarabad. This second phase of development comes about as a result of the efforts of the pupils of Astarābādī's pupils. Scholars in these locations may well have had contact with Astarābādī's Akhbarism earlier than the travels of al-Ḥurr al-ʿĀmilī, Aḥmad al-Baḥrānī and other second generation Akhbārīs. The currently available sources indicate that it was through the activities of the pupils of Astarābādī's pupils that Iraq and Isfahan gained a significant Akhbārī presence. Furthermore, the emergence of Akhbārī teaching establishments in these locations does not necessarily indicate Akhbārī dominance. There was, of course, a variety of intellectual trends within the Shīʿī intellectual elite of the period, and this variety would have been reflected within the scholarly establishments within a particular place.

By 1080, this secondary development of scholarly centres gave way to a more general acknowledgement of Astarābādī's importance and a more general popularisation of Akhbarism across the Shīʿī world. Many of the early copies of *al-Fawāʾid al-Madaniyya* that currently exist were transcribed during the three decades after Astarābādī's death,[82]

[82] See, for example, (and in order of composition): Marʿashī, MS#3028 (*Marʿashī Fihrist*, v. 8, p. 214, date deails are found on p. 217) dated 1042 AH; Sepāhsalār, MS#1053 (*Sepāhsalār Fihrist*, v. 7, p. 601) dated 1053 AH; Marʿashī MS#8276 (*Marʿashī*

and there was clearly a concerted effort to publicise his ideas both during his lifetime, and (perhaps with more vigour) after his death. Muḥammad Taqī Majlisī (Majlisī I, d. 1070) writes that:

> About 30 years ago, the great and learned Mawlānā Muḥammad Amīn Astarābādī (may God have mercy on him) began to occupy himself with the examination and study of the *akhbār* of the Sinless Imams. He studied the censure of opinion and evaluation [found in the *akhbār*] and became acquainted with the method of the companions of the holy Sinless Imams. He wrote the *Favāʾid-i Madaniyyah* and sent it to this country [that is, Iran]. Most of the people of Najaf and the Holy ʿAtabāt approved of his method and returned to the *akhbār*. The truth is that most of what Mawlānā Muḥammad Amīn said, is true.[83]

The *Lawāmiʿ*, from where this passage is taken, was completed in 1066 (though this passage may have been written earlier) and hence "about thirty years previous" would have been during the last years of Astarābādī's life. Two points can be made in the light of this citation. Firstly, it is perhaps significant that Majlisī I refers to Astarābādī "sending" *al-Fawāʾid al-Madaniyya* to Iran (rather than his pupils disseminating his ideas). This might be seen as evidence of the relatively early introduction of Astarābādī's ideas to Isfahan where Majlisī I was based, and an alternative means whereby his ideas were spread.[84] Secondly, Majlisī I refers to the people of Najaf and the ʿAtabāt as being predominantly followers of Astarābādī's ideas. If this is an accurate portrayal, Majlisī I's statement describes the spread of Akhbarism to the ʿAtabāt earlier than that mentioned by some commentators, who identify Yūsuf al-Baḥrānī as the scholar who brought about the dominance of Akhbarism in the ʿAtabāt in the twelfth *hijrī* century.[85] The earliest record of a scholar with links to Astarābādī visting the ʿAtabāt relates to Aḥmad al-Maqābī (a pupil of one of Astarābādī's pupils), and his visit occurred much later (probably some 60 years after Astarābādī's death). By 1066, however, Majlisī I was able to say that Akhbārī ideas dominated the ʿAtabāt, and hence it can be assumed that there were other means

Fihrist, v. 21, p. 237), dated 1067 AH; Mellī, MS#1440 (*Mellī Fihrist*, v. 9, p. 471) dated 1067 AH; Mashhad, MS#13977 (*Mashhad Fihrist*, v. 16, p. 436) dated 1069 AH.

83 Majlisī I, *Lawāmiʿ*, v. 1, p. 47. A full citation of Majlisī I's writings on Akhbarism found in the *Lawāmiʿ* can be found Jaʿfariyān, *Ṣafaviyyah*, v. 3, pp. 1057–161.

84 See the discussion of Akhbarism in Isfahan below, pp. 163–165.

85 See Cole, "Akhbari-Usuli Conflict", pp. 13–16.

by which Akhbarism reached southern Iraq than those recorded in the *ṭabaqāt* works.

Whilst I have found no other evidence that Astarābādī himself sent copies of *al-Fawāʾid al-Madaniyya* to Iran or the ʿAtabāt, there is a significant body of evidence, surveyed above, that his pupils, after studying with him in Mecca, returned to their own communities (many of which were in Iran) to establish modest Akhbārī teaching curricula in the *madrasas* there. These, in turn, trained cosmopolitan scholars who then publicised Astarābādī's ideas, either directly through the distribution of copies of *al-Fawāʾid al-Madaniyya*, or indirectly through the composition and teaching of their own Akhbārī works. These, the evidence suggests, were the mechanisms whereby Astarābādī's Akhbarism became recognised, both by supporters and opponents, across the whole Shīʿī community. By the death of Majlisī I, or perhaps a little earlier, *al-Fawāʾid al-Madaniyya* was well known throughout the Shīʿī world, and there was no need for teachers to introduce his ideas. Rather, from this time on, there was (from an Akhbārī perspective), the need to argue for their validity, to compose Akhbārī works in a variety of genres and thereby bring greater coherence to Akhbārī methodology than that bequeathed by Astarābādī himself. The most sophisticated expressions of Akhbarism date from this later time, two generations after Astarābādī, when Akhbarism had established itself and was the subject of sustained Uṣūlī attack.[86] In order to respond to such attacks, Akhbārī scholars required a more elaborate juristic doctrine than that put forward by Astarābādī. Hence, it was in this generation that a number of leading Shīʿī scholars, both within Safavid Iran and outside of it, produced the bulk of the surviving treatises in which the doctrines of Akhbarism are elaborated and defended. It is at this point, in the later Eleventh Century, that the criteria for an Akhbārī "school" appear to have been met.

---

86 The earliest recorded refutation of *al-Fawāʾid al-Madaniyya* is probably ʿAlī al-ʿĀmilī's *al-Shawāhid al-Makkiyya*, completed in 1055. This was printed on the margins of the original lithograph edition of *al-Fawāʾid al-Madaniyya*. On ʿAlī al-ʿĀmilī, see Madanī, *Sulāfat*, pp. 302–304.

The Development of Akhbārī Ideas Outside of Astarābādī's Scholarly Network

Astarābādī's Akhbarism first gained ground in Mecca, Shiraz, Lebanon and Bahrayn from the time of his first *ijāza* to Mājid al-Baḥrānī (1020), to around 1050. After this initial stage, the scholars trained in these first centres travelled, preaching Akhbarism in Isfahan, the shrine cities of Iraq and eventually India. However, there is also evidence that the reception of Akhbārī ideas in Isfahan was aided by the presence of what might be called a "proto-Akhbārī" position advanced by the influential scholar Mullā ʿAbd Allāh al-Tūstarī (or al-Shūshtarī, d. 1021).[87] I am not referring here to the general study of *ḥadīth*, which had never left the curriculum of Shīʿī studies. Rather, al-Tustarī's ideas, to the extent that they are available in the extant documentation, demonstrate marked similarities with Astarābādī's critique. A comprehensive comparison of their methodologies must await the publication of al-Tustarī's extant works.[88] From the available evidence, al-Tustarī proposed a *ḥadīth*-based jurisprudence, critical of those who gave extensive rein to rational exegetical procedures. His pupil, Majlisī I, terms this methodology *al-fiqh al-akhbārī*,[89] saying that al-Tustarī was the first to teach this methodology in Isfahan. This was presumably carried out in his own *madrasa*, established through an endowment from Shah ʿAbbās I after the Shah had persuaded al-Tustarī to return from his seclusion in the shrine in Mashhad. Indeed, it is from this period (after his return from Mashhad in 1006) that most of al-Tustarī's "proto-Akhbārī" views can be traced.[90] He held

---

[87] I am, here, discounting the significance of the Akhbārī *nisba*, attributed to Mīr Yūsuf ʿAlī al-Ḥusaynī "al-Akhbārī" who died after 1019, and is said to have corresponded with Nūr Allāh al-Tustarī (d. 1019), author of *Majālis al-Muʾminīn*. His recorded views concerning the Prophet's knowledge do not seem to have had any resonance with Astarābādī's own programme. See Ṭihrānī, *Dharīʿa*, v. 2, p. 93.

[88] Such as his *Jāmiʿ al-Fawāʾid* (a commentary on a commentary of al-ʿAllāma's *al-Qawāʿid*, *Mashhad Fihrist*, v. 16, p. 161, MS#13417 and MS#6736), and his *risālas* on prayer (Friday prayer, *Mashhad Fihrist*, v. 16, p. 599, MS#7055 and on prayer and purity, *Mashhad Fihrist*, v. 16, p. 397, MS#2436). The *risāla* on Friday Prayer is listed by Jaʿfariyān, *Ṣafaviyyah*, v. 2, p. 309, and argues for the individual obligation to perform Friday Prayer.

[89] See the reference in Ṭihrānī, *Ṭabaqāt*, v. 5, p. 343.

[90] Al-Tustarī's "proto-Akhbārī" ideas might be traced from his time as a pupil of Aḥmad al-Ardābīlī in Karbala (a period of 30 years according to Afandī, *Riyāḍ*, v. 3, p. 204), and therefore provide evidence of an early Akhbārī movement traced through

open and public debate with philosophers, particularly Mīr Dāmād. These debates concentrated on the use of reason as a source of religious knowledge.[91] Whilst the debates were heated, they did not prevent the emergence of the usual mutual scholarly respect. The scholars were said to be close at the time of al-Tustarī's death, with Mīr Dāmād saying the prayers at his funeral.[92] Like later Akhbārīs, al-Tustarī held that Friday prayer was individually obligatory.[93] He is said to have written no independent work in later life because he was concerned simply with disseminating *ḥadīth*. His literary activity was restricted to commentaries and marginalia. Such an attitude reminds one of Astarābādī's own view, expressed in his answers to Ḥusayn al-Ẓahīrī.[94]

Al-Tustarī, then, introduced to Isfahan ideas similar to those of Astarābādī, and therefore laid the groundwork there for the reception of Akhbarism. This is confirmed by an analysis of al-Tustarī's pupils, many of whom reacted to *al-Fawā'id al-Madaniyya* positively and developed Akhbarism further in their own works. Later Akhbārīs, when describing the period of Akhbarism immediately following Astarābādī, considered many of these pupils as preservers of Astarābādī's message. It may, however, be more accurate to consider al-Tustarī's "school" as an independent tradition of thought which fell under Astarābādī's influence. His pupils included Majlisī I, recognised as an Akhbārī by later scholars, and who himself made statements of qualified agreement with Astarābādī, the most explicit being cited above.[95] Also

al-Qaṭīfī to al-Ardābīlī to al-Tustarī (on al-Qaṭīfī's supposed Akhbarism, see Newman, "Development and Political Significance", pp. 776–886). However, the fact that al-Tustarī only turned to this Akhbārī method after his return to Isfahan from Mashhad in 1006 would indicate that it was something he developed whilst away from Isfahan, living in the Shrine of al-Imām al-Riḍā in Mashhad. It may well be that the teaching circle he founded there was revived later in the century with the arrival of al-Ḥurr al-ʿĀmilī. See above, p. 158.

[91] For an account of the debates, see Iskandar-Bayg, *Ta'rīkh-i ʿĀlam-Ārā*, v. 2, p. 1417: they discussed "*mubāḥithāt-i ʿilmī va-masā'il-i ijtihādī*". Al-Ṭihrānī describes al-Tustarī as "defending the Akhbāriyya, and Mīr Dāmād defending the ʿAqlāniyya." (Ṭihrānī, *Ṭabaqāt*, v. 5, p. 343).

[92] Iskandar-Bayg, *Ta'rīkh-i ʿĀlam-Ārā*, v. 2, pp. 1417–1418. A similar respect existed between the great Akhbārī-Uṣūlī adversaries, Yūsuf al-Baḥrānī and Muḥammad Bāqir al-Bihbihānī (see Gleave, "Akhbārī-Uṣūlī Dispute", pp. 95–96).

[93] See Afandī, *Riyāḍ*, v. 3, p. 196.

[94] See above, pp. 157–158.

[95] See also Majlisī I's *Kitāb al-Mas'ūlāt* when, in direct response to a question concerning his support for Astarābādī, he says, "The method of this slave [i.e. Majlisī I] is to act on the basis of *ḥadīths* of which I have knowledge (*ʿilm*) or *ẓann* which

amongst his pupils were scholars later claimed to be Akhbārīs such as Muḥammad Ṣāliḥ al-Māzandarānī (d. 1081 or 1086), a pupil (and son-in-law) of Majlisī I, whose commentary on al-Kulaynī's *al-Kāfī* was considered a significant Akhbārī work, though his identity as an Akhbārī is disputed.[96] Also amongst the pupils who trained in al-Tustarī's *madrasa*, was ʿAbd Allāh al-Tūnī (d. 1071), claimed by Akhbārīs as one of their own.[97] None of these scholars had direct links with Astarābādī, and yet they were all active during the period immediately following Astarābādī's death. Some of their doctrines, explored in the following chapters, do bear striking resemblance to some of those found in *al-Fawāʾid al-Madaniyya*, though they cite him rarely. Only Majlisī I of those mentioned here is known to have read Astarābādī's work. Al-Tustarī's *madrasa* is perhaps best seen as a concurrent proto-Akhbārī development, which enabled Astarābādī's ideas to take secure root in Isfahan later in the Twelfth Century.

Concerning ʿAbd Allāh al-Tūnī, it is noteworthy that he was close to Khalīl al-Qazwīnī (d. 1089), a well-known exponent of Akhbarism, who also had no (recorded) scholarly linkage with Astarābādī. We do know, however, that al-Qazwīnī made his first pilgrimage to Mecca between 1031 and 1032, and it is possible he met or studied with Astarābādī at that time. Whilst the scholars he "relates from" were philosophers and *mujtahids* (in particular, he relates from Mīr Dāmād and al-Shaykh al-Bahāʾī), his own work is firmly in the emergent Akhbārī tradition. He certainly held Astarābādī in great regard, collecting and editing his comments on the *Uṣūl al-Kāfī*, and composing his own commentaries in Persian and Arabic on *ḥadīth* collections. There are numerous humorous stories of his encounters with famous Akhbārī scholars of the day recorded in the *ṭabaqāt* works, including his fractious relationship with Fayḍ and Majlisī I. For example, al-Qazwīnī prohibited the use of tobacco, and wrote a treatise to this effect, sending it to Majlisī I. Majlisī I, who smoked, used the treatise's binding as a *qaliyān* cover and sent it back to

comes close to knowledge. I do not act on the basis of *ijmāʿ* without *ḥadīths*." Majlisī I, *Kitāb al-Masʾūlāt*, p. 692.

[96] He is claimed as an Akhbārī by Fatḥ ʿAlī Zand (*al-Fawāʾid al-Shirāziyya*, f.2A.4). For further discussion of his alleged Akhbarism, see below, pp. 239–241 and pp. 263–264.

[97] Al-Tūnī's important work of *uṣūl al-fiqh*, *al-Wāfiya*, is analysed below, pp. 238–239 and pp. 262–263.

al-Qazwīnī with an insulting note.[98] Stories such as these, recorded by later Uṣūlī *ṭabaqāt* writers, may have been designed to highlight discord within the ranks of Akhbārī scholars. They certainly provide evidence of the existence of doctrinal and personal conflicts amongst the early Akhbārī scholars. Whilst they held some doctrinal and juristic views in common, there was not (yet) a shared school identity preventing such personal rivalries and disagreements from becoming publicly known. Al-Qazwīnī's Akhbārī school in Qazwin was particularly active, producing famous Akhbārī scholars, including ʿAlī Aṣghar al-Qazwīnī (alive 1092) and Raḍī al-Dīn al-Qazwīnī (d. 1096). The school, however, seems to have developed without the direct input of Astarābādī or his pupils. It was probably through the greater distribution of copies of *al-Fawāʾid al-Madaniyya*, rather than the proselytising efforts of Astarābādī's pupils, that Akhbarism took hold in Qazwin. The Qazwin Akhbārī school flourished during Akhbarism's more developed (that is, post–1080).[99]

Two other (supposedly) Akhbārī scholars of the period between Astarābādī's death and 1100 cannot be linked to Astarābādī through *ijāzas*, tutorship or the more general transmission lines (*riwāya*). The first is Ḥusayn b. Shihāb al-Dīn al-Karakī (d. 1076), who is invariably identified as an Akhbārī. He was known as a poet, amongst whose compositions were a number in praise of the Ahl al-Bayt.[100] The work relevant to this study is his *Hidāyat al-Abrār ilā Ṭarīq al-Aʾima al-Aṭhār*, better organised than Astarābādī's *al-Fawāʾid al-Madaniyya* but making many of the same points.[101] He was born in 1012, though it is not known where. His *nisba* indicates ʿĀmilī heritage, though he may have been born to ʿĀmilī parents resident in Iran or elsewhere.[102] It is unlikely that he acquired his Akhbarism in the Jabal ʿĀmil school begun by al-Shaykh al-Ẓahīrī, and later strengthened by Zayn al-Dīn al-ʿĀmilī. This is unlikely because there is a reference to an *ijāza* given him by al-Shaykh al-Bahāʾī, who died in Isfahan in 1030 (that is, when Ḥusayn al-Karakī was 18 years old).[103] Like

---

[98] Khwānsārī, *Rawḍāt*, v. 3, p. 259.

[99] See Momen, "Uṣūlī, Akhbārī, Shaykhī, Bābī", Shahīdī, "Madrasah-yi Falsafī-i Qazvīn dar ʿAṣr-i Ṣafavī".

[100] A number of his poems are cited in Madanī, *Sulāfat*, pp. 355–367.

[101] See below, pp. 249–251.

[102] Abisaab (*Converting Persia*, p. 95, p. 107) refers to him as one of the ʿĀmilī émigrés and an ʿĀmilī descendent.

[103] The *ijāza* is quoted in the *Shudhūr al-ʿAqliyān* of Iʿjāz al-Kintūrī (d. 1286/1869) and cited by Kashmīrī in his *Nujūm al-Samāʾ*, p. 94.

many early Akhbārīs, he studied with al-Shaykh al-Bahāʾī, though did not feel contrained by al-Shaykh al-Bahāʾī's Usulism. On the other hand, there may be reason to doubt the authenticity of this report of an *ijāza* from al-Shaykh al-Bahāʾī to Ḥusayn al-Karakī. Firstly, al-Karakī was still young (though not impossibly so) to receive an *ijāza* from such eminent a scholar as al-Shaykh al-Bahāʾī. Secondly, al-Shaykh al-Bahāʾī is mentioned in *Hidāyat al-Abrār* without any explicit indication of Ḥusayn al-Karakī having studied with him.[104] In the work, al-Shaykh al-Bahāʾī is condemned for following the way of al-ʿAllāma al-Ḥillī.[105] According to al-Ḥurr, Ḥusayn al-Karakī lived in Isfahan "for a time" (*muddat*[an]), and if the *ijāza* is genuine, this must have been for around 45 years. When he completed *Hidāyat al-Abrār* in 1073,[106] he was still resident in Isfahan, but the work's vehement condemnation of the *mujtahids* may have lead to criticism which in turn may have prompted his move to Haydarabad. He moved there two years before his death at the age of 64.[107] As indicated previously, Astarābādī's dedication of the *Dānishnāmah-yi Shāhī* to the Quṭbshāhī ruler in India may have led to the establishment of a group of Akhbārī followers there.[108] If this is the case, then Ḥusayn al-Karakī may have found a sympathetic audience for his approach on his arrival in Haydarabad. According to Mīrzā Muḥammad al-Akhbārī, *Hidāyat al-Abrār* was a popular work amongst the early Akhbārīs, and as is argued in subsequent chapters, it constitutes the first systematic attempt at codifying Akhbārī principles.[109] What is

---

[104] He is referred to as "our shaykh" (*shaykhunā*), though this may be purely honorific. See, for example, Karaki, *Hidāyat*, p. 69.

[105] Karakī, *Hidāyat*, pp. 10–11.

[106] This dating is based on the colophon found on the manuscript of the work in the library of the Ḥusayniyya Shushtariyya in Najaf (MS#, 4–133) and cited in the printed edition: Karakī, *Hidāyat*, p. 312. Of course, he may have completed the work before then, and this merely refers to this copy of the work.

[107] See Ḥurr, *Amal*, v. 1, p. 71. One manuscript records the age of death as 68, though this is an isolated reference, and subsequent biographical dictionaries record him as dying at the age of 64.

[108] Both Astarābādī and his vehement critic, ʿAlī al-ʿĀmilī, dedicated works to the Qutbshahi kings. It seems highly likely that Akhbarism had some presence in India, which is probably what prompted ʿAlī al-ʿĀmilī's dedication of his *al-Shawāhid al-Makkiyya fi madāhid ḥujuj al-khayālāt al-madaniyya* to the Sulṭān ʿAbd Allāh Quṭbshāh. See Ṭihrānī, *al-Dharīʿa*, v. 14, pp. 244–245, #2391.

[109] The work cited is Mīrzā Muḥammad's *Munyat al-Murtād*, cited in Khwānsārī, *Rawḍāt*, v. 7, p. 134. Whether this means the work was sent back or even part composed in Jabal ʿĀmil is not clear. Specifically he refers to ʿĀmilī Akhbārīs, which could refer to Akhbārīs from Jabal ʿĀmil or simply ʿĀmilīs resident in Iran who study the *akhbār*.

clear is that there is no evidence of Ḥusayn al-Karakī having studied with Astarābādī or even his pupils, though the similarity of the ideas presented in the two works makes it likely that al-Karakī was aware of Astarābādī's achievements,[110] particularly likely since Majlisī I, an Isfahani contemporary of Ḥusayn al-Karakī, explicitly states that he has read *al-Fawā'id al-Madaniyya* in he wrote his *Lawāmiʿ Ṣāḥibqirānī* in 1066.[111]

A second, though minor, Akhbārī scholar who cannot be linked with Astarābādī through *ijāza* or *riwāya* is one ʿAbd al-Aẓīm b. ʿAbbās al-Astarābādī, a pupil of Shaykh al-Bahā'ī. He also was clearly not a follower of al-Bahā'ī's Uṣūlī jurisprudence, neither did he follow his other Uṣūlī teacher, Fakhr al-Dīn al-Ṭarīḥī (d. 1081), who wrote a refutation of Muḥammad Amīn al-Astarābādī.[112] ʿAbd al-Aẓīm al-Astarābādī's death date is not recorded, but he must have been a younger contemporary of Astarābādī, and therefore active as an Akhbārī in the years following Astarābādī's death. He is described as an Akhbārī (*min al-ʿulamā' al-akhbāriyyīn*) by Yūsuf al-Baḥrānī,[113] though no literary work is attributed to him. If he was an Akhbārī, it is not clear whether he came into contact with Astarābādī's ideas through regional connections (they share the *nisba*), or whether he was sufficiently late to have benefitted from the growing general recognition of Astarābādī's importance (he probably died after his teacher, that is, after 1081). His significance for the early development of Akhbarism lies in the fact that he gave an *ijāza* to the famous Qur'ān commentator, Hāshim b. Sulaymān al-Baḥrānī (d. 1107). Hāshim al-Baḥrānī, mentioned in later lists of Akhbārī scholars, authored *al-Burhān fī tafsīr al-Qur'ān* which is characterised as a work of Akhbārī *tafsīr* by Lawson.[114] Hāshim al-Baḥrānī, in turn, taught Sulaymān b. ʿAbd Allāh al-Māḥūzī (mentioned above as one who wavered between the Akhbārī and Uṣūlī positions). Sulaymān, in

[110] Karakī, *Hidāyat*, p. 172.
[111] See above, n. 83.
[112] See the citation from al-Ṭarīḥī's *ijāza* to his son, quoted in Afandī, *Riyāḍ*, v. 4, p. 335. The work is entitled (and glossed) *Jāmiʿ al-Fawā'id fī radd ʿalā al-Mawlā Muḥammad Amīn al-qā'il bi-buṭlān al-ijtihād wa'l-taqlīd*. See Ṭihrānī, *al-Dharīʿa*, v. 5, pp. 73–74 (where copies of the work are mentioned attached to other works of al-Ṭarīḥī) and v. 10, p. 186 (where two of al-Ṭarīḥī's works in refutation of Astarābādī are mentioned).
[113] Baḥrānī, *Lu'lu'a*, p. 66.
[114] Lawson, "Akhbārī *tafsīr*", pp. 187–195.

turn, taught the famous Akhbārī, ʿAbd Allāh al-Samāhījī (d. 1135), author of *Munyat al-Mumārisīn*.[115] This constitutes yet another of the many Bahrayni Akhbārī networks which emerge in the century after Astarābādī's death. Whilst ʿAbd al-Aẓīm al-Astarābādī cannot be directly linked to Astarābādī through teaching relationships, his tutorship of the Akhbārī *mufassir*, Hāshim al-Baḥrānī, gives him a minor role in the early development of Akhbarism.

Mature Akhbarism

The early period of Akhbarism after Astarābādī can be traced initially through his pupils, who probably spread Akhbarism to the Hijaz, Shiraz and Jabal ʿĀmil. Later the pupils of these pupils travelled more widely through out the Shīʿī world, disseminating Akhbarism until, sometime in the late Eleventh Century, Astarābādī's ideas and Akhbarism in general became a commonplace element of Shīʿī juristic debates. From the turn of the Twelfth *Hijrī* Century, teaching linkages with Astarābādī become less important since (it seems) there were sufficient copies of *al-Fawāʾid al-Madaniyya* in circulation for scholars to become Akhbārīs through acquaintance with his text alone. Khalīl al-Qazwīnī, Hashim al-Baḥrānī and perhaps even ʿAbd Allāh al-Tūnī represent this second phase of Akhbarism's development. From this period on, Akhbarism's growth was determined, not only by eloquent proselytes, but also by the force of Astarābādī's argument communicated through the written word.

In this period of what might be called "mature Akhbarism", the lines of thought within the Akhbārī school were defined. Amongst the authors of the important Akhbārī works of the period are:

1. Muḥammad Ṭāhir al-Qummī (d. 1098), a pupil of Majlisī I, a virulent anti-Sufi who is described as an Akhbārī both by Uṣūlīs and later Akhbārīs.[116] His association with the Safavid Shahs is well-known.
2. Sayyid Niʿmat Allāh al-Jazāʾirī (d. 1112), a pupil of Majlisī II and author of a number of important Akhbārī works, not least his literary

[115] These linkages are laid out in al-Samāhījī's *ijāza* to Nāsir al-Jarūdī al-Qaṭīfī (see Schmitdke, "The *ijāza*", pp. 70–71).

[116] See Zand, *al-Fawāʾid al-Shīrāziyya*, f.1B.14 and Khʷānsārī, *Rawḍāt*, v. 4, p. 140.

pot-pourri, *al-Anwār al-Nuʿmāniyya*. His son, Nūr al-Dīn al-Jazāʾirī, author of the famous linguistic work, *Furūq al-Lughāt*, may have also been an Akhbārī. His grandson, ʿAbd Allāh b. Nūr al-Dīn al-Jazāʾirī (alive in 1151), is described as an important Akhbārī during the period of Nādir Shāh. He wrote a commentary on Muḥsin Fayḍ's *al-Nukhba*.[117]

3. Nūr al-Dīn Muḥammad al-Kāshānī al-Akhbārī (d. 1115), a descendent of Muḥsin Fayḍ's brother and pupil of Fayḍ. He is the author of, amongst many other works, an Akhbārī *tafsīr* entitled *Kitāb al-Muʿīn*.
4. ʿAbd Allāh b. Sulaymān al-Samāhījī (d. 1135) whose list of 40 differences between Akhbārīs and Usulis (found in his *Munyat al-Mumārisīn*) has already been mentioned.
5. Yūsuf al-Baḥrānī (d. 1176), author of many works including the influential *al-Ḥadāʾiq al-Nāḍira*, the most accomplished attempt to write an Akhbārī *fiqh*, not withstanding Astarābādī's criticism of the genre (and, indeed, al-Baḥrānī's criticism of Astarābādī).
6. Muḥammad al-Akhbārī (d. 1232), a prolific author and colourful personality, who supposedly struck a (failed) bargain with Fatḥ ʿAlī Shāh Qājār to make Akhbarism the state religion of Iran.
7. Fatḥ ʿAlī Zand al-Shīrāzī (d. after 1236), author of *al-Fawāʾid al-Shīrāziyya*, a Persian language refutation of Usulism and the *mujtahids*.

Apart from these major Akhbārī figures, the biographical and bibliographical works within the Shīʿī tradition preserve the names of a large number of Akhbārī scholars from the Twelfth and Thirteenth *Hijrī* Centuries.[118]

Major *mujtahid*/Uṣūlī scholars are, however, also recorded from the period. Muḥammad Bāqir al-Sabzawārī (d. 1090) and Muḥammad al-Khʷānsārī (d. 1099) provide examples of the survival and continued influence of Usulism throughout the period. Akhbarism did not monopolise scholarly activity either in Iran or elsewhere. Al-Sabzawārī rose to high office in the Safavid state and had cordial relations with Muḥsin Fayḍ, even though he did not share Fayḍ's Akhbārī approach. There was also a developing and influential philosophical/mystical school of Shīʿī thought in which the ideas of Mullā Ṣadrā (d. 1050)

[117] ʿAbd Allāh al-Jazāʾirī, *al-Tuḥfat al-Sāniya*, analysed in Gleave, "Scriptural Sufism".

[118] A list of such references is found in the appendix, below, pp. 306–310. For the most part, their works have not survived or remain in manuscript form.

were elaborated upon and developed. For some Akhbārīs (such as Fayḍ), this tradition could be married with Akhbarism.[119] For others, philosophy was excluded along with *ijtihād* in the religious sciences. Of course, scholars explicitly identified as Akhbārīs, represent only a small proportion of the scholars of the period catalogued in a work such as al-Ṭihrānī's *Ṭabaqāt al-Aʿlām*. However, those who are identified as Uṣūlīs or *mujtahids* are also not so numerous. The evidence concerning the majority of minor scholars of the period in question is insufficient to categorise them as either Akhbārīs or Usulis, and until the extensive extent manuscript sources from the period become available, it is on contemporary depictions of the intellectual situation in the major Shīʿī centres (such as that of Majlisī I referred to above) that research must rely. These sources describe Akhbārī ideas, stemming from the thought of Astarābādī and subsequently dominating the major centres of Shīʿī academic activity for a century, dating from the mid to late Eleventh Century *Hijrī* Century. The biographical compendia, most of which admittedly date from the early Thirteenth Century, openly identify numerous scholars as Akhbārīs. The most important theological and juristic works of the period were composed by scholars claimed by later Akhbārīs as their own. Whilst there was undoubtedly a variety of intellectual currents in the *madrasas* of Isfahan, Shiraz, Qum and the ʿAtabāt, the available evidence indicates, not only that law dominated the seminary curriculum, but also that the legal methodology most utilised within that curriculum adhered to Akhbārī interpretive principles. It is to the analysis of the variety within this "mature Akhbarism" that the following chapters are devoted.

Conclusions

The rise of Akhbarism from the period of Astarābādī to its dominance of the Iraqi shrine cities under Yūsuf al-Baḥrānī has been traced here through the biographical references of scholars, teacher-pupil relationships, surviving works of theology and law and marital/family relationships. The network of relationships (depicted in Figure 1) explains how Astarābādī's Akhbarism spread across the Shīʿī world.

[119] See the appendix to Rizvi's *Mulla Sadra Shirazi*.

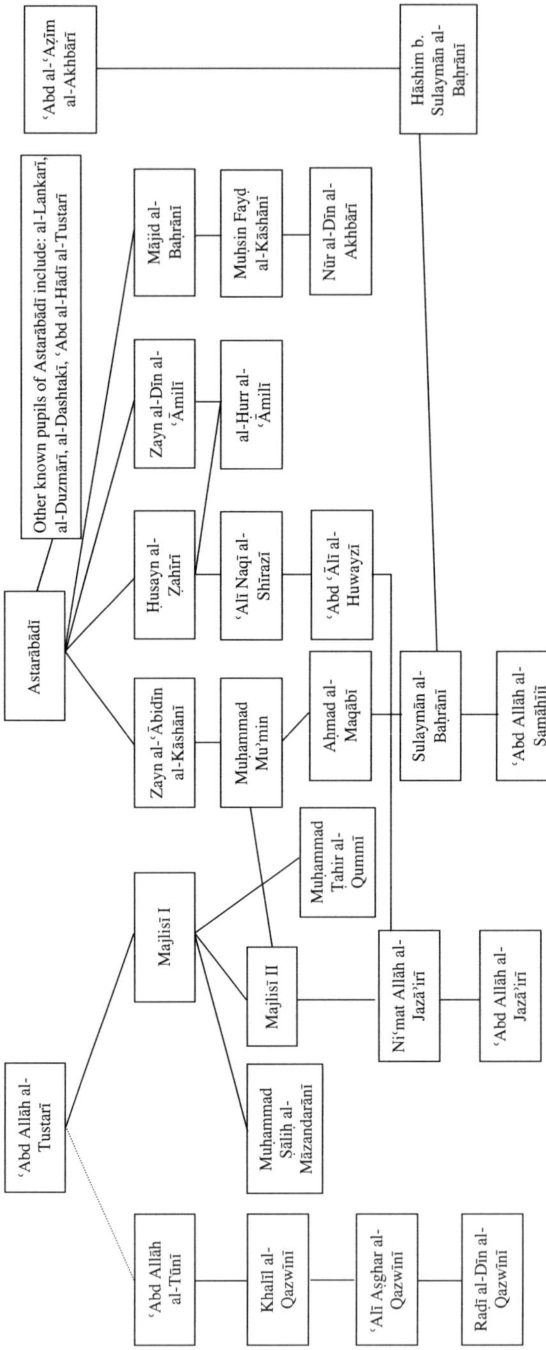

Figure 1. Scholarly linkages between the major Akhbārī scholars, from Astarābādī to Fatḥ ʿAlī Zand
—— = Ijāza Link. = ……… Probable studentship

The criteria of a *madhhab* used by Melchert, and expanded upon by Stewart, would seem best fulfilled by Akhbarism at around the time of al-Ḥurr al-ʿĀmilī.[120] A founder had been established (in the form of Astarābādī himself), standard school doctrines were formed,[121] commentaries on previous Akhbārī works (such as Muḥsin Fayḍ's *al-Nukhba*) were being composed, the school had established an exclusive institutional basis (in Iran, but also in the ʿAtabāt and Jabal ʿĀmil) and a network of distinctive Akhbārī teacher-pupil relationships was in place. Of course, there is a certain artificiality in asserting that an Akhbārī *madhhab* came into existence at one point rather than another, since this ultimately depends on the criteria for a *madhhab* employed to make such an assessment. However, the criteria developed within the Western study of Islamic law were fulfilled by the time of al-Ḥurr al-ʿĀmilī, and not before. Needless to say, the criteria are not fulfilled for the period before Astarābādī himself.[122]

There remains, however, the more difficult question of why Akhbarism proved popular at this particular point in Imāmī Shīʿī history. From an Akhbārī perspective, of course, the rise of Akhbarism was a return to the truth, ultimately brought about by God, inclining the hearts of the believers away from falsehood. Uṣūlī explanations, in which the pernicious Muḥammad Amīn al-Astarābādī dupes the community by a simplistic appeal to tradition, are equally unsatisfying. More comprehensive explanations for Akhbarism's popularity in the Eleventh and Twelfth Centuries, however, have been proposed by a number of commentators. Amongst them are attempts to link Akhbarism to the rise of particular social classes in society (Arjomand),[123] royal patronage designed to undermine clerical authority (Jawdat

---

[120] See above, p. 158.

[121] See the analysis of difference lists, below, pp. 177–215.

[122] By which I mean, that Akhbarism at this time had (1) a recognised founder in the person of Astarābādī (even if some Akhbārīs wished to simply claim that Akhbarism was the Shiʿism of the Imams), (2) a growing corpus of commentaries on standard works of law (in particular the various commentaries on Muḥasin Fayḍ's *al-Nukhba*), (3) the regular transmission of of legal knowledge through an *ijāza* system which was (relatively) exclusive to Akhbārī teachers and pupils, (4) a stable set of school doctrines in terms of *uṣūl al-fiqh*, even though one can detect some variation and dispute within Akhbarism and (5) a number of dedicated Akhbārī institutions, following an exclusively Akhbārī curriculum in Lebanon, Iraq and Iran. In sum, these criteria of a *madhhab* are fulfilled by the time of al-Ḥurr, but not before.

[123] Arjomand, *Shadow of God*, pp. 146–155.

and Babayan),[124] Akhbarism's supposed cooption of popular folk Shiʿism, metanoia brought on by societal changes in the Safavid state (Abisaab),[125] a reaction to the rise of Sufism (Jābirī and al-Wardī),[126] an unholy alliance of Gnostic Sufis and Traditionists (Moussavi)[127] and a puritan reaction against the decadence and irreligiosity of the Safavid court. Such theories provide only partial explanations. Both Arabs and Persians can be found amongst Akhbarism's leading exponents, hence casting doubt on Arjomand's theory that it represented the "Persian" clerical estate (or indeed the alternative theory, that it was primarily an "Arab" movement). Furthermore, both Akhbārīs and Uṣūlīs enjoyed state patronage (and there were those in both schools who did not), making the simple explanation of royal patronage unconvincing.[128] Similarly, Akhbarism generally does not seem to have held a doctrine akin to the supposedly "folk" Shiʿite notions of the Imams (often associated with the *ghulāt*),[129] nor is there a single Akhbārī position on the role of philosophy and mystical experience in the discovery of religious knowledge.[130] Some Akhbārīs lived ascetic lives (such as Khalīl al-Qazwīnī), and others enjoyed the pleasures brought by wealth (such as, by all accounts, the Majlisīs). Akhbarism could clearly be married with both puritan and profligate lifestyles. The most striking characteristic of the Akhbārī school—if it has sufficient coherence to deserve the term—is the multifarious intellectual interests and diverse academic careers of its various adherents. Furthermore, not enough is known of the level of its grassroot support within the Shīʿī world to determine whether or not the Akhbārī-Uṣūlī conflict was a dispute confined to the scholarly elite, and whether or not Shīʿī society was seriously divided by it.[131] A global explanation for the rise of Akhbarism, then, remains elusive.

---

124 Jawdat "al-Ḥaraka al-Akhbāriyya", pp. 2–4; Babayan, *Mystics, Monarchs, and Messiahs*, p. 412.

125 Abisaab, *Converting Persia*, pp. 105–106.

126 Jābirī, *al-Fikr al-Salafī*, p. 279 and Wardī, *Lamaḥāt*, v. 1, pp. 57–61.

127 Moussavi, *Religious Authority*, p. 98.

128 See Gleave, "The Qāḍī and the Muftī".

129 See above, p. 25.

130 See Gleave, "Scripturalist Sufism".

131 Davānī argues that Bihbihan was divided between Akhbārī and Uṣūlī factions (Davānī, *Vaḥīd-e Bihbihānī*, pp. 112–120). Qazwin was also, supposedly, divided into quarters, some dominated by Akhbārīs and some by Usulis (see, Ḥasan al-Amīn, *Mustadrak Aʿyān*, v. 2, pp. 303–304, Shahīdī, "Madrassah-yi Falsafī-i Qazvīn dar 'Aṣr-i Ṣafavī", Shahīdī, "Sahm-i Ḥawzah-yi 'Ilmiyyah-yi Qazvīn", pp. 140–143 and Momen, "Uṣūlī, Akhbārī, Shaykhī, Bābī").

As is clear from the above analysis, identification of an Akhbārī in Shīʿī biographical and historical works is normally triggered by an adoption and espousal of certain key elements of Astarābādī's programme. These include the rejection of *ijtihād* and *taqlīd* as legitimate mechanisms of legal operation and a relaxed attitude towards the processes necessary to validate the authenticity or probative force of *ḥadīths*. These, in turn, are based upon the adoption of a particular legal epistemology in which certain knowledge as to the duties God demands of his subjects is available through an examination of the revelatory sources. From these minimal doctrines, scholars had relative freedom to bolt on a variety of positions on other religious issues. High mysticism, Muʿtazilī-influenced theology, folk Sufism and illuminationist philosophy could all be married with Akhbarism. At times, the marriage was supported by explicit appeal to Akhbārī hermeneutic principles; at others there is no explicit link made between a scholar's legal Akhbarism and his other intellectual pursuits. Similarly, some Akhbārīs, in their condemnation of these non-legal religious disciplines, appealed to Akhbārī doctrines. Even within Akhbārī legal theory, there was variety and scholarly difference. In short, a jurist's allegiance to Akhbārī legal doctrine did not necessarily determine his attitude towards other areas of religious enquiry. Furthermore, as I have indicated elsewhere, Akhbārī allegiance in terms of *uṣūl al-fiqh* did not always dictate unanimity in terms of positive law (found in *furūʿ* works and collections of *fatāwā*).[132] Akhbarism's central doctrines are, then, restricted to issues of legal theory (*uṣūl al-fiqh*) and concern the effect of a particular epistemological stance on the questions of *ijtihād, taqlīd* and the efficiency of the *akhbār* as sources of legal knowledge. Akhbārī jurists managed to achieve a union between these central doctrines and a variety of legal positions concerning broad questions of state legitimacy and juridical authority, as well as legal minutiae such as the purity of wine and the legal requirements of a legitimate marriage.[133] Conflicting legal views were held equally vehemently by different Akhbārīs, and each was justified by an appeal to Akhbārī doctrine. It is a measure of the skill and sophistication of Akhbārī authors that they were able to

[132] See Gleave, "Intra-madhhab *ikhtilāf*".
[133] See Gleave, "The Qāḍī and the Muftī".

take Akhbarism in contrasting and conflicting directions, both within the law and outside of it.

Once this variety is understood, the desire for a global explanation for Akhbarism's emergence is concomitantly reduced. Ironically (given its portrayal as a strict, unbending and doctrinaire tradition in the secondary literature), Akhbarism's flexibility, and its ability to weld itself onto a variety of legal, theological and philosophical views, brought it scholarly popularity in the Eleventh Century. That a scholar's Akhbārī identity was acquired through his position on technical questions of legal theory—and not by anything else—gave Akhbārī scholars a freedom to link their Akhbarism with religious investigation outside of the legal sphere. Such freedom may not have been so readily available to supporters of the Uṣūlī position, for whom the demands of *taqlīd* to a *mujtahid*, at least theoretically, dictated the level of a scholar's involvement in other strictly non-legal religious activities. The search for global explanations of Akhbarism's popularity referred to above are premised on Akhbārīs constituting a unified, almost homogenous, body of scholars. As becomes clear in the following chapters, it is the vibrant intra-Akhbārī debate as well as the broader Akhbārī-Uṣūlī dispute (and possibly also the Shīʿī-Sunni polemic of the period) which best explain the prevalence of Akhbārī juristic scholarship amongst the Shīʿī *ʿulamāʾ* during the Eleventh and Twelfth *Hijrī* Centuries.

CHAPTER SIX

DEFINING THE AKHBĀRĪ-UṢŪLĪ CONFLICT

Designating a particular scholar as an Akhbārī should ideally be based on an established, agreed set of criteria. As has already been explained,[1] the competing conceptions of Akhbarism found in literature originating within the Shīʿī tradition, has led to conflicting accounts of Akhbarism's origins and development. This variety of definitions led also to conflicting roll calls of past Akhbārī scholars, making an assessment of the movement's key phases and doctrines problematic. Producing an account of Akhbarism's development is further complicated by the fact that a writer's conception of Akhbarism is often keyed to his personal evaluation of the contribution of the Akhbārī school. Those within the tradition critical of Akhbarism tend to see it as originating with Astarābādī. Those supportive of it tend to view it as having its origins in the dicta of the Imams themselves. This broad division, denuded of its polemical edge, is also replicated in the writings of modern commentators external to the Shīʿī tradition. The debate over the appropriate dating for the emergence of Akhbarism depends, then, upon the definition of Akhbarism adopted by an author. As I have argued, the terms *akhbāriyya* and *akhbāriyyūn* (and *akhbāriyān* in Persian), whilst present in early Imāmī literature, lacked consistency of reference, and should not be considered proper names as such. The pre-Astarābādī usage of the term *akhbāriyya*, indicated neither a set of defined doctrines, nor a common hermeneutic approach.[2] Only after the work of Muḥammad Amīn al-Astarābādī did terms such as *akhbārī, akhbāriyya* and *akhbāriyyūn* acquire a consistent meaning. A coherent Akhbārī position, it is argued, was initiated by Astarābādī, developed by his pupils and emerged as an important feature of the intellectual landscape of Safavid and early Qajar Iran, and within the same time frame amongst Shīʿī communities outside of Iran. This is not to say that this later Akhbarism had no

[1] See above, pp. 6–8.
[2] See above, p. 25.

connection with intellectual (or indeed, anti-intellectual) Imāmī Shīʿī trends before Astarābādī. Rather, I argue that the Akhbārī approach to questions of legal hermeneutics only gained coherence after (and through) the work of Astarābādī; even then the movement does not fulfil even the most minimal criteria of a "school" (*madhhab*) until at least half a century later.[3] To speak of Akhbarism before Astarābādī, and mean by this a *madhhab*, is simply an inappropriate use of the technical terminology emerging from within the Shīʿī tradition.

Astarābādī's legal theory has already been described, its defining feature being the rejection of the dominant legal epistemology (that is, the graded conception of knowledge, stretching from *ẓann* to *ʿilm*). This rejection led to a dismissal of the hermeneutic techniques which relied on this epistemology (included under the umbrella term *ijtihād*). In its place, Astarābādī proposed a binary epistemology (between *ʿilm* and *ẓann*), and although he did propose a sketch of the reasoning process which should replace Usulism, it was far from comprehensive.[4] It was only later that Akhbārī scholars put forward sophisticated alternatives to the Uṣūlī/*mujtahid* legal methodology.[5] However, even in its more mature stage, authors supporting alternative Akhbārī legal theories display a preoccupation with refuting Usulism, and Akhbarism continued to define itself through a rejection of Uṣūlī doctrine. Akhbārī scholars were unable, generally speaking, to relinquish this defensive mode of self-presentation. Given the reactionary nature of post-Astarābādī Akhbārī discourse, an account of the developing conceptions of the central points of difference between Akhbārīs and Uṣūlīs is, in effect, an account of Akhbārī reflexive thinking. Such an account can be constructed from the various lists of differences between Akhbārīs and Uṣūlīs, composed by Akhbārī scholars. This chapter is an analysis of these lists, from which Akhbarism's perception and prioritising of its own doctrines can be deduced. My concern here is to trace the conceptions of Akhbarism proposed by Akhbārī scholars themselves; or to put it another way, I aim to analyse the development of post-Astarābādī Akhbārī self-definition.

[3] The criteria of a *madhhab*/school are discussed above, p. 173.

[4] See above, pp. 66–87 for a description of Astarābādī's alternative "Akhbārī" legal theory.

[5] See below, pp. 268–296.

Defining the Akhbārī-Uṣūlī Conflict

A considerable number of lists of differences between the *akhbāriyyūn* and the *uṣūliyyūn/mujtahidūn* have been composed. Indeed the constant popularity of the format is not confined to participants in the dispute or writers within the Shīʿī tradition; external commentators have also chosen to present the dispute in the form of comparative lists.[6] Works dedicated to examining the differences between the two schools became an established genre and invariably included such a list.[7] Indeed the emergence of the numbered list genre is, perhaps, one of the most interesting literary consequences of the Akhbārī-Uṣūlī dispute, having little precedence in juristic literature. The lists examined in this chapter are generally presented as non-partisan expositions of the dispute, and are usually found in works of *ajwibat al-masāʾil* ("answers to questions"). The positions of the two schools are defined, but, whatever the allegiance of the author, explicit condemnations or affirmations of either school opinion do not dominate. Furthermore, any variety of opinion within each camp receives rather limited, occasional attention, particularly in the earlier lists. The authors prefer to present the two schools as holding unified doctrinal positions which can be easily compared and contrasted. Having said this, the lists are not merely accounts of the conflict between Akhbārīs and Uṣūlīs. When authored by Akhbārīs, they are a record of the school's self image. Development within these lists reflects not only shifts in the focus of the dispute, but also modifications in the self-definition of the Akhbārī school. As lists, these documents "sharpen the outlines of the categories" forcing the list user (in this case, the reader) to

---

[6] See for example, Falaturi, "die zwolfer-Schia", pp. 81–90; Momen, *Shīʿī Islam*, pp. 223–225, Iskandarjūʾī, "al-Muqārana", pp. 102–115.

[7] A preliminary list of 25 works, specifically dedicated to an exposition of the differences between Akhbārīs and Uṣūlīs, is given in Appendix 2, below, pp. 311–314, with a numerical indication of the "differences" where known. Of course, there may be many additional "lists of differences" found in works not dedicated to delineating the areas of dispute. All the lists examined here are found in works dealing with general Akhbārī themes. That works dedicated to examining the differences between the two schools emerge quite late (the earliest reliably dated work was written sometime before 1203, though there are possible earlier candidates) could be significant. The emergence of independent difference lists between the two schools (as opposed to lists of differences presented in the course of more general works where the focus is not merely to catalogue the differences) is, perhaps, a feature of the more mature phase of Akhbarism's development.

make decisions.[8] This, as Goody has argued, encourages hierarchical thinking in which some listed items are thought more important than others.[9] The lists, then, both define and simplify the positions within the dispute; these two linked processes (definition and simplification) function as part of the author's own argumentation for or against a particular position. Polemic and accompanying argumentation infects the listing process in the less obvious ways examined below. The encroachment of an author's school perspective into what is presented as an "objective" list includes the selection of differences, ordering (even if there is no explicit hierarchical arrangement) and the manner in which doctrinal differences are formulated and presented. In designing his list, an author has the opportunity to control the terms of the dispute and thereby push forward his own thesis. Indeed, the lists analysed below[10] both reflect and create Akhbārī and Uṣūlī "schools"; the lists could be said to act like a creed (indeed, creeds are themselves a form of list) by defining the central doctrines of one school in contradistinction to its principal opponents. In short, the lists display a characteristic common to most lists: they make a play for non-normative status by eschewing open assessments of the items described. However, the processes of selection and arrangement, together with the formulation of literary descriptions of the differences between Akhbārīs and Uṣūlīs, mean the composition of these

[8] See Goody, *Savage Mind*, p. 102. Goody sees lists as a feature of literate societies, and therefore as a consequence of a writing culture.

[9] Goody, *Savage Mind*, pp. 103–108. In a longer list such as al-Samāhījī's *Munyat al-Mumārisīn* (see below, pp. 205–211), this hierarchical thinking is clear: the statement of al-Samāhījī's view of the principal difference (*ijtihād*) is followed by detailed exposition of *ijtihād*-related doctrines (the issue of *ghayba* vs. pre-*ghayba* legitimacy, the command to obey the *ẓann* of the *mujtahid* and so on).

[10] The lists used in the following analysis consist of (in chronological order): Ḥurr (d. 1104), *al-Fawā'id al-Ṭūsiyya*, pp. 447–450 (listing 23 differences); Jazā'irī (d. 1112), *Manba'*, pp. 40–66 (9 differences with commentary); Māḥūzī (d. 1121), *Mas'ala: mā al-Farq bayn al-Akhbārī wa'l-mujtahid* (5 differences, recorded in al-Akhbārī, *Ḥirz al-Ḥawāss*, f.13a.1–13b.6/pp. 26 l.1–27 l.6); Samāhījī (d. 1135), *Munyat al-Mumārisīn* (40 differences in Newman, "Akhbārī/Uṣūlī Dispute, pt. 1", pp. 24–38). Reference is also made to the University of Tehran Manuscript of al-Samāhījī's *Munyat al-Mumārisīn* (not used by Newman; the list of differences is found on f.16a.l25–f.20b.l6.), and to citations of Samāhījī's 40 differences found in Akhbārī, *Ḥirz al-Ḥawāss*, f.13b.6–16b.14/pp. 27 l.6–33 l.14 and Zand, *al-Fawā'id al-Shīrāziyya*, f.13b.4–33a.7; 'Abd al-Nabī al-Shīrāzī (d. after 1175, presuming this citation is from the same 'Abd al-Nabī al-Shīrāzī al-Baḥraynī described in Ṭihrānī, *Ṭabaqāt*, v. 6, p. 476), *Risālat* (4 principal differences, cited in Akhbārī, *Ḥirz al-Hawāss*, f.16b.14–22b.3/pp. 33 l.14–45 l.3); Baḥrānī (d. 1175), *al-Ḥadā'iq*, v. 1, pp. 167–170 (3 differences).

lists cannot be divorced from the desire, on the part of the authors, to present Akhbarism as the more appropriate and sensible juristic methodology.

The lists considered in this chapter vary considerably in length. The longest (written by ʿAbd Allāh al-Samāhījī) consists of forty points of dispute, the shortest (by Yūsuf al-Baḥrānī) comprises only three. Mere numerical calculation should not, in itself, be taken as a measure of a writer's assessment of the dispute's importance. In shorter lists, for example, there is a tendency to include a number of disputed issues under a single heading. In longer lists, each issue receives its own discrete point.[11] Similarly al-Ḥurr's relatively long list of twenty-three differences is specifically designed to demonstrate that the differences between Akhbārīs and Uṣūlīs are substantive (*maʿnawī*), and not merely terminological (*lafẓī*). The following survey of the content of these lists reveals a reoccurring set of doctrines which define the Akhbārī position *vis-à-vis* that of the Uṣūlīs or *mujtahids*. In accordance with Astarābādī's own emphasis, the lists delineate the Akhbārī suspicion of those interpretive methods which produce uncertain results. Some lists devote individual points to these interpretive methods (making the list more expansive); other authors aim at concision, bracketing the methods under a single general heading (such as *ijtihād* or *dalīl al-ʿaql*). Indeed, epistemological concerns underlie almost every difference outlined in these lists, with the Uṣūlī-*mujtahids* emphasising the legitimacy of *ẓann*, and the Akhbārīs advocating a strict adherence to *ʿilm*.

Within these parameters, points of difference can be classified into three broad categories, related to (1) hermeneutics, (2) the status of sources and (3) the role of the jurist.[12] Under hermeneutics, the issue of *ijtihād* predominates: the Akhbārīs reject its validity, whilst the Uṣūlīs-*mujtahids*[13] advocate it. The Uṣūlī-*mujtahid* advocacy of

[11] This said, al-Baḥrānī's desire to abbreviate the dispute to three (albeit substantial) points is probably linked to his general desire to reduce the bitterness and tension between Akhbārī and Uṣūlī scholars within the Shīʿī scholarly community. See Gleave, *Inevitable Doubt*, pp. 248–253.

[12] This three-fold categorisation of principal Akhbārī doctrines is employed in my overview of Akhbārī juristic theory, discussed in Chapters 7, 8 and 9 below.

[13] The significance of the use of the terms *mujtahid* and/or Uṣūlī as the opponents of Akhbarism is discussed below, p. 208. In the following, I use the composite Uṣūlī-*Mujtahid* when talking in general terms. In discussing a specific point in a specific list, I replicate the particular author's usage.

ijtihād is not always presented as having a uniform character. As al-Samāhījī states, some *mujtahids* say *ijtihād* is obligatory individually (*ʿayn*[an]), others say it is obligatory "by choice" (*takhyīr*[an], meaning an individual, once he has chosen to employ *ijtihād*, cannot then decide to ignore its results).[14] Similarly, some say *ijtihād* is obligatory upon all Muslims (*ʿaynī*), though only a few become qualified to practise it, whilst others argue that it is obligatory for the community as a whole (*kifāʾī*—meaning if part of the community study, become qualified, and then practise *ijtihād*, the community as a whole has discharged its obligation). The majority of *mujtahids*, al-Samāhījī states, argue for the latter (*kifāʾī*) position.[15]

The treatment of *ijtihād* itself (as opposed to linked hermeneutic issues) is stated briefly and precisely in the earlier lists. The difference over the legitimacy of *ijtihād* is presented as the first difference in the list of al-Ḥurr al-ʿĀmilī (d.1104/1693), and his presentation merely states that Uṣūlīs say *ijtihād* is "permitted, nay obligatory". The Akhbārīs argue that one can only act on the basis of a text.[16] By locating this difference at the head of his list, al-Ḥurr indicates that this is the principal point of contrast between the two schools. By implication, then, the results of *ijtihād* are not based on texts,[17] but on some other foundation (namely fallible human reason). Al-Jazāʾirī (d. 1112/1700) expands this rather simplistic formulation. For him, the Akhbārīs argue that the early Shīʿī scholars forbade *ijtihād*, and this prohibition remains in force, whilst the *mujtahids* argue that the early scholars had no need of *ijtihād* because of the wealth of legal sources available to them (including, for the earliest scholars, the Imams themselves). Later, according to the *mujtahids*, most of the textual sources were lost and *ijtihād* then became a necessary interpretive tool. Al-Samāhījī (d. 1135/1723) develops this point even further, and covers the issue of *ijtihād* directly in four points. There is the bald statement of Akhbārī denial of *ijtihād* (point 1, mentioned above), followed by a discrete point on the Akhbārī rejection of any distinction between the *ghayba* and post-*ghayba* periods with respect to *ijtihād's*

[14] Newman, "Akhbārī-Uṣūlī Dispute, pt. 1", p. 24.
[15] Newman, "Akhbārī-Uṣūlī Dispute, pt. 1", p. 33.
[16] Ḥurr, *al-Fawāʾid al-Ṭūsiyya*, p. 447.
[17] The term *naṣṣ*, used by al-Ḥurr here, means more than simply "text". Rather it implies a text which conveys its meaning to the hearer or reader in a clear and unambiguous manner.

validity (point 7), followed by a denial of the *mujtahid* position that *ijtihād* is obligatory when one cannot communicate with the Imam (point 17) and finally a rejection of *ijtihād* as either an individual or collective obligation (point 27). Similarly, ʿAbd al-Nabī al-Shīrāzī's (d. after 1175/ 1762) discussion of *ijtihād*, though his presentation is not divided into separate points, delineates Akhbārī rejections, not just of *ijtihād*, but also of the details of *mujtahid* doctrine. *Ijtihād*, for Akhbārīs, is forbidden "in the occultation and in the time of the Imam's presence" (*khʷāh dar zamān-i ghaybat-i Imām bāshad yā dar zamān-i ḥuḍūr*).[18] The simple statement of Akhbārī denial of *ijtihād* in the early lists was elaborated by later list writers, and a more detailed comment on the various *mujtahid* doctrines related to *ijtihād* (and the Akhbārī rejection of them) became necessary.

Other hermeneutic issues are also well-attested as points of dispute between Akhbārīs and Uṣūlīs. All these could be said to relate to *ijtihād*, since employing it implies validation of these interpretive devices. Beginning with the earliest list, al-Ḥurr delineates nine such devices, all condemned by the Akhbārīs. These include implications which can be drawn from texts (*mafhūmāt*),[19] analogical reasoning (*qiyās*, even when the *ʿilla* is stated), *a minori ad maius* arguments (*al-awlā*), "the presumption of licitness" (*al-barāʾa al-aṣliyya*) and "the presumption of continuance" (*istiṣḥāb al-ḥāl*). Each of these is presented as a distinct point of dispute. Al-Jazāʾirī, on the other hand, brackets most of these interpretive devices together (under *al-istinbāṭāt al-ẓanniyya*—"uncertain methods of interpretation"), though he gives separate points to analogical arguments, *al-barāʾa* and *istiṣḥāb al-ḥāl* (the latter two being combined in a single point). Al-Māḥūzī's list opens with the dispute over *al-barāʾa al-aṣliyya*, and a description of how it should be distinguished from the principles of *ibāḥat al-aṣl* (that all things are permitted unless there is specific textual evidence to the contrary) and *istiṣḥāb*.[20] With respect to each device, the Akhbārīs refuse to accept its validity, since its employment leads to uncertain (*ẓannī*) results. Al-Samāhījī also mentions some of these interpretive devices, though interestingly, only towards the end of his forty points (*al-barāʾa*, for example, is not mentioned until point 32; the various

[18] Quoted in Akhbārī, *Ḥirz al-Ḥawāss*, f.19b.7–9.
[19] *mafhūm al-sharṭ*, *mafhūm al-ṣifa*, *mafhūm al-ghāya* and "the rest of the *mafhūmāt*"—on which, see Gleave, *Inevitable Doubt*, pp. 179–181.
[20] Quoted in Akhbārī, *Ḥirz al-Ḥawāss*, f.13a.2–9.

mafhūmāt are not mentioned at all). Al-Shīrāzī does not even refer to these devices as a point of dispute (they are presumably included in his catch-all phrase *al-adilla al-ẓanniyya*, but they deserve no separate points of their own). Finally, although al-Baḥrānī (d. 1186) deals with *al-barā'a al-aṣliyya, istiṣḥāb* and the *mafhūmāt* extensively in his prologues, he does not see them as requiring separate enumeration as points of dispute between Akhbārīs and Uṣūlīs.[21] There is, then, a shift in focus, from detailed delineation of disputes over interpretive devices to their inclusion under more general headings. This does not necessarily indicate that these issues were no longer disputed. Rather, there was a tendency to bracket these issues into one or two points, in order to make room for issues which had come to the fore in the Akhbārī-Uṣūlī dispute—specifically, the issue of the status of revelatory texts, and how different categories of text interact.

Astarābādī's perspective on the status of the revelatory corpus has already been outlined.[22] His view was that the legal contents of the Qur'ān could not be understood without reference to the *akhbār* of the Imams, and that the *akhbār* found in the early collections (in particular the "Four Books") were of certain authenticity (*qaṭʿiyyat al-wurūd*). These positions gave rise to a number of areas of dispute between Akhbārīs and Uṣūlīs, and they were inevitably translated into points within the lists analysed here. Al-Ḥurr gives relatively scant coverage to the issue of the status of revelatory texts. Instead he concentrates on hermeneutic matters. He does mention isolated traditions (*khabar al-wāḥid*): Uṣūlīs argued that they were a basis for action, whilst Akhbārīs argued that they could only act in this way when there was external evidence (*qarā'in*) affirming their authenticity (point 3). The passage is, however, rather poorly formed, as the Akhbārī position, even in al-Ḥurr's own writings, is that the category of *khabar al-wāḥid* is redundant.[23] Similarly al-Ḥurr's fifth point, concerning whether or not one can act on a report of uncertain provenance (*ẓannī al-sanad*), is rather misleading. Whilst Uṣūlīs would, in contrast to Akhbārīs, act on a report which they have classified as *ẓannī al-sanad*, the central point of dispute was over the definition of such terms. The Akhbārīs had a more lax set of criteria whereby a report might be classified as

---

[21] *Al-barā'a al-aṣliyya* and *istiṣḥāb* are listed as elements in *dalīl al-ʿaql* which forms part of al-Baḥrānī's second difference (Baḥrānī, *al-Ḥadā'iq*, v. 1, p. 168).
[22] See above, pp. 74–79.
[23] See below, pp. 246–248.

authentic. The problem here is that al-Ḥurr is not yet able to express the Akhbārī position in independent terminology. He is forced to use the terms that he has inherited from dominant Uṣūlī discourse (*khabar al-wāḥid, ẓannī al-sanad* etc.), a feature which also reflects his general tendency to present Akhbarism primarily as a reaction to Usulism. Al-Jazā'irī's statement of the differences between Akhbārīs and *mujtahids* also fails to cover the issue of the status of revelatory-legal sources adequately. He devotes one point to the issue, in which the *mujtahids* are said to support five sources of law (*kitāb, sunna, ijmā', dalīl al-'aql* and *istiṣḥāb*), whilst the Akhbārīs either reject the last three or have a different understanding of them.[24] Astarābādī's doctrines concerning the authenticity of the *akhbār*, the role of the four hundred *uṣūl* and the collection of the Four Books do not appear in either of these early lists—though they are known to be Akhbārī doctrines which challenged the Uṣūlī position. In al-Māḥūzī's list of differences, however, the issue of the status of revelation is given a distinct point, and receives more extensive coverage:

> For the Akhbārīs, the Four Books are sound in their entirety . . . as more than one of them has explicitly stated. The categories [of *ḥadīth*] are for them two: sound and weak. Every *ḥadīth* which al-Shaykh [al-Ṭūsī] acts upon in his writings, and in *al-Kāfī* and *al-Faqīh*[25] are likewise sound (*ṣaḥīḥ*). *Ṣaḥīḥ* for them means every *ḥadīth* in which they can have confidence, or which is attached to something which cannot be disbelieved. Such [*ḥadīths*] are many [in the Four Books] . . . The *mujtahids* divide the *ḥadīths* into four—*ṣaḥīḥ, ḍa'īf, ḥasan* and *muwaththaq*. This is the position of al-'Allāma and those who follow him, and possibly was unknown before him.[26]

Al-Samāhījī, a pupil of al-Māḥūzī, expands the coverage even further. He lists five separate points relating to the status of revelatory texts: point 2 concerns the superiority of the *sunna* over other sources of law; point 4 concerns the illegitimacy of the four-fold categorisation of *ḥadīths*; point 5 concerns the definition of "sound" (*ṣaḥīḥ*) amongst the Akhbārīs; point 29 concerns the definition of "reliable" (*thiqa*) in *ḥadīth* classification; point 39 concerns the soundness of

---

[24] Jazā'irī, *Manba'*, pp. 43–44. Any distinction between *dalīl al-'aql* and *istiṣḥāb* is rejected by later list authors as the latter is incorporated into the more general category of *dalīl al-'aql*. See below, p. 275.

[25] A reference to al-Kulaynī's *al-Kāfī* and Ibn Bābūya's *Man lā Yaḥḍuruhu al-Faqīh*. Al-Māḥūzī is, then, referencing all the "Four Books" here.

[26] Quoted in Akhbārī, *Ḥirz al-Ḥawāss*, f.13a.16–13b.4.

all Four Books of *ḥadīth*.[27] Even in the pruned and concise lists of al-Shīrāzī and al-Baḥrānī, these issues are given prominence, through the declaration that the sources of law, for Akhbārīs, are two, not four (supplemented by additional glosses).[28]

As the issue of the status of the revelatory sources eclipses purely hermeneutic concerns, so the interrelationship between sources emerges as a principal area of dispute. By the interrelationship between the revelatory sources I am referring to how one source might act as an interpreter of another (such as the *akhbār* acting as a *tafsīr* of the Qurʾān), and, furthermore, how contradictions between and within revelatory sources are solved. These are, of course, hermeneutic issues, and perhaps should be considered alongside linguistic implications (*mafhūmāt*) and analogical reasoning. However, they are hermeneutic issues which only arise once a position regarding the status of revelatory sources has been taken. The Akhbārīs, from Astarābādī onwards, famously adopted the position that the Four Books are of certain authenticity. The issue of the interrelationship between revelatory sources becomes particularly important for them following the adoption of such a doctrine. The common *mujtahid* solution of demoting one of the two contradictory sources to a lower status of authenticity is not available to Akhbārīs (as all *akhbār* are considered "sound"). Furthermore, the Akhbārīs also adopted a set of hermeneutic techniques which gave the *akhbār* a single definable meaning. Holding this restrictive interpretive position also prevents another solution to revelatory contradiction from being used, namely interpreting apparently contradictory reports in such a way that their meanings can be reconciled (the so-called *al-taʾwīlāt al-baʿīda*). Al-Ḥurr's list, however, gives the impression that the dispute between Akhbārīs and Uṣūlīs primarily concerned hermeneutics, with minor issues concerning the status of sources and contradictions between them. The only point where the interrelationship between sources is tackled concerns the interpretation of the Qurʾān by the *akhbār* (point 16). If the apparent meaning of a Qurʾanic passage (*ẓāhir al-qurʾān*) is not confirmed by a report, then the Qurʾān cannot act as a source. This is a rather indirect way of saying that the legal meaning of the

[27] Newman, "Akhbārī-Uṣūlī Dispute, pt. 1", pp. 25–26, p. 34.

[28] al-Shīrāzī, quoted in Akhbārī, *Ḥirz al-Ḥawāss*, f.17b.17–18a.2 and f.20a.8–11; Baḥrānī, *al-Ḥadāʾiq*, v. 1, p. 167.

Qur'ān is only understood after having read its interpretation by the Imams: the Qur'ān can only be understood through the *akhbār*. This is a standard Akhbārī position and it contradicts the Uṣūlī-*mujtahid* view that the Qur'ān can act as an independent source. It is mentioned in a number of the lists reviewed here, but it is the only point in al-Ḥurr's list which deals with the interrelationships between sources.[29] Similarly, al-Jazā'irī and al-Māḥūzī do not view the Akhbārī-Uṣūlī dispute as concerning the interrelationship of sources, with no specific points outlining the Akhbārī or Uṣūlī positions on this issue. This lack of emphasis is, however, remedied in al-Samāhījī's list. He devotes much space to outlining the positions of the two schools on revelatory contradiction and inter-source relationships. Al-Samāhījī's point 11 describes how the *mujtahids* prefer their own opinions on occasions of inter-source conflict, whilst Akhbārīs decide between contradictory reports on the basis of procedures revealed by the Imams themselves. Points 13 and 14 refer to interpreting reports in such a way that they no longer demand obligations or prohibitions, but only refer to recommendations or discouragement. The "*ijtihādī*" doctrine allows contradictory reports to be reconciled, whilst the Akhbārī rejection of it means that the interpreter must choose between reports. Point 16 concerns whether or not the Qur'ān can be interpreted without the aid of the *akhbār*. Point 33 concerns whether or not *al-barā'a al-aṣliyya* can be used to reconcile contradictory *akhbār* (unsurprisingly, the Akhbārīs do not consider this permitted).

Now, the issue of source interrelationship (rather than source interpretation) within these lists naturally follows on from the foregrounding of the issues of source status, and hence one would expect it to emerge as a major disputed area subsequently. This is indeed the case, as the lists of al-Ḥurr, al-Jazā'irī and al-Māḥūzī mention, but do not foreground, the Akhbārī position on the status of revelatory sources. Al-Samāhījī, probably reflecting a shift in focus within the Akhbārī-Uṣūlī dispute in the decades prior to him writing his list, gives the issue of the interrelationship between sources prominence. It took time for the differences between Akhbārīs and Uṣūlīs on these issues to dominate discussions, and hence, unsurprisingly, later lists (such

[29] See Jazā'irī, *Manba'*, pp. 52–58; Newman, "Akhbārī-Uṣūlī Dispute, pt. 1", p. 31; Baḥrānī, *al-Ḥadā'iq*, v. 1, p. 167 (constituting the second of al-Baḥrānī's three differences).

as those of al-Samāhījī and al-Baḥrānī) feature such issues much more prominently.

The final broad category of disputed points relates to the qualifications, role and authority of the jurist. Once again, these issues are marginal in al-Ḥurr's list. The principal point of relevance to these issues refers to whether a non-*mujtahid* should follow a report or the opinion of a *mujtahid* (al-Ḥurr's point 18).[30] Al-Jazā'irī and al-Māḥūzī neglect these issues entirely. As with the issue of revelatory contradiction, al-Samāhījī brings the Akhbārī position on these matters to the fore. The disputed division of the community into *mujtahid* and *muqallid* is outlined in point 6 (Akhbārīs deny this division, saying all community members are equal in following the Imam). Who is permitted to assume the role of judge and market inspector (*muḥtasib*) is discussed in point 8 (the Akhbārīs argue that a transmitter of reports is the only individual qualified; the *mujtahids* argue it should be a *mujtahid*). The definition of a "scholar" (*ʿālim*) is discussed in point 9 (Akhbārīs say it is a transmitter; *mujtahids* say it is a partial or absolute *mujtahid*). The educational qualifications of scholars are delineated in points 10 and 28 (the *mujtahids* demand proficiency in an extended list of disciplines; the Akhbārīs say only skills in the Arabic language are required for an *ʿālim*). The requirement to follow a *mujtahid* is discussed in points 12 and 30. The legitimacy or otherwise of following the rulings of a dead scholar (*taqlīd al-mayyit*) is discussed in point 15. As Newman rightly points out, al-Samāhījī's list does indicate that the role of the jurist had emerged as a major point of contention between Akhbārīs and Uṣūlīs in the late Safavid period.[31] This does not, however, mean that the Akhbārīs had no scholar-layperson hierarchy. In fact, al-Samāhījī's text indicates that the basis of a scholar's authority was, for Akhbārīs, to be found in disciplines related to the *ḥadīth* sciences, and not in knowledge accrued from a broader range of disciplines. It was the basis of scholarly authority which became central to the two parties' disagreement.

These general observations, gathered from a comparison of the various lists, reveal a shift in focus from the work of al-Ḥurr to that of al-Baḥrānī. In the earlier texts, the authors concentrate on

---

[30] The point is amplified into three points, but as is argued below, these are not separate points, but merely *reductio ad absurdum* argumentation. See below, p. 193.

[31] See Newman, "The Akhbārī-Uṣūlī Dispute, pt. 2".

hermeneutic issues, primarily *ijtihād* and the tools to be used by the scholar in the analysis of texts. In later lists, these concerns are marginalised. Though *ijtihād* maintains its iconic status, the various hermeneutic devices (*mafhūmāt, al-barāʾa al-aṣliyya, istiṣḥāb*) are bracketed under general terms and no longer deserve distinct points in the lists. The issues of the status and interrelationship of revelatory texts and the authority of the scholar within the community come to the fore. These points were not absent from early Akhbārī discourse (one finds them well attested in Astarābādī's own writings). However, they did not feature to any great extent in early lists of differences between Akhbārīs and Uṣūlīs. Al-Samāhījī's list indicates a change in the dispute's focus (and hence in the self-definition of Akhbārīs), as he defines the dispute as having a more extensive remit than in previous lists. This is not merely because his list is longer than those written previously, but also because he broadened the focus of the dispute from pure hermeneutic matters to more involved problems of juristic methodology (the status and relationship between texts) and matters of community importance (the qualifications and role of the jurist). List authors subsequent to him usually refer to their own lists as "abbreviations" of, or as drawing on, al-Samāhījī's *Munyat al-Mumārisīn*. The refocusing of the principal areas of Akhbārī-Uṣūlī dispute, as laid out in later lists, reflects the emergence and establishment of an Akhbārī school after the work of al-Ḥurr al-ʿĀmilī. In the previous chapter, this development was traced through an analysis of the *ijāza* linkages between Akhbārī scholars. Here we see a similar process occurring in the list genre, though with a delay of two decades. In the previous analysis, Akhbarism begins to fulfil the criteria of a "school" (*madhhab*) at the time of al-Ḥurr.[32] Here we see some *madhhab* concerns emerging two decades later, in al-Samāhījī's list. The lag is perhaps not surprising. That al-Ḥurr represents the first accomplished list composition points towards the emergence of a defined Akhbārī juristic position (one criterion of a *madhhab*). Al-Samāhījī's list reflects two decades of school dispute between Akhbārīs and Uṣūlīs. The points of dispute have not only multiplied over this time, but their content has also developed. The experience of the Akhbārīs as a distinct school, in constant dispute

[32] For the criteria being used here, see above, p. 160. For the dating of an Akhbārī school to the time of al-Ḥurr, see above p. 173.

with their Uṣūlī-*mujtahid* opponents, has led to "second tier" questions (such as the role of the jurist and the interrelationship between revelatory sources) dominating the discussion.

Presenting the Akhbārī-Uṣūlī Dispute

The earliest post-Astarābādī candidate for a "list of differences" is found in a chapter entitled "*fī bayān aṣl al-ikhtilāf*" ("On explaining the root of the difference") found in al-Karakī's *Hidāyat al-Abrār*, a work analysed in greater depth in the following chapters:[33]

> The reason (*sabab*) which caused the difference (*ikhtilāf*) is the clear opposition of modern scholars to the early scholars on three issues:
>
> First, a group of early scholars . . . explicitly stated that it is not permitted to prove legal rulings by *ẓann*. The modern scholars permit this.
>
> Second, the early scholars agreed . . . [with al-Ṭūsī's statement] that "only one [scholar] has the truth, and for that there is an indicator. He who opposes this indicator is a miscreant in error." The modern scholars say that the erroneous *mujtahid* has not sinned.
>
> Third, a group of the early scholars explicitly state that the reports they transmit in their books, and upon which they base their action, are all sound . . . the modern scholars say that isolated reports on their own bring only *ẓann*.[34]

For al-Karakī, then, the blame for the dispute is clearly attributed to the "modern scholars": they have strayed from the methodology of the earlier scholars (which, by implication, has a greater claim to legitimacy). This is not, then, a list of differences between Akhbārīs and Uṣūlīs/*mujtahids*, but a polemical attempt to attribute blame for the deviation of the community from the true path and the resultant dispute and division. Whilst the passage outlines al-Karakī's primary mode of attack against the *mujtahids*/Uṣūlīs, it is not (and nor does it aim to be) a list of differences, despite its portrayal as such by later Akhbārī scholars.[35] Where subsequent lists of differences can be distinguished from al-Karakī's three "causes of the dispute" is in their authors' attempts to catalogue differences and present them as purely descriptive. The above passage does, however, outline the

---

[33] See below, pp. 248–251 and pp. 276–278.
[34] Karakī, *Hidāyat*, pp. 6–7.
[35] Akhbārī, *Ḥirz al-ḥawāss*, f.7a.11–7b.6/pp. 14 l.11–15 l.6.

fundamental difference in epistemology (that is, between binary and graded approaches to knowledge, proposed by Akhbārīs and Uṣūlīs respectively). All the other differences laid out by subsequent Akhbārī writers, and examined below, are based upon a rejection of the legitimacy of the Uṣūlī category of *ẓann*.

Another early "list" is that found in al-Ḥurr al-ʿĀmilī's *al-Fawāʾid al-Ṭūsiyya*. The work generally is a rather inchoate collection of 102 sections (titled as numbered *fāʾidas*).[36] In each *fāʾida*, al-Ḥurr presents his remarks on separate religious topics, and many of these relate to the author's professed Akhbarism. Some *fāʾidas* are presented as responses to petitioners' questions, others as thoughts on particular *ḥadīths* or Qur'anic verses and yet others are explicit refutations of an opponent's position. Works of this sort (which can be termed *fawāʾid* works)[37] emerged as a popular form of religious literature amongst the Safavid Shīʿī scholars. They are, in effect, a variorum of short articles requiring less sustained effort than a work in an established genre, such as *uṣūl, fiqh* or Qurʾān/*ḥadīth* commentary. In *fāʾida* 92, al-Ḥurr lists the 23 differences between Akhbārīs and Uṣūlīs already mentioned, but he provides little explanatory comment on the validity of either school, despite his own Akhbārī allegiance. The list comes in the midst of a refutation of an unknown scholar's *al-Risāla fīʾl-ijtihād* to which al-Ḥurr is replying in the standard atomised point by point manner. The fact that the first true list of Akhbārī-Uṣūlī differences dates from the late Eleventh Century[38] further confirms the hypothesis, outlined in the last chapter, that Akhbarism only attained a "school" identity seventy years or so after Astarābādī's death (and even then in an ill-formed and ill-defined manner). Before this date, Akhbārī scholars were yet to acquire a common heritage and a common set of doctrines; Astarābādī's legacy consisted of a general unease over the use of *ijtihād* and related hermeneutic principles. Al-Ḥurr's list is evidence of the construction of Akhbārī doctrinal coherence over time, and also a contributory factor in the creation of that coherence. By cataloguing the differences, an author projects his view

[36] Al-Ṭihrānī mentions that the number of *fāʾidas* is "102 or 103", though the printed edition is clearly numbered 102. Ṭihrānī, *al-Dharīʿa*, v. 16, p. 347, #1616.

[37] See above, p. 102.

[38] The work itself is undated, though it was clearly written between al-Ḥurr's arrival in Mashhad (in 1073) and 1090. The latter is the date of the earliest manuscript of the work known to al-Ṭihrānī (*al-Dharīʿa*, v. 16, p. 347 #1616).

of the essential characteristics of Akhbarism. Over time, such lists contributed to Akhbarism's increasingly strong school identity, and for this reason one can say that the lists both reflected and created Akhbarism as a school.

Al-Ḥurr's list of differences is presented as a response to the statement (found in *al-Risāla fi'l-ijtihād*) that the difference between the Akhbārīs and Uṣūlīs is merely "terminological" (*lafẓī*). Al-Ḥurr, of course, considers the difference to be substantive (*maʿnawī*), and he demonstrates this by listing twenty-three doctrinal issues (numbered by *abjad*) on which Akhbārīs and Uṣūlīs differ. He introduces them as "all the substantive differences between the two groups". In the main, each description of a difference follows a consistent model: the statement of Uṣūlī doctrine, and a statement of its denial by the Akhbārīs. An Akhbārī alternative to the Uṣūlī doctrine is rarely provided, and no reason for the denial is given. For example, the second difference reads:

> Bā': The Uṣūlīs say that it is permitted, or rather obligatory, to act in accordance with the opinion of the *mujtahid* and no one else. The Akhbārīs say this is not permitted.[39]

This rather uninformative declaration of the invalidity of an Uṣūlī doctrine hides the Akhbārī alternative to the *mujtahid* authority system which had emerged by al-Ḥurr's time.[40] The model of description becomes so repetitive ("the Uṣūlīs say such and such . . . the Akhbārīs deny it) that al-Ḥurr al-ʿĀmilī resorts to an abbreviation: "in the same way (*ka-dhālik*), such and such . . ." meaning that the Uṣūlīs assert a doctrine and the Akhbārīs deny it:

> Ḥā': The Uṣūlīs say that it is permitted to perform an action on the basis of a report whose chain of transmission is merely of less than certain validity (*ẓannī al-sanad*). The Akhbārīs say it is not permitted.
> Wāw: A report which is of uncertain indications is the same.
> Zā': *al-aṣl* [i.e. *al-barā'a al-aṣliyya*] is the same.
> Hā': *al-Istiṣḥāb* is the same . . .

This abbreviated form of presentation continues from point 6 (*wāw*) to point 17 (*yā' ẓā'*). Including these, 15 of the 23 points of difference are expressed as mere Akhbārī denials of Uṣūlī doctrines. Most of the

[39] Ḥurr, *al-Fawā'id al-Ṭūsiyya*, p. 447.
[40] See Gleave, "The Qāḍī and the Muftī".

principal Akhbārī doctrines identified by al-Ḥurr are, then, reactionary: they are not doctrines at all, but denials of Uṣūlī doctrines without an alternative Akhbārī hermeneutic being proposed, or an exploration of any variety within the Akhbārī school being given.

A few of al-Ḥurr's remaining points of difference do describe an Akhbārī alternative to a given Uṣūlī doctrine. However, they have clearly been selected to ridicule the Uṣūlī position. Al-Ḥurr's aim in formulating these points is not merely to describe the main differences between the two positions. Rather, it is to highlight those Uṣūlī doctrines which can most easily be shown to be incoherent. Al-Ḥurr's list, on these points at least, consists of covert *reductio ad absurdum* arguments. For example, in point 17, al-Ḥurr states that when a non-*mujtahid* comes to know a ruling through reading an authentic report, then (for Akhbārīs) he should follow the ruling contained in the report. The Uṣūlīs, however, argue that he should ignore this and follow a *mujtahid's* opinion. This "point of dispute" between the schools consists in al-Ḥurr taking the Uṣūlī position to its logical conclusion and thereby demonstrating it to be counter-intuitive. This is then contrasted with the Akhbārī position, which appears obviously reasonable by comparison. The point is pressed home in points 18 to 22. Point 18 states that if the non-*mujtahid* knows of 1000 "sound *ḥadīths*" which explicitly indicate a particular legal ruling, but hears a *mujtahid's* opinion which contradicts this ruling, then (according to the Uṣūlīs) he should follow the *mujtahid*. According to the Akhbārīs, of course, he follows the *ḥadīths*. In a subsequent point, al-Ḥurr states that for Uṣūlīs, if a member of the *ʿulamāʾ* knows all *ḥadīths*, without exception, but his knowledge is deficient in a single discipline or issue, then he cannot follow the *ḥadīths* he knows, but must follow the *mujtahid*. The Akhbārīs assert he is qualified to follow the *ḥadīths* he knows. With each point, al-Ḥurr increases the intensity of the hyperbole (a single *ḥadīth*, a thousand *ḥadīths*, all *ḥadīths* that exist), demonstrating that the logical conclusion of the Uṣūlī position is the rejection of certain knowledge in favour of *ẓann*. Whilst al-Ḥurr lists these as discrete points, they are all, in fact, a single point of dispute: for the Uṣūlīs the *mujtahid's* opinion as to God's law holds more authority than the *ḥadīth*. The point is expressed repeatedly, but formulated with different conditions, making this a steady, creeping piece of argumentation, rather than a list of separate points of difference between the two parties.

On two other occasions (points 1 and 19), an alternative Akhbārī method is clearly presented. The first point reads:

> Alif: The Uṣūlīs say that *ijtihād* is permitted, nay obligatory, in deriving rulings. The Akhbārīs say that it is not permitted to act on anything other than a text.[41]

Even here though, the Akhbārī method is presented in negative terms, demonstrating a reactionary, rather than constructive, character.

Point 19 is more clearly doctrinal (as opposed to reactionary). On the issue of the correct procedure on issues for which there is no text (Qurʾanic or otherwise), the Akhbārīs argue that one should act with caution (*iḥtiyāṭ*) whilst Uṣūlīs say one should follow the *mujtahid's* ruling. Once again, al-Ḥurrr wishes to reduce the Uṣūlī position to ridicule:

> For the Uṣūlīs, it is obligatory to act in accordance with [the *mujtahid's*] *ẓann*, even if it opposes caution. Those who argue for *ijtihād* mostly rule on issues without reference to the text (*al-naṣṣ*), and then give *fatwās* upon [those issues]. If one studies [the sources], one finds that at times [answers to] those issues are recorded in the [revelatory] texts. [Sometimes] these differ from the *fatwās* [of the *mujtahid*], and [at other times] they agree.[42]

The Uṣūlīs, then, do not simply stipulate obedience to the *mujtahid's* opinion. They stipulate it even when this opinion directly contradicts the natural reading of a revelatory text.

Unlike al-Karakī's catalogue of reasons (and hence blame) for the conflict, al-Ḥurr's presentation can more accurately be described as a list of differences. However, Akhbarism is portrayed as a reactionary movement, revolving around a condemnation of Uṣūlī doctrines. Al-Ḥurr rarely presents Akhbārī alternatives, and he structures his list, and describes the differences, in such a way as to advance his own anti-Uṣūlī stance. Hence a single point of dispute is expressed three times with more and more outlandish riders.

Al-Jazāʾirī's *Manbaʿ al-Ḥayāt*, written around twenty years after al-Ḥurr's *al-Fawāʾid al-Ṭūsiyya*, represents a development in presentational technique, as well as reflecting developments in Akhbārī doctrine more generally.[43] The work as a whole is a defence of the

---

[41] Ḥurr, *al-Fawāʾid al-Ṭūsiyya*, p. 447. The use of the term *naṣṣ* here is explained above, n. 17.

[42] Ḥurr, *al-Fawāʾid al-Ṭūsiyya*, p. 449.

[43] A summary of al-Jazāʾirī's list of differences can be found in Sefatgol, *Sakhtār-i Nahād*, p. 529. Sefatgol is using the Majlis Library manuscript, #2761 (see *Majlis Fihrist*, v. 9, pt. 1, p. 152) rather than the printed text.

doctrine that the opinions of a scholar continue to have probative force after that scholar's death. This was a distinctive Akhbārī doctrine: previous to al-Jazā'irī, Astarābādī had denounced it,[44] al-Karakī had refuted it[45] and al-Ḥurr had mentioned it in his list of differences at point 21.[46] The Uṣūlīs argued that an ordinary believer's *taqlīd* must be to a living *mujtahid*. For them, when that *mujtahid* dies, his opinions no longer have probative force (summarised in the slogan *ḥukm al-mayyit ka'l-mayyit*—"the dead person's ruling is like the dead person"). The believer must then choose a new *mujtahid*. Al-Jazā'irī takes as his text the *risāla* of Shahīd II (d. 966) on following a living *mujtahid*.[47] One element of Shahīd II's argument is the claim that the believer who errs due to his ignorance of the law is not excused his error (he is not excused because he should have referred to a *mujtahid*). In reply to this, al-Jazā'irī first outlines his own position—namely that the believer is charged with obedience to the Imams (through the *akhbār*), and not to a *mujtahid*. As part of his reply, he lists eleven differences between Akhbārīs and Uṣūlīs in order to demonstrate that legal knowledge is generally available and not restricted to the *mujtahid*. He describes each difference in turn, according to a regular model:

[1] he numbers and names the area of dispute;
[2] he outlines the *mujtahid* opinion;[48]
[3] he outlines the Akhbārī opinion (phrased as a rejection of the *mujtahid* opinion);
[4] he comments on (and argues for) his own position on the topic under discussion, introducing this last section with "I say..." (*aqūlu*).

---

[44] Astarābādī, *al-Fawā'id*, p. 299.

[45] Karakī, *Hidāyat*, pp. 303–304.

[46] Ḥurr, *al-Fawā'id al-Ṭūsiyya*, p. 449. Here the difference is part of al-Ḥurr's polemic against the Uṣūlī position that the *mujtahid's* opinion is always preferable to a non-*mujtahid's* reading of *ḥadīth*. The point refers to the supposedly illogical Uṣūlī doctrine that if a dead *mujtahid's* opinion agrees with 1000 *ḥadīth*s, and a living *mujtahid's* opinion does not, then the living *mujtahid's* opinion is preferable.

[47] See Ṭihrānī, *al-Dharī'a*, v. 4, p. 392, #1736. See above p. 188.

[48] In the description of the seventh area of dispute (concerning whether or not the early Shī'ī scholars forbade or allowed *ijtihād*), al-Jazā'irī places the Akhbārī position before that of the *mujtahids*. In this point, al-Jazā'irī also deviates from his usual pattern by dispensing with the general description of the disputed area at the outset. See Jazā'irī, *Manba'*, pp. 54–57. In the eleventh area of dispute (concerning the meaning of a sound *ḥadīth*), the *mujtahid* position is not introduced as such, but by reference to past scholars claimed as *mujtahids* (particularly Muḥammad b. Idrīs al-Ḥillī). See Jazā'irī, *Manba'*, p. 82.

For example:

> [1.] The fourth issue concerns the probative force of *qiyās al-awlawiyya* and *manṣūṣ al-ʿilla*.
>
> [2] The *mujtahids*—may God be pleased with them—say that both have probative force. They make them the reason (*manāṭ*) underlying many of the legal rulings (*al-aḥkām*) in such a way that, on some occasions, they give them priority over *akhbār* which have a *sanad* which is classified as unsound under the new method of [*ḥadīth*] classification.
>
> [3] As for the Akhbārīs—may God sanctify their spirits—they deny probative force to both of them. They say that deducing rulings from either [*qiyās al-awlawiyya* or *manṣūṣ al-ʿilla*] came about because of [the influence of] the Sunni method [of deriving rulings], since [for the Sunnis], at times, the texts are unable [to provide rulings].
>
> [4] I say that the true position here is that of the Akhbārīs, for there are many *akhbār* which declare both [types of analogical reasoning] to be invalid in principle...[49]

The last part of each section, in which al-Jazāʾirī argues for his own position, is always extensive (in the above cited passage, it runs to three pages in the printed edition). The issue at stake in the point cited above is the validity of two hermeneutic mechanisms. The first, *qiyās al-awlawiyya*, refers to the *e minori ad maius* argument in which the prohibition of a minor case is taken to indicate a similar prohibition of a major case. Hence, the prohibition on addressing one's parents with a rebuke (*uff*, Q17.23) is also a prohibition against hitting them. The dispute over this form of argumentation was not concerned with its validity (most Sunni jurists argued it had probative force, though the Ẓāhirīs famously rejected it), but over its classification: is it best considered an analogy (that is, the transfer of the ruling from the minor case to the major case), or is it merely a linguistic implication (that is, prohibiting a minor offence implies that anything more serious than that minor offence is also prohibited)?[50] The legitimacy of the procedure was also generally accepted by pre-Astarābādī Shīʿī jurists, though it was classified as a linguistic inference in order to avoid contravention of the uncompromising revelatory statements of the Imams condemning *qiyās*.[51] Similarly, there was dispute over

[49] Jazāʾirī, *Manbaʿ*, p. 45. The *masʾala* here ends at the foot of p. 47.
[50] See Zysow, "Economy of Certainty", pp. 157–162.
[51] See Gleave, "Qiyās".

qiyās manṣūṣ al-ʿilla—cases where the reason (*ʿilla*) for a ruling was explicitly stated (*manṣūṣa*) in a text, thereby (supposedly) enabling a simple transfer of the ruling from recorded to novel cases. Once again, the dispute amongst Sunni jurists concerned classification of this reasoning process as analogy (*qiyās*) or linguistic inference (*mafhūm*).[52] As with *qiyās al-awlawiyya*, the Shīʿī jurists preferred to consider this a linguistic inference, and thereby legitimise it as a valid element in Shīʿī *uṣūl al-fiqh*.[53]

Al-Jazāʾirī supports the Akhbārī rejection of both procedures, considering them types of *qiyās*.[54] He argues for this position by citing *akhbār* in which the Imams denounce *qiyās* (either directly or through their rejection of rulings resulting from *qiyās*). This is followed by a list of invalid analogies attributed to Abū Ḥanīfa. Then, in reference to *qiyās al-awlawiyya*, al-Jazāʾirī states that "the Qurʾān was only revealed in the language of the Arabs, that is, how they spoke in their everyday discourse." They would not have understood hitting one's parents to be prohibited from a text in which saying *uff* to them was prohibited. Similarly, with respect to *manṣūṣ al-ʿilla*, Imam Mūsā al-Kāẓim is reported as having said "one should not ask God why something is *ḥalāl* and why another thing is *ḥarām*." Al-Jazāʾirī rationalises the prohibition on *qiyās manṣūṣ al-ʿilla* thus: whilst it might appear as if the ratio (*ʿilla*) for a particular ruling is available in a particular case (that is, it is written within the text under consideration), it is not known whether or not this ratio is valid for all cases which appear to fall under the same category. One cannot deduce from the presence of words indicating causality in a report (*liʾann, li* etc.) that a similar causal process will occur in all instances similar to (or more accurately, presumed to be similar to) the case in question.[55]

Now, two comments can be made here. Firstly, whilst it is clear that al-Jazāʾirī is arguing for the Akhbārī position from his exposition (in section [4] above, beginning with *aqūlu*), his description of the difference of opinion (that is, parts [1]–[3] above) is non-partisan.

---

[52] See Zysow, "Economy of Certainty", pp. 323–329.

[53] See Gleave, "Qiyāṣ",

[54] That both procedures were not rejected by all Akhbārīs. Gleave, *Inevitable Doubt*, pp. 103–105 and pp. 151–152 in relation to Yūsuf al-Baḥrānī.

[55] The reasoning here is outlined in Jazāʾirī, *Manbaʿ*, p. 47.

From the description alone, the author's school loyalty cannot be determined. This is true of all al-Jazāʾirī's "points of dispute": they are presented as neutral descriptions of the dispute, and by doing this, al-Jazāʾirī claims some sort of independent legitimacy for his list. Argumentation is not absent from the text as a whole; however, it is removed from the presentation of the dispute, and hived off into section [4]. In short, his brief portrayal of the *mujtahid* position in section [2] above (and in the equivalent section of all of his eleven points of dispute) does not appear obviously skewed by his personal school affiliation. Here we see a development in the presentation of the dispute from al-Ḥurr's list. Al-Ḥurr composed a series of points (points 17 to 22) which formed a *reductio ad absurdum* argument of the *mujtahid* position, and thereby covertly introduced his own position into his description of the dispute. Al-Jazāʾirī, perhaps more confident of Akhbarism's ability to withstand *mujtahid* criticism, felt able to describe the *mujtahid* position without explicit prejudice.

Secondly, whilst al-Ḥurr's list is primarily a list of Akhbārī denials, in al-Jazāʾirī's text the Akhbārī alternative is described and, at times, justified also. In the above cited passage, section [3], the Akhbārīs are said to justify the rejection of these two types of analogical reasoning because they derive from Sunni juristic thinking. This was a common motif in Akhbārī literature and accords with the implicit (and sometimes explicit) assumption that any legal process originating in Sunni discourse has no place in Shīʿī legal discussions. In section [4], al-Jazāʾirī suggests an alternative interpretive procedure which demands that the language of the revelatory texts be understood according to the everyday discourse of the Arabs (*bi-lisān al-ʿarab wa-bi-mā kāna yajrī baynahum fī maḥāwarātihim*). A further justification for the Akhbārī position is also supplied: namely, even when God gives a reason for a ruling, one cannot assume that the ruling is transferred to all similar cases. Only on those occasions when God explicitly says that a ruling can be transferred, or when he is announcing a general prohibition, can we assume that the ruling is transferred. On each occasion that al-Jazāʾirī describes a point of dispute (sections [1]–[3] in the above model), there is a clear attempt to prevent the description from being infected by polemic against one or other of the opinions expressed. Al-Jazāʾirī reserves his own opinion and argumentation for section [4].

Al-Jazāʾirī's provision of alternative Akhbārī doctrines (rather than simple Akhbārī denials) is perhaps the most significant development from earlier lists. Admittedly, in ten of al-Jazāʾirī's eleven points, the *mujtahid* view is described before the Akhbārī opinion, and the portrayal of Akhbarism as a reactionary movement is thereby maintained. However, al-Jazāʾirī is keen to flesh out Akhbārī alternatives to the *mujtahid* position (as was seen in the citation from the fourth point of dispute mentioned above). The fifth point constitutes another example of this, and concerns "taking legal rulings from the Qurʾān" (*akhdh al-aḥkām min al-Qurʾān*). Al-Jazāʾirī describes the *mujtahid* position as permitting rules to be taken directly from the Qurʾān, interpreting the Qurʾān according to a set of hermeneutic procedures (*amārāt al-istinbāṭ*) and rejecting those reports which contradict the meaning of the Qurʾanic text (as mediated through the employment of these hermeneutic procedures). The Akhbārī position is described thus:

> As for the Akhbārīs—may God sanctify their tombs—they argue that, for us, all the Qurʾān is of unclear reference (*mutashābih*). Hence it is only permitted for us to take rulings from it when there is an indication within the *akhbār* as to its meaning.[56]

The Akhbārīs are not only presented as rejecting the *mujtahid* position. An alternative, Akhbārī hermeneutic is outlined: the Qurʾān can only be understood in the light of the rulings of the Imams. This is a consistent feature of al-Jazāʾirī's presentation. For al-Ḥurr, Akhbarism's most significant features lay in its denial of Uṣūlī methodology. For al-Jazāʾirī, it consists of an alternative hermeneutic. Whilst the alternative does indeed clash with Usulism, it is described in its own terms. One sees here the early development of a distinctive Akhbārī position with its own canon of independent and coherent principles which should replace dominant Usulism.

This development can also be seen in al-Jazāʾirī's tendency to discuss the interrelationship between points. The linking of *qiyās al-awlawiyya* and *manṣūṣ al-ʿilla* into one point, as described above, is an example of this. For al-Ḥurr, these two types of analogy consisted of separate areas of dispute between Akhbārīs and Uṣūlīs (namely, points 13 and 14 in his list).[57] Another example is al-Jazāʾirī's eighth area of dispute, concerning *al-barāʾa al-aṣliyya* (the presumption of licitness). The *mujtahids* argue that in the absence of an explicit prohibition in a revelatory text, the assessment on any occasion (be

[56] Jazāʾirī, *Manbaʿ*, p. 48.
[57] Ḥurr, *al-Fawāʾid al-Ṭūsiyya*, p. 448.

it concerning a substance or an action) is of the permitted nature of the item under scrutiny. The Akhbārī position is described thus:

> The Akhbārīs—may God have mercy on them—say that the meaning of the *akhbār* is that God has a specific ruling for everything, even to the [number of] lashings and the length of the whip [used in *ḥudūd* punishment]. Every occasion, be it minor or major,[58] has had a ruling placed upon it by the Imams. Some of them have reached us and the rest remain [unknown]. So, whenever a text reaches us, saying that an action is obligatory, forbidden, recommended, discouraged or licit, we know this from its contents. Whenever a text has not reached us, then we must suspend judgement (*tawaqquf*) concerning [the matter] until its ruling becomes clear, when we then act accordingly. If it remains hidden from us, then *tawaqquf* is obligatory for us. Hence the fundamental nature of a thing is not that it is forbidden [merely that its assessment is unknown]. We have heard that some modern Akhbārīs forbid the wearing of clothes other than those used at the time of the Imams, because there is no permission from the Lawgiver to [wear them]. Similarly, [they forbid] the consumption of meals which were not eaten at that time, because there is no specific text. They stipulate that there must be a specific indicator [in the revelatory texts] for every instance.[59]

Al-Jazā'irī's willingness to present a coherent Akhbārī alternative to an Uṣūlī doctrine is, once again, in evidence here, but more than this: the Akhbārī position on *al-barā'a al-aṣliyya* is explicitly linked with general Akhbārī positions on knowledge of the law, the sufficiency of the *akhbār* to provide guidance and the need to suspend judgement in cases of revelatory silence. These points had also been made by al-Ḥurr, but in discrete differences. What is significant here is that through an exploration of the interdependence of Akhbārī doctrines, the coherence of the Akhbārī position is further strengthened in the reader's mind. Another point of interest is al-Jazā'irī's hint at different Akhbārī positions: some "modern" Akhbārīs hold a view on meals not prepared at the time of the Imams, and by implication, others do not.

Finally, it should be noted that al-Jazā'irī's support for the Akhbārī opinion is not unequivocal, notwithstanding his personal renown as an Akhbārī. In his own assessment of the above dispute concerning *al-barā'a al-aṣliyya*, al-Jazā'irī states that it is the *mujtahid* position which is the stronger (*al-ẓāhir anna al-aqwā hunā qawl al-*

---

[58] Reading *ʿaẓā'im* for *ʿaḍā'im* here: Jazā'irī, *Manbaʿ*, p. 58.

[59] Jazā'irī, *Manbaʿ*, p. 58.

mujtahidīn), in that the *akhbār* do indeed indicate that the Imams ordered a presumption of licitness when the precise assessment of an action is unknown from the revelatory texts. This is probably a criticism of those Akhbārīs who only permit meals and clothes which were current at the time of the Imams. However, the *mujtahids* "go too far" (*afraṭū*) when they declare a thing to be licit when there is only a weak report in which it is prohibited. In such circumstances, al-Jazā'irī argues, the so-called weak report is sufficient evidence to establish a prohibition. Similarly, the Akhbārīs "go too far" (*afraṭū*) when they argue that any action which is not explicitly designated by a text is forbidden. Al-Jazā'irī argues that only in cases of total revelatory silence can a presumption of licitness be made. He takes similarly "moderate" opinions on the questions of the probative force of *khabar al-wāḥid* (the ninth area of dispute)[60] and the obligatory nature of caution on occasions of revelatory contradiction (the tenth area of dispute).[61] He can do this, and remain an Akhbārī (he supports the Akhbārī position more often in his commentary on points of dispute), by selecting for description the most extreme of the various Akhbārī and *mujtahid* views on a particular topic, and then presenting himself as a moderate. Not all Akhbārīs demanded an explicit textual indicator for every assessment, and not all *mujtahids* argued for licitness in the face of so-called weak *ḥadīths*,[62] but it serves al-Jazā'irī's purpose to portray the positions in this way.

Al-Jazā'irī's desire to present himself as a "moderate" in the dispute could stem from his tutorship under Muḥammad Bāqir al-Majlisī,[63] who famously stated that he pursued a "middle way" (*bayna ifrāṭ va tafrīṭ*) between the two schools. Al-Jazā'irī is concerned that the dispute might generate declarations of unbelief. On the definition of "the sound report" (*al-ḥadīth al-ṣaḥīḥ*), al-Jazā'irī outlines the different opinions and concludes:

> The truth is with the Akhbārīs [on this point]. In sum, it is clear to one who follows the opinions of the Akhbārīs and *mujtahids* that there is both excess and negligence (*ifrāṭ*[an] *wa-tafrīṭ*[an]) in them. The Akhbārīs have extended the tongue of insult to the *mujtahids*, and attributed error

---

[60] Jazā'irī, *Manba*ʿ, p. 65.
[61] Jazā'irī, *Manba*ʿ, p. 76.
[62] For further discussion on these issues, see below, p. 204.
[63] Majlisī II, *Ajwibah*, p. 4. The most explicit statement by Muḥammad Bāqir al-Majlisī is found in these answers to questions set by Khalīl al-Qazwīnī.

> and misguidance to them. This is an insult which is inappropriately placed, because the *mujtahids*—may God sanctify their spirits—have always exerted effort in their search for legal rulings, and in bringing understanding to any obscurity. The truth lies in a middle way between the two opinions and a path between the two paths ... This is the middle path which a group of our modern scholars have followed, and that is the way of caution which does not lead the follower astray.[64]

Al-Jazāʾirī's list, then, highlights his view of the main areas of dispute, and at times he exaggerates both Akhbārī and *mujtahid* opinions in order to present himself as taking a middle way. However, this cannot hide the fact that on most issues relating to the dispute, he adopts an Akhbārī position, and his mode of argumentation is decidedly Akhbārī (for example, the citation of *ḥadīths* and the preference for the opinions of pre-ʿAllāma scholars).

Notwithstanding these general observations, al-Jazāʾirī's list develops al-Ḥurr's presentation of the dispute in a number of important ways. Firstly, there is a formal distinction between describing the dispute and evaluating the views of the two schools, and this is achieved through the institution of a separate section in which the assessment is made. Al-Ḥurr blurs this distinction in order to expose the *mujtahids* to ridicule through rhetorical phrasing of certain differences. Secondly, al-Jazāʾirī describes the Akhbārī position as an alternative hermeneutic, and not simply as a rejection of Usulism. Thirdly, in the course of this more elaborate exposition of the Akhbārī position, al-Jazāʾirī also explores the coherence of the Akhbārī position by showing the interconnected nature of their doctrines. Fourthly, there are occasional references to different opinions within the Akhbārī school, in particular between what might be termed "moderate" and "extreme" positions. Al-Jazāʾirī attempts to associate himself with the former in his commentary sections.

Al-Jazāʾirī's distinction between the description of the opinions of Akhbārīs and Uṣūlīs, and the presentation of argumentation for one of these positions is developed in Sulaymān al-Māḥūzī's list of differences, recorded in one of his *ajwiba* works (containing answers to questions set by a fellow scholar).[65] As has already been mentioned,

[64] Jazāʾirī, *Manbaʿ*, p. 83.

[65] The text is found cited by al-Akhbārī in his *Ḥirz al-Ḥawāss* (see above, n. 00). Al-Samāhījī, one of al-Māḥūzī's pupils, also cites this work in his famous list of 40 differences (see below). The wording cited there differs from al-Akhbārī's text

al-Māḥūzī's position within the Akhbārī-Uṣūlī dispute is uncertain.[66] In his outline of the differences between the two schools (numbering six in all), he does occasionally express his opinion, though, unlike al-Jazā'irī, he never enters into explicit argumentation. On each occasion he expresses an opinion, it accords with that of the Akhbārīs, and hence this can be validly characterised as an "Akhbārī" presentation of the dispute. For example, on the question of *istiṣḥāb al-ḥāl*, he outlines the Akhbārī position (which is shared with some *mujtahids*), after which he says "this is the stronger position in my opinion" (*huwa al-aqwā ʿindī*). He also mentions the question of contradictory *akhbār*, and the solution proffered by the two schools: the Akhbārīs argue for a set number of legitimate techniques, sanctioned by the Imams; the Uṣūlīs argue that the full weight of interpretation should be brought to bear on the contradictory *ḥadīths* in order to prevent a contradiction. Al-Māḥūzī states that "following the path of the Akhbārīs concerning preference [between *ḥadīths*] is probably stronger."[67] Apart from these asides, the dispute is described in non-partisan terms. The focus of the discussion is the dispute over *al-barā'a al-aṣliyya* (that is, the assumption that an action is licit when there is no clear indication in the revelatory texts that it is forbidden). The Akhbārī position is described in a nuanced manner. The Akhbārīs do not permit the believer to act on the basis of *al-barā'a al-aṣliyya* in the following circumstances:

1. *al-barā'a* cannot be used to pronounce an action licit when the action is only obliquely mentioned in revelation. The example given concerns the prohibition on touching the actual text of the Qur'ān (as opposed to its covers) whilst in a state of ritual impurity.[68]

---

slightly—as al-Akhbārī himself states, he has summarized al-Māḥūzī's answer. I have been unable to locate the original.

[66] See above, p. 159 (see also Baḥrānī, *Lu'lu'a*, p. 10, where it is stated that he was a supporter of *ijtihād*, but "returned to something close to the Akhbārī position" in later life).

[67] Newman, "Akhbārī/Uṣūlī Dispute, pt. 1", p. 28 (translation, pp. 42–43). This passage is found in al-Samāhījī's forty points, but is missing from al-Akhbārī's summary of al-Māḥūzī's answer. It is possible that there were two answers to the same (or similar) questions concerning the difference between Akhbārīs and Uṣūlīs.

[68] *ḥurma fī'l-wujūdī*: Akhbārīs argue that there is sufficient revelatory indication that this action is possibly prohibited (since minor ritual impurity prohibits one from touching various other items), and therefore should be treated as prohibited on the basis of caution, and the presumption of licitness cannot apply here. Akhbārī, *Ḥirz al-Ḥawāss*, f.13a.3/p. 26 l.3.

2. *al-barāʾa* cannot be used to pronounce a ruling of licitness when the ruling pertains to a situation obliquely mentioned in revelation. The example concerns whether or not a bodily discharge from an orifice other than the anus or the penis causes ritual impurity.[69]

3. *al-barāʾa* cannot be used to express a preference for one or other of two (or more) contradictory reports.[70]

However, *al-barāʾa* can be used legitimately, according to the Akhbārīs, to move an action which could be obligatory (due to its similarity with other actions) to a lesser category (namely, recommended and permitted). The example given concerns the late evening prayer called *ṣalāt al-witr*: this is not obligatory, but recommended. This is not because of any underlying licit assessment of all actions (*al-aṣl*), but because there are many *ḥadīths* which indicate that, on certain matters, God has chosen to hide the ruling from his servants, or the people are unable to know the ruling for some other reason. In such circumstances, the ruling "placed" (*mawḍūʿ*) on the action cannot be obligation.

The Akhbārī position, then, is carefully described by al-Māḥūzī here. For Akhbārīs, *al-barāʾa al-aṣliyya* can be used as an interpretive principle to reduce a suspected obligation to a lesser assessment; however it cannot be used to declare a suspected prohibition permitted. How such a principle works in the delineation of the law is explored in Chapter 9 below, but the significant point here is the care with which al-Māḥūzī describe the Akhbārī position on *al-barāʾa al-aṣliyya*. Similarly, the Akhbārī position on declaring a preference between contradictory reports (*tarjīḥ taʿāruḍ al-akhbār*) is given a detailed description, and related to the principles of interpretation laid out by al-Kulaynī in the introduction to his *al-Kāfī*.[71] The same care is taken over the Akhbārī positions on other issues of *uṣūl al-fiqh: ijmāʿ* (including the problems of knowing the participants in any

---

[69] *ḥukm waṣfī*: Akhbārīs argue that there is sufficient revelatory evidence to indicate that all bodily discharges (flatulence, urine, faeces, blood, lachrymal substance etc.) violate a state of purity, and the presumption of licitness for those unmentioned in the revelatory text is not valid.

[70] Akhbārī, *Ḥirz al-Ḥawāss*, f.13a.6/p. 26 l.6. That is *al-barāʾa* cannot be used as evidence that a report which indicates the licitness of an action predominates over a report which indicates another ruling. It should be noted that the exact wording of al-Māḥūzī's description here is replicated by his pupil, al-Samāhījī, in his list (see Newman, "Akhbārī-Uṣūlī Dispute, pt. 1", pp. 34–35).

[71] On these principles, see Gleave, *Inevitable Doubt*, pp. 119–121.

consensus, and whether or not the Imam might be one of them), the division of all rulings into permitted, prohibited and uncertain, the issue of the validity of all the reports in the "Four Books" and *istiṣḥāb al-ḥāl*. There are indications in the list of differences of opinion amongst the Akhbārīs on certain points. For example, on the issue of *taʾkhīr al-bayān* (the delay of explanation, in which God reveals a command, but not the means of fulfilling that command), some Akhbārīs (*jamāʿa minhum*) say it is permitted (and by implication, others are noncommittal or opposed to it).[72] There are also instances where some *mujtahids* agree with the Akhbārī position. "Some *mujtahids*" (*baʿḍ al-mujtahidīn*) are said to agree with the Akhbārī position and disregard the *ijmāʿ* of the modern scholars (meaning those after al-ʿAllāma al-Ḥillī). Together they argue that it is not a valid proof because there is no way of knowing if the Imam's opinion is included in the consensus, unless there is a report to that effect (and in such cases, the report is the proof rather than the simple fact of a consensus).

Al-Māḥūzī's list, then, develops further the depiction of the Akhbārī-Uṣūlī conflict. The explanation of the Akhbārī position on each point is detailed and non-partisan. The expression of personal comment and explicit argumentation for or against a particular position is minimal. Akhbārī positions are presented as alternative solutions to particular hermeneutic problems; in fact, the Akhbārī position is given prominence, and appears to control the general presentation. The *mujtahid* position is expressed, in the main, as a reaction to the Akhbārī stance (that is, a reversal of al-Ḥurr's presentation technique). Al-Māḥūzī's pupil, ʿAbd Allāh al-Samāhījī, continues to develop these characteristics. Al-Samāhījī produced what was probably the best known of the lists of differences between Akhbārīs and Uṣūlīs. His list of forty differences has been edited by Newman,[73] and was cited by most later Shīʿī commentators on the dispute.[74] It

[72] See Gleave, *Inevitable Doubt*, pp. 235–236.

[73] See Newman's articles, "The Akhbārī-Uṣūlī Dispute, pt. 1" and "The Akhbārī-Uṣūlī Dispute, pt. 2". The latter article is Newman's own analysis of the list.

[74] See, for example, Baḥrānī, *al-Ḥadāʾiq*, v. 1, p. 167; Akhbārī, *Ḥirz al-Ḥawāss*, f.13b.5–16b.14; Zand, *al-Fawāʾid al-Shīrāziyya*, 13b.4–77b.9 (with extensive commentary from Zand himself and citations of other lists of differences, with points numbered differently, and no point after point 34 being listed); Khwānsārī, *Rawḍāt*, v. 1, pp. 136–140 (an abbreviated version which has been extensively analysed by Newman, "Akhbārī-Uṣūlī Dispute, pt. 2"); Dizfūlī, *Fārūq al-Ḥaqq*, f.240b.6.

gained, then, a canonical status, and later lists were, on the whole, commentaries, abbreviations or developments of al-Samāhījī's list. A few general comments can be made about al-Samāhījī's presentation. The *mujtahid* position is invariably mentioned first, and the Akhbārī position preserves its reactionary character which had, to an extent, been marginalised in al-Māḥūzī's list. However, the Akhbārī position receives extensive explanatory comment from al-Samāhījī, and the view of the Akhbārīs is outlined in great detail. The Akhbārī view is expressed first, before that of the *mujtahids*, in only six of the forty points, coalescing at the end of al-Samāhījī's text.[75] This deviation from the normal pattern follows an unusually formulated point (point 33), discussed further below. There is a single instance of different opinions amongst the Akhbārīs (the only explicit reference being the implication that there were different opinions over *ta'khīr al-bayān*, in line with al-Māḥūzī's reference above). There is a certain amount of expansion (or perhaps, mere padding) in the list, probably in an attempt to reach the memorable number of 40 points. For example, al-Ḥurr and al-Jazā'irī restrict the Akhbārī denunciation of *ijtihād* to a single statement. Al-Samāhījī devotes 11 points in all to the dispute over *ijtihād*[76] and does not seem embarrassed over any repetition. For example:

> [al-Ḥurr]:
> Alif: The Uṣūlīs say that *ijtihād* is permitted, nay obligatory, in deriving rulings (*al-iḥkām*). The Akhbārīs say it is not permitted to act on anything other than a text.[77]
>
> [al-Jazā'irī]:
> Issue 7: The Akhbārīs say that the early Shī'ī Akhbārī scholars ... forbade *ijtihād* and *taqlīd* ... the *mujtahids* say that the early scholars had no need of *ijtihād* ... because the 400 *uṣūl* were still extant at that time.[78]

---

[75] Points 32, 33, 34, 35, 38 and 39. Point 34 is actually a reference to "a group of Akhbārīs" and their opinion concerning *ta'khīr al-bayān*. Newman, "Akhbārī-Uṣūlī Dispute, pt. 1", pp. 34–36 (translated on pp. 50–51).

[76] Points 1, 6, 7, 9, 12, 14, 17, 18, 23, 26 and 28, though many other points are also directly related to the question of *ijtihād*. See Newman, "Akhbārī-Uṣūlīs Dispute, pt. 1", p. 24, pp. 26–27, p. 28, p. 29 and pp. 31–32.

[77] Ḥurr, *al-Fawā'id al-Ṭūsiyya*, p. 447.

[78] Jazā'irī, *Manba'*, pp. 54–55.

[al-Samāhījī]:[79]
1. The *mujtahids* say that *ijtihād* is obligatory, individually or optionally . . . the Akhbārīs say it is forbidden . . .[80]

7. The *mujtahids* say that seeking knowledge during the occultation is through *ijtihād*, and during the time of the [Imam's] presence, by taking [knowledge] from the Sinless One . . . The Akhbārīs do not distinguish between the occultation and the time of presence . . .[81]

17. The *mujtahids* permit *ijtihād* in legal rulings when one is not able to gain knowledge of the Sinless One's opinion . . . The Akhbārīs do not distinguish in this matter [that is, between being able to, and not being able to, gain knowledge of the Imam's opinion][82]

That the rejection of *ijtihād* formed the crux of the Akhbārī-Uṣūlī dispute is clear, and indeed many points mentioned by al-Ḥurr and al-Jazāʾirī can be traced back to the Akhbārī rejection of *ijtihād*. The rejections of various interpretive devices by Akhbārīs (*istiṣḥāb, al-barāʾa al-aṣliyya, mafhūm al-sharṭ, mafhūm al-ṣifa*) are given discrete points in al-Ḥurr's list. They are all techniques used by the *mujtahid* in reaching his opinion of God's law. The same could be said of *qiyās al-awlawiyya* and *manṣūṣ al-ʿilla*, which are given a separate point in al-Jazāʾirī's list. A rejection of *ijtihād* necessarily entails a rejection of these devices. What is significant, though, is that both these earlier lists restrict the explicit mention of *ijtihād* to a single point, whilst al-Samāhījī devotes a significant number of points, not only to the Akhbārīs' straightforward rejection of *ijtihād*, but also to the consequences of this rejection (see for example, points 7 and 17 cited above). This is partly because al-Samāhījī aims at comprehensiveness, whilst earlier lists aimed as concision. Furthermore, al-Samāhījī wished to delineate a memorable number of 40 points of dispute, and expanding one point into three or more points was a convenient way of doing so (the above mentioned padding). On the specific issue of *ijtihād*, al-Samāhījī's expansion of the number of points explicitly referencing this element of the dispute, is combined with making the general rejection of *ijtihād* his first point. This enables him to portray the dispute as primarily about a single issue (*ijtihād*), with all

---

[79] These three points are merely a sample of al-Samāhījī's points which directly relate to the issue of *ijtihād*'s permissibility. A full list is given above, n. 76.
[80] Newman, "Akhbārī-Uṣūlī Dispute, pt. 1", p. 24.
[81] Newman, "Akhbārī-Uṣūlī Dispute, pt. 1", p. 26.
[82] Newman, "Akhbārī-Uṣūlī Dispute, pt. 1", p. 29.

other points (that is, the subsequent thirty nine) being ramifications of this rejection.

Al-Samāhījī's use of the term *mujtahid* (as opposed to Uṣūlī) for the Akhbārīs' opponents could be said to intensify the centrality of *ijtihād* to the dispute. The variation in usage has been noted by Newman who states that al-Samāhījī's use of the term *mujtahid* "suggests that the term Uṣūlī may not have been uniformly used in the late Safawid period."[83] The inconsistent use of terms by Akhbārīs to describe their opponents inhibits any general conclusions concerning the use of the terms *uṣūliyyūn* and *mujtahidūn*. For any unambiguous employment of either term, there are numerous exceptions. In general terms, however, *uṣūliyyūn* (or *uṣūliyya*) is used to describe those who participate in the maintenance of the discipline of *uṣūl al-fiqh* in its established form, be they Sunni or Shīʿī. A commitment to this discipline in its established form invariably leads to a validation of *ijtihād*. The legitimacy of *ijtihād* is an accepted element of *uṣūl* discourse, and a chapter on *ijtihād* (*bāb al-ijtihād*) is an established element of the *uṣūl* genre. However, in so far as the Akhbārī polemic revolves around a rejection of the discipline of *uṣūl al-fiqh*, their opponents are *uṣūliyyūn*, and the use of this term indicates a rejection of *uṣūl al-fiqh* as a valid discipline within the academy, rather than a specific complaint against the doctrine of *ijtihād*. The term *mujtahid*, however, has two distinct uses in these Akhbārī texts: one who supports the doctrine of *ijtihād* (usually expressed in the plural as *mujtahidūn*) and one who has attained the necessary educational qualifications to give independent judgements. The former is an expression of a doctrinal position; the latter is a description of a scholar's status and reflects a particular conception of the learned hierarchy. It is the second usage that is employed by al-Samāhījī in his thirty-first point:

> [32] The *mujtahid* and Akhbārī are combined in one way and differ, one from each other, in another. This demonstrates that between them is a general-particular distinction based on a quality (*ʿumūm*^{an} *wa khuṣūṣ*^{an} *min wajh*).[84]

As already noted, this deviation from al-Samāhījī's standard presentation introduces the points in which the Akhbārī doctrine is mentioned first (in points prior to this, the *mujtahid* doctrine occupies

[83] Newman, "Akhbārī-Uṣūlī Dispute, pt. 1", p. 38, n. 6.
[84] Newman, "Akhbārī-Uṣūlī Dispute, pt. 1", p. 34 (translation, p. 49).

the introductory position). By point 31, al-Samāhījī is struggling to maintain the *mujtahid*-Akhbārī model, and is forced to compromise the method of presentation, using a more discursive style.

Whilst it would seem natural that one who has reached the rank of *ijtihād* would also advocate its employment in the derivation of rulings, al-Samāhījī proposes a distinct category of scholars who qualify as *mujtahids*, but who do not support *ijtihād*. He terms such scholars "*mujtahid-muḥaddith*". Examples of *mujtahid-muḥaddiths* are given, and include Astarābādī, Khalīl al-Qazwīnī, Muḥsin Fayḍ al-Kāshānī, Muḥammad Ṭāhir al-Qummī and al-Ḥurr al-ʿĀmilī.[85] "*Mujtahid*" is a general category, of which a *mujtahid-muḥaddith* is a subclass: the former is made up of scholars who are both qualified and support *ijtihād*, and those who are qualified but choose to reject the validity of *ijtihād*. Both are to be distinguished from a *muḥaddith*, who may be learned in *ḥadīth*, but does not have the qualifications of a *mujtahid*. Hence the Akhbārī scholars listed as "*mujtahid-muḥaddith*" are specific types of *mujtahids: mujtahids* who deny the legitimacy of *ijtihād*. This is what is meant by the phrase borrowed from Arabic logic analysis: *ʿumūm*[an] *wa khuṣūṣ*[an] *min wajh*.[86] The word *mujtahid*

[85] The identify of "ʿAbd Allāh al-Yazdī" mentioned by al-Samāhījī is unclear, as noted by Newman ("Akhbārī-Uṣūlī Dispute, pt. 1", p. 49, n. 35). Since all manuscripts of al-Samāhiji's *Munyat al-Mumārisīn* seem to state "al-Yazdī" (apart from those consulted by Newman, see also University of Tehran, #1160, f.19b.11 and the citation in Fatḥ ʿAlī Zand, *al-Fawāʾid al-Shīrāziyya*, f.30b.2), I am not convinced it could be a scribal error. It is more likely that this scholar is the eighteenth student of Muḥammad Bāqir al-Majilsī, ʿAbd Allāh b. al-Ḥusayn al-Yazdī, mention in al-Majlisī's *Biḥār* (v. 102, p. 93; his *ijāza* from Majlisī I is cited in *Biḥār*, v. 107, p. 150 and Majlisī II, *Ijāzāt al-Ḥadīth*, pp. 77–78). Al-Yazdī was alive in 1013, and the chronology would seem to fit with the list given by al-Samāhījī.

[86] This type of general-specific distinction is normally distinguished in Arabic logic from *ʿumūm wa khuṣūṣ muṭlaq* (a general-specific distinction based on species). For example, describing an animal (the general category) as a white animal (a specific category) is based on a quality of the animal, and the relationship between the two categories is termed *ʿumūm wa khuṣūṣ min wajh*. Describing a human being (a specific category) as an animal (a general category) is based on the former being a subclass of the latter, and the relationships between the two categories is described as *ʿumūm wa khuṣūṣ muṭlaq*. The categories *mujtahid* and Akhbārī, according to al-Samāhījī, can have a relationship analogous to the animal-white animal relationship: *mujtahid* is a general category for those with specific qualifications, and within that category of scholars, there are Akhbārīs who deny the legitimacy of *ijtihād* (called *mujtahid-muḥaddith*). Newman's translation of *ʿumūm wa khuṣūṣ min wajh* ("generality and particularity in one respect", "Akhbārī-Uṣūlī Dispute, pt. 1", p. 49) does not do justice to the logical point al-Samāhījī is making here, and leads him to, I believe, misunderstand al-Samāhījī's point in his commentary (Newman, "Akhbārī-Uṣūlī Dispute, pt. 2", p. 259).

used by al-Samāhījī in the opening sentence of point 31 is, then, merely a measure of an individual's qualifications, and not an indication of support for *ijtihād*. This distinguishes it from its use in the plural elsewhere in the 40 points (*mujtahidūn*).

The difference between a *mujtahid* (be he a supporter or opponent of *ijtihād*—that is, be he a *mujtahid* or a *mujtahid-muḥaddith*) and the *muḥaddith* is also mentioned, and relates to the acquisition of certain educational qualifications (*jāmiʿ al-sharāʾiṭ*) in the former and the absence of these (and their replacement with knowledge of the *ḥadīth* alone) in the latter. The *muḥaddith* may have greater knowledge of *ḥadīths* than the *mujtahid*, but since he has not studied all the relevant sciences, he is not classified as a *mujtahid*. Indeed, he may not wish to be called so: it is not that a *muḥaddith* is striving to be *mujtahid*, but has not yet completed his training—it is quite possible that he has not studied the requisite sciences because he does not consider them valuable. Al-Samāhījī refers to "pupils of al-Ḥurr al-ʿĀmilī in Mashhad whom we met, and others" as falling into this category. They are Akhbārīs, but unlike the scholars mentioned above, they are not qualified *mujtahids*. According to al-Samāhījī, within the cadre of Akhbārī scholars there are those who are *mujtahids*, but reject *ijtihād*, and those who are *muḥaddith*s who lack the necessary qualifications to be considered a *mujtahid*. That al-Ḥurr is in the former category and al-Ḥurr's pupils in the latter indicates that these two groups of Akhbārīs are not necessarily in conflict or competition. Instead, al-Samāhījī's characterisation represents an embryonic hierarchy amongst the Akhbārīs between the *mujtahid-muḥaddiths* and the *muḥaddiths*.[87] For al-Samāhījī, then, a *mujtahid* can refer to a supporter of *ijtihād* (as it does when al-Samāhījī states "the *mujtahids* believe x"), but it can also be used to describe a person sufficiently qualified to employ *ijtihād* (and this latter description need not exclude the term's application to a learned Akhbārī).

[87] This analysis, then, does not accord with Newman's view that the *muḥaddith-mujtahids* and the *muḥaddith*s are "moderate" and "pure" Akhbārīs respectively (Newman, "Akhbārī-Uṣūlī Dipute, pt. 2", p. 261). Al-Samāhījī's point 34 does not indicate this. Instead it indicates that there were Akhbārī *mujtahids* (in the sense of being sufficiently learned) who rejected *ijtihād* (a white animal is still an animal); and there were Akhbārī scholars who had not reached this level (and need not reach this level to be involved in the promulgation of the law). The absence of certain names from the list of *muḥaddith-mujtahids* should not be given any great significance ("Akhbārī-Uṣūlī Dispute, pt. 2", p. 260).

In sum, then, the early lists analysed here (al-Ḥurr's list and early Akhbārī writings more broadly) show a terminological ambivalence between the use of the terms *uṣūliyyūn* or *mujtahidūn* when describing the Akhbārīs' opponents. There is a general tendency to use the former when portraying Akhbarism's rejection of *uṣūl al-fiqh* (both Sunni and Shīʿī) as a discipline worthy of a place in the Shīʿī curriculum. The latter is used to portray Akhbarism's specific criticisms of the *ijtihād* doctrine (supported by Shīʿī *mujtahids*). However, sufficient counter examples exist to prevent a hard and fast delineation of the distinction between these two uses. Later lists, however, do show a much greater consistency in their use of terminology. Opponents of Akhbārīs are invariably described as *mujtahids*, and the term Uṣūlī is less frequently employed. The receding use of the term Uṣūlī by Akhbārīs, as the following chapter demonstrates, shadows a development in the Akhbārī position. Akhbārī writers shift from unreservedly condemning the discipline of *uṣūl al-fiqh*, to proposing an alternative version of *uṣūl al-fiqh*.[88] For some Akhbārīs, an outline of this alternative can be presented in a form approximating to the established *uṣūl* genre. An example of this is Yūsuf al-Baḥrānī's introduction to his *al-Ḥadāʾiq al-Nāḍira*,[89] which follows the format of an *uṣūl* text. In it, al-Baḥrānī consistently refers to *mujtahids* as the opponents of the Akhbārīs,[90] plays down the reactionary nature of Akhbarism and attempts to outline a coherent, alternative Akhbārī *uṣūl al-fiqh*. His list of three differences between Akhbārīs and *mujtahids* comes at the end of his prologues. The three differences are presented as a summary

[88] Stewart's characterisation of Akhbarism as an anti-*madhhab* movement is based, in part, on its rejection of *uṣūl al-fiqh* as a worthy discipline. Stewart concentrates on the Akhbārī rejection of consensus (see Stewart, *Islamic Legal Orthodoxy*, pp. 175–208), and one could question whether the rejection of *ijmāʿ* was the defining Akhbārī doctrine. I would argue that the rejection of *ijtihād* holds a more central position in the Akhbārī polemic, and that the Akhbārī-Uṣūlī/*mujtahid* dispute on *ijmāʿ* was of secondary importance. Furthermore, whilst the characterisation of Akhbarism as "anti-*madhhab*" might be considered accurate for the early stages of Akhbarism, the evidence from these lists, and from other Akhbārī works (analysed below, pp. 235–237) indicates that Akhbarism can be characterised as a *madhhab* with greater accuracy after al-Ḥurr's contribution to the debate.

[89] I analyse al-Baḥrānī's *uṣūl al-fiqh* in detail in Gleave, *Inevitable Doubt*.

[90] See Baḥrānī, *al-Ḥadāʾiq*, v. 1, p. 15, p. 166, p. 167 and p. 168. Al-Bahrānī uses the term *uṣūliyyūn* (v. 1, p. 26, p. 27, p. 44, p. 52 and p. 56) but always contrasts it with *muḥaddithūn*. Here *uṣūliyyūn* appears to be used for writers of *uṣūl al-fiqh* rather than supporters of *ijtihād* (though, obviously, there will be overlap between these two categories).

of al-Samāhījī's forty points of dispute and comprise: (1) the division of the sources into four or two, (2) the legitimacy or otherwise of the division of "things" into three categories (3) the legitimacy or otherwise of interpreting the Qur'ān directly.[91] For al-Baḥrānī, the dispute is not insignificant, but, then again, debate on these three issues need not lead to intra-community bitterness.

There is a move from presenting the Akhbārīs as simply rejecting the *mujtahid*/Uṣūlī position (as in al-Ḥurr's list), to a more comprehensive exploration of Akhbārī alternatives with justificatory remarks (as in al-Samāhījī's list and within the context of a full exposition of Akhbārī *uṣūl al-fiqh* in al-Baḥrānī's prologues). This also reflects the incorporation of the discipline of *uṣūl al-fiqh* into Akhbārī literature, as the Akhbārī writers use the lists as opportunities to present (and at times define) their position with increasing sophistication over time.

The potential confusion caused by the use of the disparate set of terms (*muḥaddith*, Akhbārī, Uṣūlī, *mujtahid*) is further explored by Akhbārīs writing after al-Samāhījī. Al-Baḥrānī, for example, always speaks of *muḥaddith*-Uṣūlī differences and *mujtahid*-Akhbārī differences. He does not, however, explain the relationships between, say, the *muḥaddith* and the Akhbārī, or the *mujtahid* and the Uṣūlī. Later, the Akhbārī Fatḥ ʿAlī Zand devotes two sections of his *al-Fawā'id al-Shīrāziyya* to defining these terms and his discussion results in a convoluted typology:

> Between the Akhbārī and the *muḥaddith* there is a general-particular distinction based on a quality. Between the Uṣūlī and the *ijtihādī* there is a general-particular distinction based on one being a subspecies of the other. Some say that between a *mujtahid* and a *muḥaddith* is total mutual exclusion (*tabāyun-i kullī*), but you should know that [the term] *mujtahid* in the technical usage of the ancient Imāmī scholars, the modern experts and the recent [scholars] is "one who searches the clear book of God, and the *sunna* the final Prophet and the *ḥadīths* of the people of his house [that is, the Imams]". He derives rulings in accordance with the principles

[91] These are analysed in Gleave, *Inevitable Doubt*, pp. 249–250. There I say that the absence of *ijtihād* in this list is an attempt to reduce tension between Akhbārīs and Uṣūlīs by omitting the most contentious doctrine. As a corrective, one could argue that al-Baḥrānī, by omitting *ijtihād*, is rather linked to the general marginalisation of pure hermeneutic issues, and the shift of focus to the interrelationship between sources.

> of *uṣūl* which have been recorded in the Book and the *Sunna* . . . This is the position of the *muḥaddith-mujtahid* jurists (*fuqahā-yi muḥaddithīn-i mujtahidīn*) of the *madhhab* of the *ahl al-bayt*.[92]

Firstly, then, a *muḥaddith* may be described as an Akhbārī. However, describing a scholar as "*muḥaddith*" does not necessarily mean that he is an Akhbārī. Secondly, the *ijtihādī* (that is, a scholar who has reached the level of *ijtihād* and supports its use in the derivation of legal rules) is a subspecies of the wider category of Uṣūlīs. Uṣūlīs generally support the use of *ijtihād*, though not all are qualified to employ it.[93] An *ijtihādī* scholar is, by definition, an Uṣūlī, but not vice versa. Finally, there is nothing which excludes a *muḥaddith* from also being a *mujtahid*. In this, Fatḥ ʿAlī Zand displays apparent agreement with al-Samāhījī. However, there is a difference between these two Akhbārī scholars. For al-Samāhījī, "*mujtahid*" means one who has reached a certain educational level, and this was open to both those who supported and rejected *ijtihād*. For Fatḥ ʿAlī Zand, the *mujtahid* is merely one who derives rules according to transmitted principles from the Qurʾān and the *Sunna*. Nonetheless, for both scholars there is no contradiction in being a *muḥaddith* and a *mujtahid*, and furthermore, a *muḥaddith* may also be an Akhbārī. The logical relationships between these categories, as envisaged by Fatḥ ʿAlī Zand, are outlined in figure 2, and represent Fatḥ ʿAlī Zand's rather superficial categorisation scheme, supposedly devised to clarify the relationship between the term Akhbārī, Uṣūlī, *mujtahid* and *muḥaddith*. The need to define these categories and describe where they overlap demonstrates a more nuanced position than that of Astarābādī, two centuries earlier. Mutually exclusive categories (such as Akhbārī and Uṣūlī) are subdivided and intersect with other categories (*ijtihādī, mujtahid, muḥaddith*) to produce an intricate map of the intellectual allegiances of the Shīʿī jurists. Akhbārī writers, by the time of Fatḥ ʿAlī Zand, are describing the dispute in increasingly complex terms, reflecting the increasing sophistication and variety within the Akhbārī school.

---

92 Zand, *al-Fawāʾid al-Shīrāziyya*, f.8a.19–8b.9.

93 "*miyān-i Uṣūlī va ijtihādī ʿumūm va khuṣūṣ-i muṭlaq ast*", Zand, *al-Fawāʾid al-Shīrāziyya*, f.8a.20. This is analogous to the animal-man relationship outlined in n. 86 above.

(a) Relationship between Akhbārī and *muḥaddith*

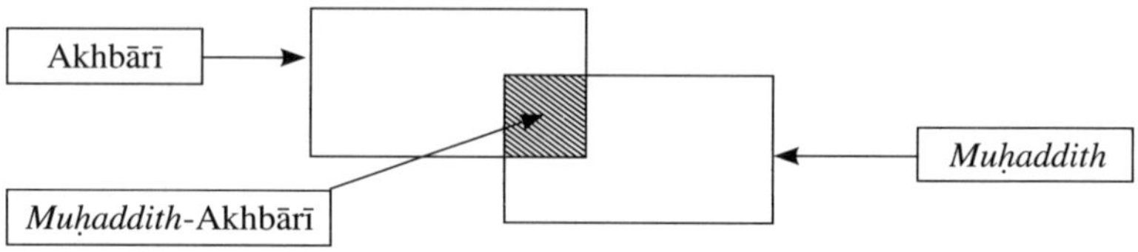

(b) Relationship between Uṣūlī and *ijtihādī*

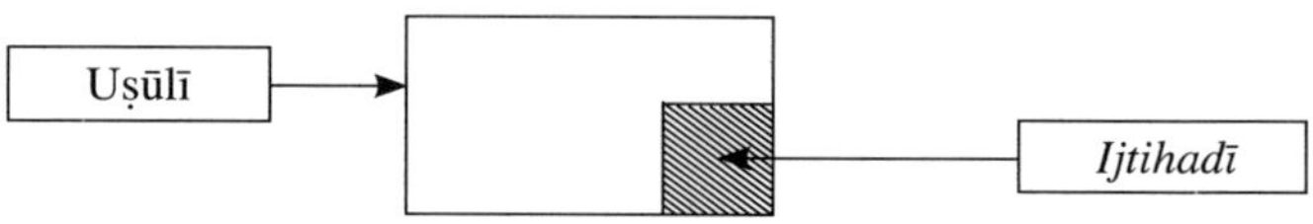

(c) Relationship between *Muḥaddith, Mujtahid* and *Muḥaddith-Mujtahid*

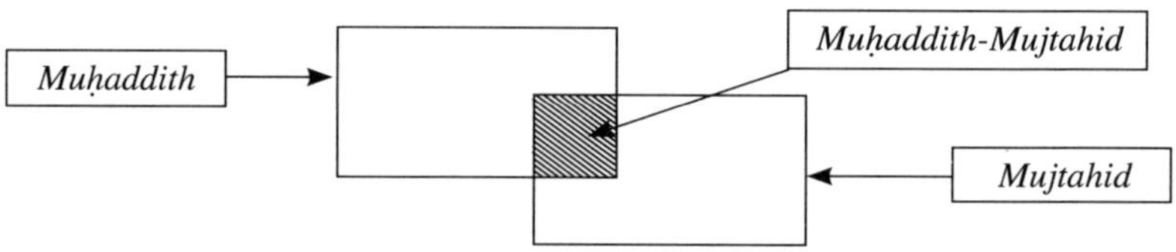

(d) Composite of Figs (a), (b) and (c).

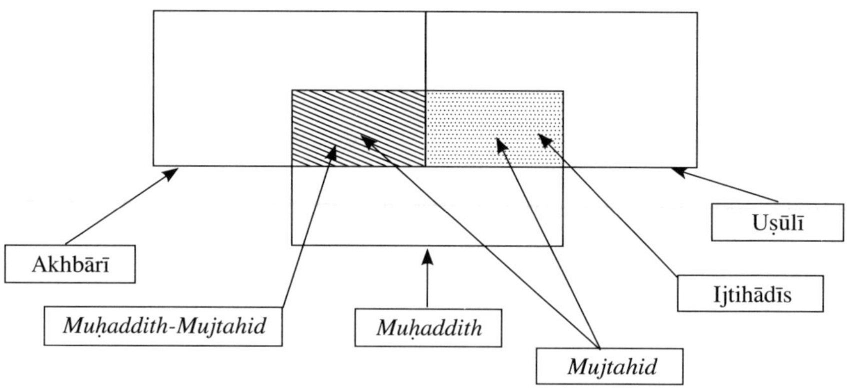

Figure 2. Relationships between Akhbārī, *muḥaddith*, Uṣūlī, *Mujtahid* and *ijtihādī* in Fatḥ ʿAlī Zand's *al-Fawāʾid al-Shīrāziyya*

Conclusions

The composition of a list of differences between one's own school and its principal opponents is, then, an act of self-definition. The popularity of the list genre amongst Akhbārīs is evidence that establishing a clear distinction between themselves and their opponents dominated their internal discussions. Early Akhbārī scholarship (between, say, Astarābādī and al-Ḥurr) was primarily reactionary—that is, the debate revolved around the faults of the Uṣūlī position, rather than the establishment of a coherent Akhbārī alternative. The early lists reflect this preoccupation, and, accordingly, list the differences between the two schools as mere Akhbārī refutations of Uṣūlī doctrines. However, as Akhbarism began to take hold, attention turned to the construction of an Akhbārī position, and the later lists, in turn, reflect this concern. Later authors do not simply list Akhbārī denials of Uṣūlī doctrines; they present, and occasionally argue for, alternative juristic mechanisms to solve the same hermeneutic problems. In al-Samāhījī's list, for example, the *mujtahid* position is regularly reduced to a single line. The Akhbārī position, on the other hand, receives expansive commentary, as not only the doctrine itself, but also the ramifications of the doctrine are explored. Furthermore, from al-Samāhījī onwards, there is an increased willingness to admit to variety within the Akhbārī school, reflecting a certain confidence acquired through more established school status, and (perhaps) a greater level of internal Akhbārī dispute and hierarchy. This latter point is also an indication of a firmer "school" foundation, since Akhbārīs were able to dispute amongst themselves without it leading to sectarian fissures within the school. The lists, then, provide an account of Akhbārī school formation which keys with the evidence of *ijāza* linkages presented in the previous chapter, though with a two decade delay. By the time of al-Ḥurr, and to a large extent, as a result of his efforts, the Akhbārī tradition of scholarship had entered the Shīʿī intellectual mainstream. It took, however, another two decades for Akhbārīs to delineate a set of doctrines, in list form, in which a coherent alternative to the position of their Uṣūlī opponents was presented.

CHAPTER SEVEN

AKHBĀRĪ QUR'ANIC INTERPRETATION

Biographical sources, *ijāza* documents and early Akhbārī-Uṣūlī difference lists tend to portray the *akhbāriyya* as a relatively united community of scholars. In biographical works (and occasionally in *ijāzas* also), the simple epithet Akhbārī was usually applied without qualification, giving the impression that the Akhbārī identity of a scholar was an unproblematic description.[1] In the difference lists, internal Akhbārī doctrinal variation emerges quite late. Not before al-Samāhījī's *Munyat al-Mumārisīn* is there a significant recognition of intra-Akhbārī disputes, and these are further developed in later lists.[2] In the earlier difference lists, Akhbarism's portrayal is rarely textured: disagreements, or more subtly, the variety of approaches to Astarābādī's legacy are, on the whole, ignored. In later texts, variety within the Akhbārī school is admitted. This should not be taken as implying that internal disputes emerged late. It is clear that there was much Akhbārī doctrinal variety from the earliest post-Astarābādī period. This includes both the diversity in legal doctrine outlined in this and the next two chapters, but also diversity amongst Akhbārīs writing in other disciplines (in particular theology and *falsafa*).[3] Akhbārīs not only modified Astarābādī's views, but also introduced novel opinions, which they considered more authentically Akhbārī than those of their opponents. These novel opinions were developed either in response to strong counter argumentation by Uṣūlīs and others, or as a result of perceived weaknesses within Astarābādī's own jurisprudence. New Akhbārī views vied with older (Astarābādī-inspired) Akhbārī doctrines for dominance, creating lively debate within the nascent Akhbārī school. Furthermore, new arguments were devised to support

---

[1] See above, pp. 40–60.

[2] See above, pp. 205–211.

[3] On theological and philosophical diversity in post-Astarābādī Akhbarism, see Gleave "Scripturalist Sufism". Indeed, the diversity of Akhbārī approaches to theology and philosophy, as mentioned above, is such that specific theological and philosophical doctrines cannot be considered central elements in determining the Akhbārī identity of particular scholars.

established Akhbārī doctrine, and in many cases, these were also introduced in order to fend off Uṣūlī attacks. In the previous chapter, based on evidence from the difference lists, I argued that Akhbarism was considered (by both Akhbārīs and others) as holding distinctive doctrines in three areas of legal theory: the status of revelatory texts, the interpretation of these texts and scholarly authority. This chapter reviews the development and variety of Akhbārī doctrines in relation to the legal interpretation of the Qur'ān.[4]

The available sources of legal knowledge (or "the sources of legal indicators"—*madārik, uṣūl, adilla*) were clearly a central Akhbārī concern in the post-Astarābādī period. The standard expression of Akhbārī doctrine indicates an acceptance of only two sources of law (*kitāb* and *sunna*), rather than the four sources of the Uṣūlīs (*kitāb, sunna, ijmāʿ* and *dalīl al-ʿaql*). Akhbārīs put forward a number of arguments against consensus (*ijmāʿ*) and reason (*dalīl al-ʿaql*—conceived of as both pure reason and hermeneutic techniques derived from reason), focussing on the uncertainty (that is, *ẓann* and the lack of *ʿilm*) of rulings derived through these two so-called "sources". These objections are part of the Akhbārī exposition of a particular exegetical method, and are properly the concern of Akhbārī hermeneutics.[5] It should be noted, however, that Astarābādī's own use of rational argumentation in his theological works indicates that the Akhbārī rejection of *ʿaql* is limited to legal matters—or put another way, whilst Akhbārīs agreed that *ʿaql* had extremely limited use as a source of law, they differed over whether it could be used as a source in other disciplines.[6] Arguments for *kitāb* and *sunna* as the only reliable sources, of course, rely on the certainty of the rulings derived from these sources, and the examination of Akhbārī statements on the operation of these sources in this and the next two chapters reveals significant intra-Akhbārī debate. With regard to the Qur'ān, the debate concerned its interrelationship with the *akhbār* (that is, the relationship of *kitāb* and *sunna*). Some Akhbārīs rejected *kitāb* as an independent source of legal knowledge, others accepted it as a source, but only in a qualified manner.

---

[4] Interpretation and scholarly authority are discussed in Chapter 9, and also in Gleave, "The Qāḍī and the Muftī".

[5] See below, pp. 205–211.

[6] See above, p. 114 and Gleave "Scripturalist Sufism".

The Unavailability of Qur'anic Meaning

The Akhbārī views on the Qur'ān as a distinct source of law has led some commentators to view Akhbarism as a continuation of earlier Shīʿī doctrines concerning the integrity of the ʿUthmanic codex. It is argued that since Akhbārīs held an equivocal view on the probative force of the Qur'ān, they can be linked with early Shīʿī "extremists" (who considered the Qur'anic text defective). There is, however, little evidence that this "extremist" doctrine was an element within Astarābādī's own argumentation, and the hackneyed Shīʿī debate about the integrity of the Qur'anic text is clearly peripheral to post-Astarābādī Akhbārī discussions of the sources of law. Astarābādī's arguments against employing the Qur'ān as a direct source revolved around the inaccessibility of its intended meaning (*murād*), not the integrity of the text itself. A particular verse may have an available meaning (that is, its *ẓāhir* may be obtainable), but this may not be God's "*murād*". Even when God's intended meaning is discovered, the legal significance of the verse still remains undiscovered. In Muslim jurisprudence generally, the lack of fit between intended meaning and legal significance is most radically demonstrated in the doctrine of abrogation (*naskh*), but it can also be seen in the interplay between general and specific rulings (*ʿāmm/khāṣṣ*) or restricted and unrestricted meaning (*muṭlaq/muqayyad*).[7] Astarābādī uses these juristic categories to argue that the Qur'ān cannot be used as a legal source when factors such as these, which affect the legal relevance of a particular verse, are not known with any certainty. It would be foolhardy, Astarābādī argues, to delineate God's law when there is uncertainty over the probative force (*ḥujjiyya*) of many Qur'anic passages.[8] For Astarābādī, then, the Qur'ān is not an independent source and it has no direct role to play in legal argumentation. It requires interpretation through a source which has indubitable probative force—namely the *akhbār*. This argument naturally trespasses on hermeneutic questions, demonstrating the interconnected nature of the revelatory sources and their interpretation for Akhbārī legal theory. The Qur'ān is excluded as an independent legal source, not because of any doubt as to its authenticity, but because of the inaccessibility

[7] See Weiss, *God's Law*, pp. 389–396.
[8] Astarābādī, *al-Fawā'id*, pp. 279–281 and pp. 336–338.

of its meaning (or at least, much of its meaning). With this as the dominant Akhbārī view, Shīʿī questions regarding the identity of the Qur'anic text with divine revelation recede, almost to the point of redundancy. The Akhbārī Muḥsin Fayḍ, for example, discusses the deficiency of the available Qur'anic text in the introduction to his Qur'anic commentary, *al-Ṣāfī*. He cites well-known reports from the Imams concerning the deficiency of the Qur'anic text. However, he concludes, God left us both the Qur'ān and the family of the Prophet Muḥammad as a guide, and the deficiency of the former is irrelevant because the latter (through the *akhbār*) provide sufficient guidance.[9] It may well be true that some Akhbārīs, Muḥsin Fayḍ included, held the view that the available Qur'anic text was deficient, but this was not a distinctive Akhbārī doctrine. A number of Shīʿī authors of the period appear to have held a similar position.[10] Akhbārī doctrine is distinctive, however, concerning the manner in which the potentially debilitating effects of this deficiency are circumvented. The *akhbār* are the sole interpreters of the Qur'ān, and potential Qur'anic deficiency is side-stepped. These two discussions (Qur'anic integrity and the Qur'ān's role as a source of law) were interrelated; the view that at least some passages of the Qur'ān are directly interpretable naturally (though not inevitably) keys with the view that the available Qur'anic text is authentic. However, the theological dispute over the integrity of the text does not figure in Akhbārī legal argumentation of the Qur'ān's role. For most Akhbārīs, it does not matter whether the Qur'ān is corrupt or inviolate: it remains redundant (or at least restricted) as a legal source because its meaning and legal significance are not immediate.[11]

On the probative force of the Qur'ān more specifically, a number of the later difference lists mention an internal Akhbārī dispute concerning the status of the Qur'ān as a source of legal knowledge. Al-Jazā'irī mentions, as his third difference between Akhbārīs and Uṣūlīs, the Uṣūlī doctrine that there are five (!) sources of the law (*kitāb, sunna, ijmāʿ, dalīl al-ʿaql* and *al-istiṣḥāb*). Akhbārīs, he argues, reject *dalīl al-ʿaql* and *ijmāʿ*, and only allow *kitāb* to be a source when its meaning

[9] Fayḍ, *Ṣāfī*, v. 1, p. 55.
[10] See Lawson, "Note", pp. 5–10 and the scholars referenced there. For an alternative view, in which Shīʿī attitudes to the Qur'ān were viewed as closer to Sunni "orthodoxy" see, Modarressi, "Early Debates", pp. 28–35.
[11] See below, pp. 238–239.

is delineated in the *akhbār* "because, [they argue, the Qur'ān] is ambiguous (*li-kawnihi mutashābih*). God spoke it to the Prophet and the People of his House [that is, the Imams]... and no ruling can be known from [the Qur'ān] unless [it is known] through [the Imams'] words. In sum, [for Akhbārīs, legal] indications are restricted to the *sunna* and nothing else".[12] Al-Jazā'irī makes no mention here of any internal Akhbārī dispute, preferring instead to distinguish his own position from the Akhbārīs.[13] He argues that there are elements of the Qur'ān—the so-called *muḥkamāt*—which can be interpreted directly. Al-Samāhījī, writing twenty years later, refers to internal Akhbārī difference on this matter. He states that the *mujtahids* accept four sources of law, whilst the Akhbārīs accept only two: "*kitāb* and *sunna*—however, some [Akhbārīs] restrict themselves to the *sunna* alone because *kitāb* is not known to us. [For them] it is not permitted to interpret [the Qur'ān] except through [the Imams]".[14] Al-Baḥrānī, writing around 50 years later than al-Samāhījī, is yet more specific: there are two Akhbārī opinions:[15] those who argue that the Qur'ān can never be used as a source (because it is *mutashābih*) and those who argue that one can interpret the Qur'ān directly.[16] Al-Baḥrānī posits a third position, a compromise between the two extremes and which he personally advocates: some of the verses of the Qur'ān (be they "*muḥkamāt*" or some other category) can be understood directly, though other elements cannot. The difference between al-Samāhījī's and al-Baḥrānī's accounts here concerns not only the number of different Akhbārī positions, but also the classification of the dispute: al-Baḥrānī considers it a hermeneutic issue (that is, a question of *istidlāl*) while al-Samāhījī considers it a question of source validation. Al-Baḥrānī does not name scholars who held the latter two positions, and only

---

[12] Jazā'irī, *Manba'*, p. 44.

[13] He does distinguish between Akhbārī positions on the Qur'ān in his commentary on Ibn Bābūya's *al-Tawḥīd* (Jazā'irī, *Nūr al-Barāhīn*, v. 1, pp. 189–190.

[14] See Newman, "Akhbārī-Uṣūlī Dispute, pt. 1", p. 25 (alternative translation, p. 39).

[15] They are mentioned, not only in his list of Akhbārī-Uṣūlī differences, but also in his section on the Qur'ān as a source of law. See Baḥrānī, *al-Ḥadā'iq*, v. 1, p. 26 where he mentions two opinions (that the Qur'ān cannot be interpreted without the *akhbār*, and that it can be interpreted directly), and claims a third as his own.

[16] Al-Baḥrānī's own position can be roughly equated with the second of these, though he terms the verses which can be understood directly as "general and unrestricted" (*mujmal wa 'āmm*). See Baḥrānī, *al-Ḥadā'iq*, v. 1, p. 29 and Gleave, *Inevitable Doubt*, p. 54.

Astarābādī is named as holding the view that the Qurʾān can never act a source. Fatḥ ʿAlī Zand, writing two generations after al-Baḥrānī, also mentions two opinions amongst the *akhbāriyān*:

> They [the Akhbārīs] agree that for those verses of the Qurʾān which are known to be "*muḥkam*", neither an interpretation (*tafsīr*) nor an interpreter (*mufassir*) is needed. However, there is a need for a *tafsīr* for *ẓāhir* meaning. They differ over whether the *ẓāhir* [meanings] of the Qurʾān have probative force when there is no explanatory comment from the Imam (*ḥujjiyyat-i ẓawāhir bidūna mubayyan-i maʿṣūm*).[17]

This is a yet more sophisticated depiction of the internal Akhbārī dispute and turns on the difference between the hermeneutic categories *muḥkam* and *ẓāhir* (on which, see below). It should be noted at the outset, however, that, according to Fatḥ ʿAlī Zand, all Akhbārīs agree that the "*muḥkam*" verses need no *tafsīr*. All Akhbārīs hold the view that at least some verses of the Qurʾān do not require the Imams' *tafsīr*. Al-Jazāʾirī's position (which also appears to be al-Baḥrānī's view, though without the use of the technical term *muḥkam*) was depicted by these authors as a deviation from the dominant Akhbārī stance. By the time of Fatḥ ʿAlī Zand, the availability of at least a portion of Qurʾanic meaning (the *muḥkamāt*, or some other locution) had become Akhbārī orthodoxy. If the difference lists are taken as the only source, the doctrinal history might be constructed thus: first, Astarābādī and the Akhbārīs argue that no verse of the Qurʾān can be understood; later, al-Jazāʾirī dissents from this view, arguing that the *muḥkamāt* can be understood, but this view is distinguished from that of the *akhbāriyya*; later still al-Samāhījī records a division between Akhbārīs who accept both *kitāb* and *sunna* and those who accept the latter only; these are re-iterated by al-Baḥrānī and a compromise Akhbārī solution (similar to al-Jazāʾirī's non-Akhbārī position) is suggested; al-Baḥrānī's compromise position (that the *muḥkamāt* verses can be understood directly) becomes agreed Akhbārī doctrine by the time of Fatḥ ʿAlī Zand, and he then projects this back onto the earlier Akhbārī period, portraying the acceptance of *muḥkamāt* interpretation as unanimous Akhbārī doctrine. By doing so, he reduces the difference between Akhbārīs to the technical question of whether or not the *ẓāhir* meaning of the Qurʾān has probative force on the occasions of the Imams' silence.

---

17 Fatḥ ʿAlī Zand, *al-Fawāʾid al-Shīrāziyya*, f.23b.14–15.

This possible version of the doctrine's history, based on these difference lists, does not, however, stand up to a comparison with Akhbārī juristic works. The irreducibly personal nature of the Akhbārī-Uṣūlī dispute's depiction in these difference lists has already been mentioned and is evident here also. Al-Jazā'irī portrays his own position as novel and outside of Akhbarism, even though the doctrine of the interpretability of the *muḥkamāt* was outlined much earlier. Muḥsin Fayḍ (see below) heavily criticised Astarābādī's absolute prohibition on Qur'anic interpretation from an early period, whilst simultaneously claiming Akhbārī allegiance. Al-Jazā'irī argues, not only that there is a single Akhbārī position, but also that it is extreme:

> As for [the Akhbārīs'] unreserved rejection of the Qur'ān as having probative force for legal rulings, it is unfounded because in the Qur'ān there are *muḥkam* [verses] and verses whose indication is clear (*ẓāhir al-dalāla*). That it is permitted to take legal rulings from [the Qur'ān] is proven.[18]

Of course, this also is a rather disingenuous depiction. He wishes to be considered a moderate. In order to do this, he needs to avoid identifying himself with previous Akhbārīs. The most effective means of doing this is to present Akhbārīs as both unified and extreme, and his own views as reasonable and moderate. In fact, al-Jazā'irī merely selects one of (at least) two available Akhbārī opinions. Similarly, al-Baḥrānī posits the existence of an Akhbārī opinion that the whole of the Qur'ān can be interpreted directly (and an exegete is "equal" to the Imams in his interpretive endeavours). Such an Akhbārī opinion is not present in the works of Akhbārī legal theory from Astarābādī onwards.[19] Al-Baḥrānī's categorisation of Akhbārī opinions is not descriptive but heuristic. This invented "Akhbārī" position resonates with that of the Ẓāhiriyya, some of whom did argue for direct Qur'anic interpretation without the intervention of the *sunna*.[20] The common, but inaccurate, bracketing of the Shī'ī Akhbārīs with the Sunni Ẓāhirīs perhaps makes the category more plausible to al-Baḥrānī's readers.[21]

---

[18] Jazā'irī, *Manba'*, p. 44.

[19] This assessment may not hold true for all *tafsīr*s written by Akhbārīs, though the only possible exception I have found is the *Tafsīr al-Mu'īn* of Nūr al-Dīn al-Akhbārī, the nephew of Muḥsin Fayḍ, referred to below, pp. 223–224.

[20] See Abṭaḥī, "Naqsh-i sunnat" and As'adī, "Qur'ān-i 'Mubīn'", p. 3.

[21] This was a common term in anti-Akhbārī works—see Khwānsārī, *Rawḍāt*, v. 3, p. 290 where Ẓāhirī and Ḥashwī doctrine is "identical with the claims of the Akhbārīs" (*'ayn maqālat al-Akhbāriyyīn*).

However, it is merely a literary device: in order to portray his own position as moderate (a common topos in juristic writings), al-Baḥrānī presents two "extreme" positions and locates himself between them.[22] Similarly, Fatḥ ʿAlī Zand names scholars who are supposedly members of the two trends within Akhbarism:

> The opinion of [Muḥammad] Amīn al-Astarābādī, Muḥaddith-i [Ḥurr-i] ʿĀmilī, ʿAllāma Sayyid Hāshim[-i Baḥrānī] and [Muḥammad Taqī] Majlisī-i Awwal [Majlisī I] is that [*ẓawāhir al-Qurʾān*] have no probative force. The position of Fāḍil-i [Fayḍ-i] Kāshānī and Muḥaqqiq-i [Khalīl-i] Qazwīnī is the opposite.[23]

As mentioned above and elaborated below, Astarābādī argued neither that the *muḥkamāt*, nor the *ẓawāhir al-qurʾān* had probative force (as Fatḥ ʿAlī Zand claims). Rather Astarābādī argued that the whole Qurʾān was unavailable for interpretation (that is, the doctrine of *taʿmiya* or "blindness"). This doctrine was the subject of intense criticism by Muḥsin Fayḍ from within the Akhbārī school on the basis that the meanings of *muḥkam* verses are readily available. In short, Fatḥ ʿAlī Zand is also presenting a description of the dispute which fails to accord with a reading of the available earlier sources. He describes it in this way to minimise the perception of intra-Akhbārī dispute, reducing the dispute to a technical matter of the preponderance or otherwise of the *ẓawāhir al-qurʾān*. What emerges from this is that difference lists cannot act as one's sole source in any description of the development of Akhbārī doctrine.

Assessing Fatḥ ʿAlī Zand's depiction of the different Akhbārī positions on the question of the Qurʾān as a source of legal knowledge does, however, enable us to understand Akhbarism's end-point on this doctrine, even if it took nearly two centuries to reach it. The final Akhbārī doctrine, as described by Zand, relies upon a particular view of how utterance (*lafẓ*) and meaning (*maʿnā*) interrelate. Al-Karakī provides a succinct, technical definition of the relevant terms:

> If an utterance cannot be interpreted in a manner other than that which is understood from it linguistically (*lughat*[an]), then it is "*naṣṣ*". If this is not the case, then the more acceptable meaning (*al-rājiḥ*) is called "*ẓāhir*",

[22] See above, pp. 201–202. Majlisī I portrays himself as a moderate, as does his son, Muḥammad Bāqir al-Majlisī; Majlisī II's pupil, al-Jazāʾirī, also portrays himself as a moderate.

[23] Fatḥ ʿAlī Zand, *al-Fawāʾid al-Shīrāziyya*, f.23b.15–17.

> and the less acceptable meaning (*al-marjūḥ*) is called "*mu'awwal*". If there is equality [between two or more possible meanings—that is, none can be said to be more acceptable than the others], then [the utterance] is termed "*mujmal*".[24] If [an utterance] combines the first two categories [named *ẓāhir* and *naṣṣ*] then it is "*muḥkam*". If it combines the latter two properties it is termed "*mutashābih*".[25]

A *naṣṣ* utterance (in this case, a *naṣṣ* verse) is one which has a meaning of such clarity that no alternative interpretation is possible or permitted. Furthermore, this indisputable meaning must be its linguistic (*lughawī*) meaning. A *muḥkam* utterance is not only *naṣṣ* (that is, it not only fulfils these two criteria), but also its single linguistic meaning is the meaning every reader would immediately recognise (that is, its "obvious", "apparent" meaning, its *ẓāhir*). This, according to most Muslim exegetical theory, is what God means when he says in Q3.7:

> It is He [Allah] who has sent down to you the Book. In it are *muḥkamāt* verses, which are the mother of the Book, and others which are *mutashābihāt*...

As is well-known, this verse provided Muslim hermeneutics with two basic categories: clear (*muḥkam*) and ambiguous (*mutashābih*), the definitions of which have inspired debate and discussion in subsequent legal theory. That the later, established technical meanings of *muḥkam* and *mutashābih* are projected onto the Qur'anic usage is obvious, and the verse has merely become a peg for these theories.

In later theory, the single obvious meaning of a *muḥkam* verse is unaffected by the broader context of the utterance and unconnected with any implication the hearer might discern. Utterances are, however, rarely so unambiguous. They infrequently have single possible meanings: they can be interpreted in different ways depending on the context (the identity of the speaker, the audience, the occasion, the preceding and subsequent utterances etc). Such occasions (termed *mutashābihāt*, and certainly more numerous than the *muḥkamāt*) require a mechanism whereby one of the competing interpretations might be selected. Muslim legal theorists have often argued that in these circumstances the *ẓāhir* (apparent) meaning of an utterance takes

---

[24] Reading *mujmal* for *jumal* in the printed edition.

[25] Karakī, *Hidāyat*, p. 243. See also Karakī, *Hidāyat*, p. 155: *muḥkam* "is when *naṣṣ* is explicit in terms of its meaning, such that it [cannot be] interpreted in a way other than what is understood from it."

precedence, and is the interpreter's default solution to the ambiguity. There must be sufficient evidence (*qarā'in*) to dislodge it and move to an interpreted ("*mu'awwal*") meaning. In *mujmal* cases, then, the presumption is that the *ẓāhir* meaning is the intended meaning, and evidence must be rallied to demonstrate that the intended meaning is *mu'awwal*. Al-Karakī's technical schema is not novel; it mirrors one of the various categorisation schemes established in the tradition of *uṣūl al-fiqh*.

The significance of the detail of this categorisation scheme developed by Fatḥ ʿAlī Zand for the intra-Akhbārī debate is clear. According to Fatḥ ʿAlī Zand, all Akhbārīs agree that when a Qur'anic verse is *muḥkam*—that is, it has one possible meaning, and this is its linguistic meaning (that is, the verse is *naṣṣ*), and it is apparent to the reader what this meaning is (that is, the meaning is the *ẓāhir* one)—then the Qur'ān can be understood directly and without the need for any *tafsīr* from the Imams. Whilst the number of verses which fulfil these criteria may be few in number, Akhbārīs (according to Fatḥ ʿAlī Zand) were never entirely opposed to believers gaining knowledge from the Qur'ān directly. They merely argued for a highly restrictive set of occasions when this can be performed legitimately. Debate between Akhbārīs, however, supposedly centres on whether the *ẓāhir* meanings of Qur'anic verses can act as evidence for or against a proposed legal ruling (and have probative force—*ḥujjiyyat*) on those occasions when the Imams have made no statement concerning the meaning of a supposedly relevant verse. To sum up, when an utterance (that is, a verse) could have a number of possible meanings, and there is no contextual indicator which might enable the interpreter to decide between these meanings, most writers of Shīʿī *uṣūl al-fiqh* argue that the *ẓāhir* is assumed to be the intended meaning of the utterance. Contextual indicators could include statements of the Prophet or the Imams, in which one of the possible meanings of a verse is identified as God's intended meaning. The question over which Akhbārīs differed was the procedure for cases when there were no such contextual indicators (that is, the Imams are silent). Some Akhbārīs revert to the established principle of *uṣūl al-fiqh*, and argue that the *ẓāhir* becomes the presumed intended meaning. Others argue that the silence of the Imams on a verse leads to permanent ignorance (or suspension of judgement—*tawaqquf*) as to the verse's intended meaning. Any attempt to force a meaning upon the verse (be it the *ẓāhir* or any other) is, this latter party argues, a violation of the text's integrity.

Whilst Fatḥ ʿAlī Zand's depiction of the intra-Akhbārī dispute is certainly more complex than those previous to him, it is merely the final formulation of the dispute, back-projected onto an earlier period of the school's development. In fact, the development of Akhbārī doctrine on the probative force of the Qurʾān (or lack of it) followed a different trajectory.

Reviewing post-Astarābādī Akhbārī writings does indicate that there was doctrinal variety on the question of the probative force of the Qurʾān, and in this sense the later difference lists have identified a contentious doctrine. There are at least two sets of Akhbārī opinions, and within each set there was some variation. On the one hand, there were Akhbārīs who considered the *akhbār* essential to understanding any Qurʾanic passage. As explained above, Astarābādī certainly held this position, as did the so-called Akhbārī *mufassirūn* (the Qurʾanic commentators al-Huwayzī and Hāshim al-Baḥrānī). On the other hand, there were those who accepted that some Qur'anic verses could be understood directly. Whether these were termed *muḥkamāt* or not does not prevent the adherents of these positions being bracketed together. Muḥsin Fayḍ, al-Jazāʾirī and Yūsuf al-Baḥrānī held this view and, for reasons outlined in previous chapters, should all be considered part of the Akhbārī school.[26] As is argued below, the debate was not merely about the probative force of the Qurʾān. It concerned also the extent to which Akhbārī devotion to the *akhbār* could be maintained in the face of Uṣūlī counter-argumentation. More broadly, the issue lying behind this intra-Akhbārī dispute was the extent to which established *uṣūl* categories (such as *muḥkam*) should influence internal Akhbārī discourse. Some (Astarābādī, al-Huwayzī, Hāshim al-Baḥrānī and others) argued that these categories had no role in Akhbārī discourse. Others (Fayḍ, al-Jazāʾirī and al-Baḥrānī amongst them) argued that acceptance of these categories was foundational to any sensible discussion about the legal reasoning process. One sees a similar division amongst Akhbārīs when they discuss the status of the *akhbār* as sources of legal knowledge, outlined in the next chapter.[27]

The arguments used by Astarābādī for the indispensable role of the *akhbār* in the interpretation of the Qurʾān have already been

---

[26] On al-Baḥrānī, see Gleave, *Inevitable Doubt*, pp. 47–55.

[27] See below, pp. 245–267.

described.[28] This is sometimes termed the *taʿmiya* position, since the meaning of the Qur'ān is "enigmatic" (or "hidden") for the ordinary reader.[29] Continuing this position is al-Karakī, who despite his succinct definition of *muḥkam* quoted above, clearly believes that the *akhbār* are indispensible. Furthermore, he implies that there is no way of knowing whether a verse is *muḥkam* or not without the *tafsīr* of the Imams. Hence, the *akhbār* are crucial not only for interpreting *mutashābih* verses, but also for deciding which verses are *mutashābih* or *muḥkam*. For al-Karakī, we interpret the Qur'ān directly only when the Imams have given us permission to do so, and this permission is not only a general permission for all verses in a certain category (for example, *muḥkam*). Rather, this permission is granted for specific Qur'anic verses, which once specified as open to direct interpretation become *muḥkam*. It is, for al-Karakī, the Imams who categorise a verse, not our own (flawed) understanding of the range of possible meanings a verse might have. After citing many *akhbār* which, he claims, demonstrate this position, al-Karakī states:

> It can be understood from the *akhbār* that the *tafsīr* of the Qur'ān is not permitted, as is the derivation of speculative rulings from its *ẓāhir* meanings. Unless they are referred to the words of the Imams to find out their status, [Qur'anic verses] are of only probable indicative value (*ẓannī al-dalāla*): whether they are abrogated or not, whether they are general or unrestricted or not, whether they are to be interpreted or not. One who does this has erred, for the Imams are explicit in saying that the Qur'ān is only understood by the one to whom it is addressed, and that the Qur'ān was revealed in a *taʿmiya* manner in relation to the minds of common folk.[30] The knowledge of what is an abrogator and what is abrogated, and what remains as having a *ẓāhir* meaning, and what does not, is particular to [the Imams].[31]

This rather prolix statement, both in wording and content, owes its form to Astarābādī's *al-Fawā'id*. The current it reflects is found

---

[28] See above, pp. 73–75 and p. 223. Though note his statement that "most" (or "mostly"—*fī'l-akthar*) of the Qur'ān is mysterious to the "minds of the *populus*" (Astarābādī, al-*Fawā'id*, p. 179)

[29] See Astarābādī, *al-Fawā'id*, p. 270; Karakī, *Hidāyat*, p. 162.

[30] Al-Karakī lifts the phrase "revealed in a *taʿmiya* manner..." directly from Astarābādī's *al-Fawā'id* (pp. 104, 179, 270 and 483). However, al-Karakī's formulation removes the qualified "mostly" or "in the main" found in some of Astarābādī's formulations (see above, n. 28) indicating that al-Karakī's position reflects a hardening of Astarābādī's own view.

[31] Karakī, *Hidāyat*, p. 162.

within early Akhbarism, and also includes the authors of the so-called Akhbārī *tafsīr* works, al-Huwayzī and Hāshim al-Baḥrānī. There is no need to repeat in detail Lawson's description of "Akhbārī" *tafsīr*.[32] In such works, Qur'anic verses are accompanied by lists of *akhbār*. Authorial comment is minimal. Of course, whilst the authors of these *tafsīrs* may present their work as free of any personal interpretive interference, preference is manifested through the selection, ordering and abbreviation of reports. Furthermore, whilst some selected reports refer directly to the verse under examination (through an explicit statement of the Imam), others were selected by the author because they spoke to the subject of the verse. This, as al-Baḥrānī perceptively points out in relation to the practice of comparing reports with the Qur'ān, implies that verses cannot be totally incomprehensible, otherwise there would be no basis for the selection of these reports and their comparison with the Qur'anic text.[33] The technique of Akhbārī *tafsīr* has, as its *raison d'être*, the view that the Qur'ān cannot be understood without the *akhbār* of the Imams. Since the established *uṣūl* and *tafsīr* disciplines consider Q3.7 (the *muḥkamāt* and *mutashābihāt* verse) to demonstrate the opposite, the interpretation of this verse by these Akhbārī writers is a convenient starting point for their thoughts on the meaning and relevance of the *muḥkam* category. Both al-Huwayzī and Hāshim al-Baḥrānī clearly consider the established position as entailing a *petitio principii*: one cannot read Q3.7 as proving that one can understand the Qur'ān directly, unless one has already demonstrated that one can read the Qur'ān directly. In place of this logical fallacy, they each offer a selection of *akhbār* on the verse. Explicit authorial comment is absent in both commentaries. The whole verse reads:

> [Q3.7] It is He [Allāh] who has sent down to you the Book. In it are *muḥkamāt*, which are the mother of the book, and others which are

---

[32] See Lawson, "Akhbārī *Tafsīr*". See also his "Note".

[33] See Gleave, *Inevitable Doubt*, p. 50. Al-Baḥrānī's solution to this logical problem is to indicate that the topic of the verse may be clear, but its legal relevance is unknown until one has consulted the *akhbār* of the Imams. Presumably, then, selection of the *akhbār* can be made, since the topic of a verse can be known directly, though the nature of the ruling concerning that topic is unavailable. This, I take it, is what is meant by his reference to the Imams' *tafsīr* being an "account of God's intended meaning". The topic of the verse is sufficiently discernable to form the basis of a comparison between the Qur'ān and the *akhbār*. Its legal meaning is not. See Baḥrānī, *al-Ḥadā'iq*, v. 1, p. 30.

> *mutashābihāt*. As for those whose hearts have *zaygh* [perversity] in them, they follow that which is *mutashābih* in [the Qur'ān], seeking *fitna* [discord] and its *ta'wīl* [interpretation]. No one knows its meaning other than God and *al-rāsikhūn fī'l-ʿilm* [those well grounded in knowledge] who say, "we believe in it. All of it is from God."[34]

The verse itself is grammatically ambiguous: the referent of "it" in "No one knows *its* meaning" (and also in "we believe in *it*") could be "that which is *mutashābih* in the Qur'ān" or the whole Qur'ān. The ambiguity is important since its clarification determines whether the verse supports or denies the availability of the meaning of at least some (*muḥkamāt*) verses of the Qur'ān.

The *akhbār*, cited in response to this text, by Akhbārī *mufassirs* can be divided into two categories:

1. "Exegetical" reports—that is, reports which cite the verse, or parts of the verse, and an interpretation of the meaning or reference of particular words or phrases is offered. For example, both authors cite the report in which the phrase *al-rāsikhūn fī'l-ʿilm* is quoted, followed by the statement of an Imam (either Imam al-Bāqir or Imam al-Ṣādiq), "The Prophet was the best of *al-rāsikhūn fī'l-ʿilm*. God taught him all that he revealed and how to interpret it."[35]
2. "Clarificatory" reports—in which the verse is not cited, but, almost *en passant*, a word or phrase from within the verse is used by the Imam in such a way that its meaning is clear. For example, Imam ʿAlī is reported to have said, "those, other than ourselves, who claim to be wellgrounded in knowledge (*al-rāsikhūn fī'l-ʿilm*) are liars, and rebels against us."[36] The presumption is that the Imam's use of the phrase corresponds to, and is meant to be, a *tafsīr* of the Qur'anic verse.

The exegetical reports can be sub-divided into those of lexical clarification (Imam al-Ṣādiq said, "the meaning of *fitna* here is unbelief (*kufr*)"[37] and "*zaygh* means doubt (*shakk*)"[38]), and those of referent identification (Imam al-Ṣādiq stated that "the *āyāt muḥkamāt*" are "the Prince of Believers and the Imams"[39]). In Hāshim al-Baḥrānī's selection,

[34] The terms in italics refer to words and phrases, the (supposedly) precise meaning of which are identified by the reports in both selections.

[35] Huwayzī, *Nūr al-thaqalayn*, v. 1, p. 316, #35; Hāshim al-Baḥrānī, *al-Burhān*, v. 1, p. 597, #4.1599.

[36] Huwayzī, *Nūr al-thaqalayn*, v. 1, p. 315, #26.

[37] Huwayzī, *Nūr al-thaqalayn*, v. 1, p. 312, #17.

[38] Hāshim al-Baḥrānī, *al-Burhān*, v. 1, p. 599, #16.1611.

[39] Hāshim al-Baḥrānī, *al-Burhān*, v. 1, p. 597, #2.1597. Hence, it is not verses which are meant by *āyāt*, but the Imams (who are, themselves, "signs").

exegetical reports predominate: the majority of his reports quote or explicitly refer to the verse. He also cites reports (usually comprising purely lexical comments) from a Shīʿī scholarly authority, which are not attributed to an Imam.[40] Al-Huwayzī's approach is rather more lax. When an Imam uses a phrase or word found in the verse (for example, *muḥkam, al-rāsikhūn* etc.), it is taken as a reference to this particular verse. With this less restrictive attitude, al-Huwayzī manages to find thirty "relevant" reports, whilst Hāshim al-Baḥrānī cites only sixteen. Though the exegetical technique is the same for both authors, the emphasis is quite different. Hāshim al-Baḥrānī focuses on the verse itself, whilst al-Huwayzī attempts to explicate the meaning of the verse within the context of the *akhbār* corpus. Obviously the latter approach allows for greater individual creativity, as selecting which reports are relevant is, itself, a matter of authorial preference and this gains greater scope in al-Huwayzī's approach.

In terms of content, a number of points can be made concerning the *akhbār* cited by these authors. Firstly, the reports do not speak with unanimity concerning what (or who) is the referent of the phrase *āyāt muḥkamāt*. In some *ḥadīths*, it is identified as the Imams themselves—that is, they are the "decisive signs" which form the foundation of the book (*umm al-kitāb*). The book itself, by implication, is *mutashābih* (ambiguous) without their input. The *mutashābihāt* are at times identified as the enemies of the Imams. In some reports the *muḥkamāt* are identified as abrogating verses (and the *mutashābihāt* are the abrogated). In other reports, the *muḥkam* and the *nāsikh* are treated as separate categories.[41] In yet other reports, the *muḥkamāt* are a subcategory of the abrogating verses (*al-nāsikhāt*).[42] The *muḥkamāt* are often glossed as "what we believe and act upon" and the *mutashābihāt* as "what we believe in but do not act upon". In this association, the link between *naskh al-ḥukm dūna al-tilāwa* (verses abrogated but still recited) and *mutashābih* is obvious.[43] There are,

---

[40] See, for example, Hāshim al-Baḥrānī, *al-Burhān*, v. 1, p. 599, #16.1611.

[41] See Huwayzī, *Nūr al-thaqalayn*, p. 318, #45.

[42] Hāshim al-Baḥrānī, *al-Burhān*, v. 1, p. 597, #1.1597 and Huwayzī, *Nūr al-Thaqalayn*, v. 1, p. 312, #19. It is interesting here that in this report the abrogated verses consititute a section of the *mutashābihāt* (i.e. a verse can be *mutashābih* for reasons other than abrogation), and the *muḥkamāt* are amongst the abrogating verses (ie there are abrogating verses which are not *muḥkam*, but there are not *muḥkam* verses which are not abrogating).

[43] See Huwayzī, *Nūr al-thaqalayn*, p. 318, #44.

then, contradictory identifications of the *muḥkamāt* and *mutashābihāt* in the *akhbār*. This equivocality is replicated with other Qur'anic phrases, notably *al-rāsikhūn fi'l-ʿilm* (identified as the Prophet and the Imams, or solely the latter). In short, whilst the technique of citing only *akhbār* and entering no authorial comment is clearly an attempt to prevent arbitrary personal opinion playing a role in the exegetical process, no certainty as to textual meaning is necessarily thereby achieved: the reports cited may be contradictory. What is achieved, though, is an acceptable range of interpretations, each element of which is sanctioned by one of the Imams. On this particular occasion, this is less true for Hāshim al-Baḥrānī's account than for that of al-Huwayzī, but neither account leaves the reader with an unambiguous idea of the verse's reference. Despite this, the emphasis throughout is on the uninterpretability of the Qur'ān without the *tafsīr* of the Imams. This is reflected not only in the form "verse+*akhbār*", but also in the content.

The above-mentioned uncertainty over the referent of "its" in the phrase "No one knows its meaning..." receives decisive treatment in these Akhbārī *tafsīrs*. It is clearly understood to be the Qur'ān as a whole, and not "that which is *mutashābih* within the Qur'ān". It is the entire Qur'ān which requires interpretation through the Imams' words, and not merely its *mutashābih* parts:

> The Qur'ān has both the particular and the general, the *muḥkam* and the *mutashābih*, the abrogator and the abrogated—those well-founded in knowledge know it all.[44]

There is an indication that some elements of the Qur'ān are not meaningless without the *akhbār* (or rather, they have meaning in themselves and this happens to coincide with the intended meaning). Imam ʿAlī, for example, designates parts of "God's speech" which "either an ignoramus or a learned man can know",[45] but their *ta'wīl* is understood by the Prophet and Imams alone ("God only revealed to [the Prophet] that which has a *ta'wīl* which is known in its completeness to his *awṣiyā'* [the Imams]"[46]). It is clear throughout that

[44] Hāshim al-Baḥrānī, *al-Burhān*, v. 1, p. 597, #4.1599.

[45] Huwayzī, *Nūr al-thaqalayn*, v. 1, p. 312, #18, though it is not clear whether "God's speech" is a reference to the Qur'ān or a more general category.

[46] Huwayzī, *Nūr al-thaqalayn*, v. 1, p. 317, #35. There is a distinction here between verses having meanings which are available, but the referent of which is unknown (or the intended meaning is unknown), and those verses which have no meaning at

the technical meanings of *muḥkam* and *mutashābih*, specific to the disciplines of *uṣūl al-fiqh* and *tafsīr*, are not included within the range of possibilities here. There may be a range of possible referents for these terms, but none of them fit the disciplinary meaning.

The Limited Availability of Qur'anic Meaning

The *tafsīrs* of al-Huwayzī and Hāshim al-Baḥrānī can be usefully compared with Lawson's other example of Akhbārī exegesis, Muḥsin Fayḍ al-Kāshānī's *al-Ṣāfī*. Here, there is a greater willingness to enter into authorial comment on the meaning of the verse, rather than the simple citation of the *akhbār*. The *āyāt muḥkamāt* are glossed through the common technique of interpolation within the Qur'anic text:

> *Āyāt muḥkamāt*: [verses] whose expressions [are such that they] dictate that they are free of ambiguity.[47]

Similarly, *mutashābihāt* are defined as "subject to different interpretations, and the intended meaning [of the verse] cannot be made clear except with investigation". The significant point here is that the glosses accord with the usual disciplinary (that is, technical) definitions of *muḥkam* and *mutashābih*. Muḥsin Fayḍ, then, is clearly willing to allow his interpretation of the Qur'anic verse to be explicitly influenced by established traditions of learning, in particular, *uṣūl al-fiqh*. Furthermore, whilst a possible referent of "it" in the verse (in "No one knows *its* meaning..." etc.) is acknowledged as the Qur'ān as a whole, this is followed by reports which infer that the referent is "that which is *mutashābih* from the Qur'ān". Whilst Fayḍ does not explicitly state his preferred interpretation, he does at least acknowledge the possibility that the "it" in "No one knows *its* meaning" is not the Qur'ān as a whole, but only its *mutashābih* elements. We have here a hint of an

---

all (i.e. utterances with no references according to the rules of grammar and the placement of meanings on a particular arrangement of sounds). However, the distinction is not precisely drawn by al-Huwayzī or by the other Akhbārīs here. The position appears to be that a meaning of the Qur'anic text is normally available, but it is not known whether this is the intended meaning without the *akhbār*. The Qur'anic text contains words, phrases and sentences which have meaning in themselves, but for Astarābādī and his followers, this has no relationship at all to the meaning intended by the author (namely, God). I discuss this further below, pp. 235–237.

[47] Fayḍ, *Ṣāfī*, v. 1, p. 318.

alternative Akhbārī position—namely that there are some verses of the Qur᾿ān which can be understood directly, without the *tafsīr* of the Imams, and these are termed the *muḥkamāt*.

A similar hint can be found in the *Tafsīr al-Muʿīn* of Nūr al-Dīn Muḥammad al-Kāshānī al-Akhbārī, nephew and pupil of Muḥsin Fayḍ. He is recognised as an Akhbārī, following his uncle, and was also a pupil of Majlisī II and the Akhbārī Muḥammad Ṭāhir al-Qummī. Nūr al-Dīn's unusual exegetical technique is laid out in his introduction, and comprises a division of four different types of interpretation, each designated by different letters in the rest of the work:

> *mīm*: an interpretation using the actual words of the Imams.
>
> *ʿayn*: an interpretation using the meaning of the Imams' words, but not the precise wording.
>
> *yā᾿*: an interpretation taken from ʿAlī b. Ibrāhīm al-Qummī's *tafsīr*, the meaning of which is attributed to the Imam, though not using the Imams' words.
>
> *nūn*: interpretations of the *mufassirūn* (principally, his uncle's *al-Ṣāfī*).[48]

The admittance of different interpretations is itself a concession to more conventional *tafsīr* styles, in which polyvalency is accepted as an inevitable (perhaps intended) consequence of Qur᾿anic interpretation. Furthermore, the inclusion of the last of these types (*nūn* above) is a further acceptance that there are valid understandings of the Qur᾿anic text which are not explicitly delineated by the *akhbār*. Nūr al-Dīn's own understanding of the terms *muḥkamāt* and *mutashābihāt* (in Q3.7) is thoroughly disciplinary: *muḥkam* means "permitting a single interpretation alone" and *mutashābih* "permitting different interpretations".[49] Similarly, the identity of "it" in the phrase "we believe in *it*..." is explicitly identified as the *mutashābih* elements of the Qur᾿ān, not the book as a whole.[50] There are, for Nūr al-Dīn, elements of the Qur᾿ān which can be understood directly; the *mufassirūn* have performed this service, and Q3.7, properly understood, supports such a category, labelling them *muḥkamāt*. We have, then, two types of Akhbārī *tafsīr*—one which permits no interpretation other than provid-

[48] Nūr al-Dīn, *al-Muʿīn*, v. 1, p. 9.
[49] Nūr al-Dīn, *al-Muʿīn*, v. 1, p. 135.
[50] Nūr al-Dīn, *al-Muʿīn*, v. 1, p. 146.

ing *akhbār* as exegesis (al-Huwayzī and Hāshim al-Baḥrānī), and the other which permits direct interpretation of the so-called *muḥkamāt* (Muḥsin Fayḍ and his nephew Nūr al-Dīn). The two styles equate to the two broad currents within Akhbārī jurisprudence.

Fayḍ delivers a much more explicit elaboration of this alternative Akhbārī position in his *al-Uṣūl al-Aṣīla*. Here there is a chapter dedicated to the question of the availability of the *muḥkamāt*'s meaning.[51] He cites Astarābādī's *al-Fawāʾid al-Madaniyya* and sets about refuting it. If the Qur'ān's meaning was totally inaccessible, then the famous *thaqalayn ḥadīth* would be inaccurate, he argues. The Prophet would have left us a single "weight" (*thaqal*), namely his family, and the Qurʾān would become irrelevant to religious knowledge.[52] Furthermore, if we are not permitted to interpret the Qurʾān because we may not be able to discern its meaning, then surely the same can be said of the Imams' words. Astarābādī contradicts himself (Fayḍ argues): he presents verses which deny the validity of *ẓann*, but how can the meaning of these verses be understood without the *akhbār*? Astarābādī's position is invalid, he argues, and requires modification:

> The *akhbār* which prohibit interpreting the Qur'ān without texts and reports [from the Imams] must be interpreted as referring to the *mutashābihāt* of [the Qurʾān], not its *muḥkamāt*.[53]

However, even the interpretation of these *muḥkamāt* is not unrestricted. Only those with particular status can interpret them: namely, "those from the Shīʿa who are blessed in their obedience to [the Imams], who follow their way . . ."[54] The identity of these *cognoscenti* (*al-ʿārifūn*) is, of course, related to Akhbārī debates concerning scholarly authority, discussed below. However, it is sufficient to note here that Muḥsin Fayḍ's use of *ʿārifūn* is clearly a nod to his mystical system, and although he may allow direct interpretation of the Qurʾān, it is restricted, not only by subject but also by agent. Only the *muḥkamāt* can be interpreted, though the procedure for determining inclusion in this category seems remarkably pliable. Furthermore, the interpretive agent here is part of an exclusive category, the boundaries of which also remain vague. It is quite possible that this position, developed by

---

[51] Fayḍ, *al-Uṣūl*, pp. 32–40.
[52] Fayḍ, *al-Uṣūl*, p. 36.
[53] Fayḍ, *al-Uṣūl*, p. 37.
[54] Fayḍ, *al-Uṣūl*, p. 38.

Muḥsin Fayḍ in opposition to Astarābādī, emerged due to the force of Uṣūlī argumentation. Uṣūlīs, of course, argued that the Qur'ān can be interpreted, though only by a suitably qualified *mujtahid*. As has already been mentioned, subsequent Akhbārīs, beginning with Fayḍ's own nephew Nūr al-Dīn, and followed by al-Jazā'irī, Yusuf al-Baḥrānī and later by Muḥammad al-Akhbārī[55] developed this position within Akhbarism, making it a substantial challenge to the original position outlined by Astarābādī.

The effectiveness of Fayḍ's challenge to Astarābādī can be seen in the rather confused position of al-Ḥurr al-ʿĀmilī. Fatḥ ʿAlī Zand, writing much later, identifies al-Ḥurr as one who followed Astarābādī's position, though as we have already seen, Zand's depiction should not necessarily be taken as authoritative. In a section of his *al-Fawā'id al-Ṭūsiyya*, al-Ḥurr replicates Astarābādī's arguments.[56] His piece is presented as a rebuttal of a "modern" opponent, who argues that one can act on the "apparent meaning of the Qur'ān" (*ẓawāhir al-Qur'ān*). Al-Ḥurr reviews the evidence from the *akhbār* of the Imams, and points first towards the ambiguity in the meaning of the term *muḥkam* in Q3.7 (that is, the debate over the referent of "its"). He continues: the Imams, abrogating verses and other candidates for the referent of *muḥkamāt* are to be found in the *akhbār*, but none of these equate to the technical definition of *muḥkam* developed by later writers of *uṣūl al-fiqh*. The disciplinary definition, he concludes, has nothing to do with the meaning of the term in Q3.7. Even if one accepts it as a technical tool:

> The texts from them [the Imams] concerning this are both well-attested and explicit in demonstrating that no one knows [what is] *muḥkam* or *mutashābih*, abrogator or abrogated, the interpreted meaning [*ta'wīl*, of a verse] and the like except the Imams themselves. One does not need a text to see that the *ʿulamā'* have had doubt (*shakk*) over whether a verse is *muḥkam* or *mutashābih*.[57]

---

55 Muḥammad al-Akhbārī ignores the internal Akhbārī differences in his *Fatḥ al-Bāb*, saying that the Akhbārīs are simply one of the two groups of Shi'a who "consider the *muḥkamāt* of *kitāb* and *sunna*, and the texts of the *akhbār* of the pure Imams sufficient for action and for giving *fatwās*. They only use methods of preference which have been recorded, and when there is no text they do *tawaqquf*." Akhbārī, *Fatḥ al-Bāb*, f.154.12–14 (p. 107, l.12–14). See also, his *Maʿāwil*, f.24a.

56 Ḥurr, *al-Fawā'id al-Ṭūsiyya*, pp. 163–195.

57 Ḥurr, *al-Fawā'id al-Ṭūsiyya*, p. 171.

Though *muḥkam* may be a legitimate category when describing utterances in general, this does not mean that one can then recognise *muḥkam* verses easily. Indeed, the fact that there is dispute over whether or not a particular verse is *muḥkam* is evidence of the redundancy of the category when applied to Qur'anic verses. The view is developed through the citation of all the relevant *akhbār* in al-Ḥurr's *Wasā'il al-Shī'a*.[58] However, in his *Wasā'il*, one also sees a more developed (perhaps modified) version of the position found in *al-Fawā'id al-Ṭūsiyya*. Al-Ḥurr muses on the following challenge to his position: in the *akhbār*, the Imams do, occasionally, say, "Have you not heard such and such a verse?" In such circumstances, the Imam is asking the audience why they have not understood a verse of the Qur'ān. Of course, the Imam would be behaving unreasonably if he was to criticise the audience for not understanding a verse when it is actually impossible for them to understand any verse due to the dependency of Qur'anic meaning upon the *akhbār*. It would be a case of "asking the impossible" (*taklīf mā lā yuṭāq*). Al-Ḥurr's answer to this challenge is a carefully worded statement concerning the exegetical procedure alluded to by the Imams' in such reports:

> As for what is related from the Imams in some *akhbār*, "Have you not heard the statement of the Most High One?", and the like, then its meaning is this:
>
> The one who hears a verse, in which the *ẓāhir* meaning indicates a particular speculative ruling is not permitted to opt for [a ruling] which contradicts this [speculative ruling]. This is because it is possible that the verse's intended meaning is the *ẓāhir* (*iḥtimāl irāda ẓāhirihā*). For this reason, [contradicting the *ẓāhir*] is forbidden. However, he is also forbidden from declaring that the *ẓāhir* is intended here, because there is a possibility that it is abrogated (*naskh*) or particularised (*takhṣīṣ*) or subject to some other interpretation (*ta'wīl*) . . .[59]

Al-Ḥurr's position in his *Wasā'il* is, therefore, more complex than a simple assertion on the basis of *akhbār*, that the meaning of the Qur'ān is unknowable. He argues that whilst this is the case, the Imams, at times, expected their audience to understand the Qur'ān in advance of the Imams' own *tafsīr*. What is meant by this, al-Ḥurr argues, is that it is permitted to assume that the *ẓāhir* is the intended meaning of a

[58] Ḥurr, *Wasā'il*, v. 18, pp. 129–152.
[59] Ḥurr, *Wasā'il*, v. 18, p. 151.

verse if there is no available *tafsīr* of the Imams. At the same time, however, it is not permitted to declare that the *ẓāhir* is the meaning of the verse in such circumstances (and by implication, to refute those who argue differently). Rather, in such circumstances, the rule is to suspend judgement and act with caution. Caution dictates that the individual behave as if the *ẓāhir* meaning is the meaning of the verse until there is evidence from the *akhbār* that this is not the case. The line between this developed position of al-Ḥurr (who is described by Fatḥ ʿAlī Zand as denying the probative force of the *ẓawāhir al-qurʾān*) and his supposed opponents (such as Muḥsin Fayḍ, who deny it) is a fine one indeed. The fact that al-Ḥurr was forced to adopt this modification of Astarābādī's views, having defended them in *al-Fawāʾid al-Ṭūsiyya*, is, no doubt, evidence of the success of Muḥsin Fayḍ's attack. Astarābādī's formulation of the relationships between the Qurʾān and the *akhbār* was, by al-Ḥurr's time, already considered rather simplistic.

Marginal Akhbārīs

In sum, the debate between these two intra-Akhbārī currents concerns the extent to which the linguistic categories found in works of *uṣūl al-fiqh* should influence Akhbārī discourse. If one accepts the category of *muḥkam* (and surrenders to the influence of established *uṣūl* categories), then a commitment to the direct availability of the Qurʾanic meaning naturally follows. By definition, *muḥkam* verses require no extra-textual aid, and the speaker's intention is entirely clear. On the other hand, if one rejects the *muḥkam* category, along with the general interpretive paradigm of the established *uṣūl* genre, then it is easier to maintain Astarābādī's position that the Qurʾān remains incomprehensible without the *akhbār*. In general, one sees Astarābādī's position dominate the early decades of Akhbarism, though the success of Muḥsin Fayḍ's challenge is evident first in al-Ḥurr's rather delicate modification of Astarābādī's views. From al-Jazāʾirī onwards, Akhbārīs do seem to have accepted that the *muḥkamāt* of the Qurʾān can be understood directly, and that the Imams' *tafsīr* was only necessary for the *mutashābihāt*. In al-Samāhījī's list of Akhbārī-Uṣūlī differences, Akhbārīs who accept only the *sunna* (that is, Astarābādī and those who followed his position) are portrayed as a deviation from the majority Akhbārī opinion. The presentation of the relative size of the two groups was, of course, flexible depending on the exegetical aims of an individual Akhbārī author. Fatḥ ʿAlī Zand's formulation of the intra-Akhbārī dispute reflects the victory of Muḥsin Fayḍ's position regarding the *muḥkamāt* amongst the Akhbārīs. This was then back-projected by Zand onto the earlier period of Akhbarism, and the real lines of debate between Astarābādī and Fayḍ are blurred by Zand, as he refers to the acceptance or non-acceptance of the *ẓāhir* on issues where the Imams are silent. The range of Akhbārī views on the role of the Qurʾān as a source of law, and the interplay between them, was more complex than the portrayal in Akhbārī difference lists.

The range of possible Akhbārī opinions just outlined can now be used to position scholars contemporaneous with the rise and establishment of Akhbarism, but whose inclusion within the school was debateable (namely, ʿAbd Allāh al-Tūnī, Muḥammad Ṣāliḥ al-Māzandarānī and Muḥammad Bāqir al-Majlisī).[60]

Al-Tūnī's major contribution to Shīʿī scholarship is his *al-Wāfiya*, a work which, in terms of its structure, conforms to the established characteristics of the *uṣūl* genre. It has, however, been claimed by both Akhbārī and Uṣūlī writers as a work within their respective school traditions. Al-Tūnī himself was a product of al-Tustarī's "proto-Akhbārī" *madrasa* in Isfahan, and though he makes no direct reference to Astarābādī, it seems likely he was aware of Astarābādī's ideas.[61] On the specific issues of the Qurʾān as an independent source of law and the authenticity of the *akhbār*, his views are closer to the Akhbārī mainstream than they are to standard Usulism. His originality lies in the arguments he devises for these positions. On the issue of the Qurʾān as a source of law, he tackles the issue of *taḥrīf* (and the *qiraʾāt*).[62] The Shīʿa, he says, are divided on whether the Qurʾān is corrupted or not. "The difference of opinion, however, makes no difference" (*lā athr li-hādhā al-ikhtilāf*). The Shīʿa have agreed (*taḥaqquq al-ijmāʿ*), and some *akhbār* demonstrate clearly (*fī baʿḍ al-akhbār*

---

[60] See above, p. 165, p. 167 and p. 169. Unfortunately, al-Māḥūzī's writings are not yet available in sufficient quantity to make a clear statement on his school allegiance. The same can be said of ʿAbd Allāh al-Yazdī. However, al-Māḥūzī's general approach can be deduced from his own adoption of Akhbārī positions in his difference list (analysed above, pp. 202–204).

[61] The editor of the introduction to the modern edition of *al-Wāfiya* considers al-Tūnī to be refuting Astarābādī (or his followers) in the work, though little evidence is given for this conclusion (Tūnī, *al-Wāfiya*, p. 32 (editors introduction)).

[62] Tūnī, *al-Wāfiya*, pp. 147–149.

taṣrīḥ[un]), that we must base our action on the Qur'ān we have (*mā fī aydaynā*) until the return of the Mahdī. He also accepts that there are various readings (numbered at seven or ten) of the Qur'ān (that is, the *qirā'āt*). However, these are attributable to reciters (*min qibl al-ruwāt*), not any plurality in God's revelation. Whilst al-ʿAllāma may have expressed a preference for the recitations of ʿĀsim and Ḥamza, al-Tūnī has his own view:

> The preferred view is to refer in [this matter] to the *tafsīr* of the holders of the *dhikr*, the preservers of the Qur'ān [that is, the Imams] whenever possible. And if it is not possible, then one should suspend judgement (*tawaqquf*).[63]

The use of the Imams as the decisive elements in a choice between *qirā'āt*, and the implicit criticism of al-ʿAllāma al-Ḥillī's position, hints at Akhbārī sympathies. This is made more explicit in a section on *ẓawāhir al-Qur'ān*.[64] The Qur'ān was revealed in a specific time and place, and the contextual evidence (*qarā'in*) would have made the intended meaning of the speaker (*al-maqṣūd*) clear to the addressees. If we still have access to these *qarā'in*, then it is permitted to interpret the Qur'ān. "We, however, do not have knowledge of whether the *ẓawāhir* have probative force [or not]" because we have lost access to most *qarā'in*. Furthermore, most *akhbār* (al-Tūnī estimates between two thirds and four fifths) record interpretations which "contradict the apparent meaning [of the Qur'ān] which one would understand grammatically".[65] This alone indicates that the *ẓāhir* does not have a convincing claim to pre-eminence over other possible meanings of a Qur'anic verse. Finally, al-Tūnī lists the reports which indicate that knowledge of the Qur'ān (*ʿilm al-Qur'ān*) is the preserve of the Prophet and the Imams, and therefore only they can interpret it for us. Once again, although the presentation may be within the established structure of a work of *uṣūl al-fiqh*, al-Tūnī's view has clear Akhbārī resonances.

Muḥammad Ṣāliḥ al-Māzandarānī, like al-Tūnī, was a product of al-Tustarī's Isfahan *madrasa*, and like al-Tūnī, is claimed by

---

[63] Tūnī, *al-Wāfiya*, p. 149.
[64] Tūnī, *al-Wāfiya*, pp. 136–140.
[65] The phase *yukhālif al-ẓāhir al-ladhī yufham bi-hiṣb al-waḍʿ al-lughawī* (Tūnī, *al-Wāfiya*, p. 136) uses, of course, the notion of language (both grammar and the meaning of words) being designed by an individual (and therefore fixed), on which, see below, p. 209.

Akhbārīs as one of their own.[66] His major work, a commentary on part of al-Kulaynī's *al-Kāfī*, indicates an interest in *ḥadīth* generally, though not necessarily Akhbārī sympathies. His other major composition, a commentary on Shaykh Ḥasan b. Shahīd II's classic work of Shīʿī legal theory *Maʿālim al-dīn* indicates a respect for the discipline of *uṣūl al-fiqh*. His presentation in both works is, of course, constrained by the subject text (in these cases, the works of al-Kulaynī and Shaykh Ḥasan), though not to the extent that his own opinion is undiscernable.[67] On the issue of Qur'anic interpretation, it is clear that the *muḥkamāt* are interpretable by the ordinary, educated reader. The Qur'ān, he explains, has two sorts of verses, *muḥkamāt* and *mutashābihāt*. The former "allow one interpretation only." The *mutashābihāt* can also be understood with interpretive effort, though at first encounter, they allow "many interpretations and it is not clear what the [speaker's] intention might be" (*wujūḥ*[an] *mutaʿaddada lā yattaḍiḥ al-maqṣūd minhā*).[68] Interpretive effort requires one to compare the *mutashābih* verse with the *tafsīr* of the Imams. This, al-Māzandarānī claims, is based on a reading (or rather, the only possible reading) of the Imams' report on Q3.7. The report of the Imams "No one understands the Qur'ān other than the one[s] to whom it was uttered" is cited by some Akhbārīs to prove that the meaning of the Qur'ān is unavailable without the Imams' *tafsīr*. Al-Māzandarānī, in his commentary on this report, argues that the report does not refer to the whole Qur'ān, but to the *mutashābihāt* alone, and that the latter must be compared with the opinion of the Imams.[69] The general tenor of these scattered discussions accords with the Akhbārī opinion of Muḥsin Fayḍ, which (as we have seen) became the dominant opinion

[66] See above, p. 165.

[67] Commentaries are not always an ideal source for determining a scholars' own views, since the commentator is, to an extent, preoccupied with explaining the meaning of the subject text. Having said that, commentaries do give the commentator the opportunity to re-interpret a text, and thereby control how the text's author is portrayed. The subject text's author may have views which chime with the commentator's, or the commentator may wish to openly disassociate himself from the subject text's author. In either case, one's own understanding of the subject text may be irrelevant. In al-Māzandarānī's case, one finds scattered comments throughout the two works where the boundary between explicating the subject text's (and hence the subject author's) meaning and al-Māzandarānī's own opinions are blurred.

[68] Māzandarānī, *Sharḥ*, v. 1, p. 137. See also Māzandarānī, *Sharḥ*, v. 2, p. 133; Al-Māzandarānī explores the different meaning of *muḥkam* in Māzandarānī, *Sharḥ*, v. 2, pp. 310–311, though he does not offer a definitive preference.

[69] Māzandarānī, *Sharḥ*, v. 12, p. 438.

of the Akhbārī school after al-Jazā'irī. It is, then, not so surprising that later authors (such as Fatḥ ʿAlī Zand) are willing to incorporate al-Māzandarānī into the cadre of Akhbārīs.

Al-Tūnī and al-Māzandarānī held different positions on the question of the Qur'ān as a source of law and legal knowledge. Their views, however, fit into the range of Akhbārī opinion on this particular issue. Muḥammad Bāqir al-Majlisī's views on juristic matters are more difficult to ascertain. He did not write a work of *uṣūl al-fiqh*, but was instead renowned as a *ḥadīth* collector who popularised Twelver Shiʿism and played a significant role in the consolidation of the authority of the *ʿulamā'* in the late Safavid period.[70] There is no convenient *tafsīr* through which his hermeneutic method might be known. His position on the *muḥkamāt* verses, however, can be deduced from his commentaries on relevant *ḥadīths*. Majlisī II cites a *ḥadīth* in which the fourth Imam glosses the various phrases within Q3.7 (the *muḥkamāt* verse).[71] Majlisī II's commentary on this *ḥadīth* is expansive and in it he describes his general approach to the interpretation of texts.[72] The approach is, he claims, based on the *akhbār* themselves, with this particular *ḥadīth* from Imam al-Bāqir providing a key reference text. On the interpretation of the Qur'ān, he divides *muḥkam* verses into two types: *muḥkam* in terms of an utterance's words (*min jihat al-lafẓ*) and *muḥkam* in terms of meaning (*min jihat al-maʿnā*). These two categories can be glossed as referring to linguistic ambiguity (arising from homonymy or ellipsis) and indicatory ambiguity (arising from a lack of knowledge about what is referred to by a specific word).[73] When an utterance (*lafẓ*) is unambiguous on both these accounts, then it is "unconditionally unambiguous" (*muḥkam ʿalā al-iṭlāq*). When it is ambiguous on both accounts it is "unconditionally ambiguous" (*mutashābih ʿalā al-iṭlāq*). At other times, it may be ambiguous in one of these two (linguistic or indicatory) and unambiguous in

[70] See, for example, Momen, *Shīʿī Islam*, pp. 114–117.

[71] The full *ḥadīth* can be found in Kulaynī, *al-Kāfī*, v. 2, pp. 28–33 (#1), where there is an extensive commentary on the whole verse. Majlisī II's commentary on this *ḥadīth* is, then, a commentary on a *ḥadīth* which is itself a commentary on a Qur'anic verse.

[72] Majlisī II, *Biḥār*, v. 66, pp. 85–149.

[73] The example given is the attributes of God, which are *mutashābih* because we have not experienced power or knowledge in such a way that we can comprehend how God's possession of these qualities might operate ("The qualities are not comprehensible to us, for a form of something one has not felt cannot be obtained in one's self", Majlisī II, *Biḥār*, v. 66, p. 91).

the other (*muḥkam min wajh^{in} mutashābih min wajh^{in}*). The last category refers to verses which may be unambiguous in their own terms (*muḥkam min al-lafẓ*), but uncertain with respect to indication (*dalāla*), or vice versa. This division is not particularly original (it being a common element in previous studies of both this verse and Arabic rhetoric generally), but the fact that Majlisī II borrows it is the first hint of his overall position. Majlisī II's interesting hermeneutic approach involves various combinations of these categories. It is sufficient here to point out that, although he gives no examples of verses which are "*muḥkam ʿalā al-iṭlāq*", he clearly considers it a category into which at least some Quranic verses can be placed, and therefore the Qur'ān can be understood directly.[74]

For Majlisī II, parts of the Qur'ān are unambiguous both in terms of what is said and in terms of what is meant. These verses can be understood directly by any competent language speaker. In this sense Majlisī II's position is not dissimilar to that of Muḥsin Fayḍ. However, he differs from Muḥsin Fayḍ in two significant respects. Firstly, the mechanisms he recommends for retrieving the meaning of the *mutashābih* verses do not consist of citation of *akhbār* alone, as was the case for Muḥsin Fayḍ. Instead, Majlisī II recommends a rhetorical analysis as a means of clarifying *mutashābih* verses. For Majlisī II, a range of mechanisms are available to the interpreter when faced with a *mutashābih* verse. A verse may be *mutashābih* because God is speaking with brevity (*ikhtiṣār al-kalām*), or expansively (*basṭ al-kalām*) or he may be utilising word order (*naẓm al-kalām*) to affect (aesthetically or otherwise) his listeners. These categories of ambiguity emerge from the ways in which words combine (*al-kalām al-murakkab*). There are other types of ambiguity brought about from the words themselves (*al-alfāẓ al-mufrada*, such as homonymy or uncommon derivations). The exact categorisation scheme need not concern us here. Instead, it should be noted that the means whereby

---

[74] This is further confirmed by his definition of *muḥkam*, given in his commentary on al-Kulaynī's *al-Kāfī*: "*Muḥkam* is certain (*muttaqin*) linguistic meaning. In technical terms it is used in an unrestricted manner to apply to anything which has a clear meaning (*maʿnā*), and for anything which is preserved from being either abrogated or particularised (*takhṣīṣ*), or both, and for something which has a straightforward word order (*naẓm*), free of deviation. It is something which can only be interpreted in one way. Anything which does not conform to these categories is termed *mutashābih*". (Majlisī II, *Mir'āt al-ʿuqūl*, v. 1, p. 210). See also, Majlisī II, *Biḥār*, v. 2, p. 185 where *muḥkam* is glossed as "permitting only one interpretation".

uncertainty over the meaning of a verse is resolved is through reference to the standard rules of rhetoric, developed within both the profane and "transmitted" Islamic sciences. His proposed hermeneutic technique here is, then, entirely dependent upon established disciplines, and these sciences can be employed to solve the ambiguities within the Qur'anic text. This sets him apart from at least some Akhbārīs.

The second manner in which Majlisī II's position differs from that of Muḥsin Fayḍ and his Akhbārī followers is in its distinction between *tafsīr/ta'wīl* (interpretation) and *ma'nā* (meaning). In the account described above, even the more nuanced positions of Muḥsin Fayḍ and his follows do not distinguish between the meaning of what is said in the Qur'ān and its significance for religious knowledge. The Akhbārīs disagreed over how the meaning of a verse could be obtained: for Astarābādī et al, a verse's meaning is entirely understood from the *akhbār*; for Muḥsin Fayḍ et al., some verses could be directly understood (*muḥkamāt*) and the rest understood through the *akhbār*. Majlisī II, however, presents his view that the Qur'ān has meaning, and can be read and understood perfectly well without the *akhbār*. This meaning is discovered through the usual techniques employed by the *mufassirūn*. Such an analysis will lead to comprehension of meaning, but not an understanding of the significance of a Qur'anic verse within the context of the true Imāmī faith. To gain the significance of the verses, the reader needs the *akhbār* of the Imams. Hence Majlisī II says that the *'ulamā'* do not "stand alone in [their search for] Qur'anic knowledge. They need the Imam, designated by God, in order to interpret (*tafsīr*) it."[75] Through the rules of rhetoric and language, one gains access to a Qur'anic passage's meaning, and this may be understood at both the level of a meaning inherent in the words used or the grammatical constructions employed. This process will not, however, enable the reader to understand that phrases such as *ahl al-dhikr* and *al-rāsikhūn fī'l-'ilm* refer to the Prophet and the Imams. This position, in which the Qur'ān has meaning (but no correct interpretation) without the *akhbār*, represents a more nuanced understanding of the problematic nature of meaning and its interpretation in the Qur'ān than that found within the Akhbārī tradition. For Majlisī II, the *akhbār* are still essential for understanding the significance of the Qur'ān. They are not, however, essential for understanding its

[75] Majlisī II, *Biḥār*, v. 66, p. 93.

meaning. On the evidence currently available concerning the range of Akhbārī opinions on this issue, such a view was not found amongst Majlisī II's Akhbārī contemporaries.

Conclusions

The range of Akhbārī approaches to the interpretation of the Qur'anic text was settled quite early on in the dispute. Two alternative Akhbārī positions (headed by Astarābādī and Muḥsin Fayḍ respectively) became established. The range of opinions found within Akhbārī works of *tafsīr* and *uṣūl al-fiqh* are not, however, a perfect fit with the dispute's depiction in difference lists. In the difference lists—those by al-Samāhījī, al-Baḥrānī and Fatḥ ʿAlī Zand—the intra-Akhbārī dispute over the interpretation of the Qur'ān is described, but each presentation is slightly skewed in order to serve each writer's aims. The difference lists indicate that there was a dispute, but are less concerned with delineating either the precise nature of that dispute or its significance for the general Akhbārī juristic system. By examining the positions of renowned Akhbārī scholars in detail, a range of internal Akhbārī opinion emerges, and with this range one is able to position scholars of disputed Akhbārī identity. At times, these "marginal" Akhbārīs were forced to develop more sophisticated doctrinal formulations. They expressed themselves more carefully than those working within the well-established boundaries of the Akhbārī school. Perhaps they were also attempting to bring some intellectual respectability to the school, or were convinced by the Akhbārīs' more stinging criticisms of Usulism. In their responses to the question of Qur'anic interpretation, the "marginal" Akhbārīs demonstrate their indebtedness to the Akhbārī approach, even when, when, as is the case with Majlisī II, their positions cannot be categorised as within the internal Akhbārī *ikhtilāf*.

CHAPTER EIGHT

SUNNA AND THE *AKHBĀR* IN AKHBĀRĪ JURISPRUDENCE

Whilst the Akhbārīs debated the probative force of the Qur'ān, they displayed almost total unanimity concerning the authenticity and hence probative force of the *akhbār*. Most Akhbārīs accorded the status of a legal source to the Qur'ān. However, practically speaking, most Akhbārīs also asserted that it was near useless as a legal source. A more important question for Akhbārīs, however, was the authenticity of the *sunna*—by which they meant reports (*akhbār*) attributed to the Prophet and the Imams.[1] Astarābādī's position, as outlined in an earlier chapter, aimed to maximise the number of *akhbār* assessed as authentic (or "sound", *ṣaḥīḥ*),[2] and this became the established (indeed, it could be argued, *defining*) doctrine of subsequent Akhbarism. The Akhbārī position regarding the authenticity of the *akhbār* was also given prominence in subsequent lists of Akhbārī-Uṣūlī differences, and was clearly an important area of dispute between the two "schools".

Akhbārī writers subsequent to Astarābādī discussed at length the manner in which the *akhbār* might be proven to be authentic. Through reducing the jurist's epistemological requirements (from unqualified *ʿilm* to *al-ʿilm al-ʿādī*), Astarābādī was able to accept that the authenticity of some *akhbār* could be doubted by a hardened sceptic. He simultaneously argued, however, that the *akhbār* were the only legitimate legal source as the believer could gain "customary certainty" (*al-ʿilm al-ʿādī*) that they were *saḥīḥ*. In developing the arguments for the reliability of the *akhbār*, later Akhbārīs pursued two principal

[1] Astarābādī does appear to have distinguished between Prophetic reports and the *akhbār* of the Imams (Astarābādī, *al-Fawā'id*, p. 281). The former, he states, are like the Qur'ān in the sense that their meaning is unavailable without the Imams' *tafsīr*. Al-Ḥurr argues similarly in his *Wasā'il* (Ḥurr, *Wasā'il*, v. 18, pp. 152–154), though as with the Qur'ān, he moderates his position to the *ẓawāhir* of the *sunna* only. The distinction does not seem to have played a major role in later Akhbārī discussions.

[2] See above, pp. 70–72.

approaches: one historical and the other *isnād*-critical. For the first, there was a development of the historical account of how the 400 *uṣūl* formed the primary source for the "Four Books". Astarābādī had already developed this argument, and subsequent Akhbārīs aimed to establish this account as the dominant narrative. By forcing Uṣūlīs to object to this version of the collation of *ḥadīths*, these authors presented the Uṣūlī position as, not only a threat to Shīʿī orthodoxy, but also as being counter-intuitive and contrary to the available evidence. It was argued that the collectors of the "Four Books" had screened the *akhbār* before inclusion. For the second defence, there was an eagerness to outline methods of *ḥadīth* authentification which challenged the dominant methodology of the Uṣūlīs. By devising alternative methods of *ḥadīth* classification, Akhbārī writers tried to demonstrate, not only that the Four Books were authentic *in toto*, but also that individual *ḥadīths*, maligned by the Uṣūlīs, could be reclassified and form a sure foundation for legal judgement. The two arguments were not mutually exclusive, and both can be found in a single author's *oeuvre*. They do, however, represent two different Akhbārī tactics in the anti-Uṣūlī polemic. The former is an attempt to demonstrate that the science of *ḥadīth* classification is superfluous since a particular historical account of *ḥadīth* collation guarantees authenticity. The latter is a proposed reform of the science, enabling more *akhbār* to be considered sound, and therefore widening the material available for deducing legal rulings. As has already been noted, this ambivalent Akhbārī attitude to the established legal sciences was the underlying cause of intra-Akhbārī debate concerning the Qurʾān's probative force.

Arguments for the Authenticity of the Akhbār

Al-Ḥurr al-ʿĀmilī is, perhaps, the Akhbārī most faithful to Astarābādī's approach. In the conclusion to his *ḥadīth* collection, *Wasāʾil al-Shīʿa*, al-Ḥurr presents the reader with twelve sections (each named a *fāʾida*) relating to the sources he has used in collating the *Wasāʾil*. He argues that the sources he has used—which extend well beyond the famous Four Books[3]—are authentic (*saḥīḥ*) and therefore form reliable bases for

[3] In *Fāʾida* 4, al-Ḥurr lists the works which he consulted directly. The first four are the famous Four Books of al-Kulaynī, Ibn Bābūya and al-Shaykh al-Ṭūsī (Ḥurr,

legal judgement. It is noteworthy that it is the books themselves which he considers sound. Individual *ḥadīths* are reliable because they are found within sound books, not through any independent investigation. The books are sound because their authors record that they collected *ḥadīths* from reliable sources (*al-uṣūl wa'l-kutub al-mu'tamida*), and they say as much in the introductions to their *ḥadīth* collections.[4] In the ninth *fā'ida*, al-Ḥurr embarks upon a detailed (*tafṣīl*^{an}) exposition of the evidence which demonstrates that the *ḥadīth* books he used to compile the *Wasā'il* are all entirely sound. The *fā'ida* is divided into twenty-two pieces of evidence (*qarā'in*), and though the general presentation is rather repetitive, al-Ḥurr's intention is clearly to establish his version of the history of Shī'ī *ḥadīth* collection. The argument can be summarised as follows: we know that the companions of the Imams used to record the sayings of the Imams during specially arranged sessions (*tadwīn fī majālis al-a'imma*). This was the practice for nearly 300 years during the Imams' presence, and was carried out on the orders of the Imams themselves. The aim was to record those sayings which might act as a guide for the Shī'a in the future.[5] These collections were reviewed by the Imams themselves after they had been written down. They survived and were used by the authors of the Four Books ("even the Uṣūliyyūn admit this much").[6] We know that these authors, and the others from whom al-Ḥurr cites, were diligent in their task and did not record material they thought dubious.[7] There is, then, little point in scrutinising the *isnāds* found within

Wasā'il, v. 30, p. 153), but these are followed by a list totalling 82 books (and he says there are more unmentioned ones) which al-Ḥurr has consulted and for which he is claiming *ṣiḥḥa*. This is followed by a list of 96 known sources from which the authors of the aforementioned 82 works made their collections (Ḥurr, *Wasā'il*, v. 30, pp. 161–165, and once again, al-Ḥurr says this list is not exhaustive). Finally, al-Ḥurr mentions that there are many other sources which are not mentioned by name in the 82, and these number 6600 in total (Ḥurr, *Wasā'il*, v. 30, p. 165).

[4] In the sixth *fā'ida* (Ḥurr, *Wasā'il*, v. 30, pp. 191–218) various authors, including the authors of the Four Books, are quoted to this effect.

[5] *ta'līf kull mā yaḥtāj ilayhi min aḥkām al-dīn li-ta'mala bihā al-Shī'a* (Ḥurr, *Wasā'il*, v. 30, p. 252).

[6] Ḥurr, *Wasā'il*, v. 30, p. 252.

[7] Al-Ḥurr asks, rhetorically: "If it is known that biographers and historians do not relate from books they consider unreliable when they are able to relate from books they consider reliable, then what should one think of the chief transmitter [Ibn Bābūya], the truthful one of Islam [al-Kulaynī] and the leader of the saved sect [al-Shaykh al-Ṭūsī]?" (Ḥurr, *Wasā'il*, v. 30, p. 253). He expresses his amazement that anyone would slander these authors by implying that they had not completed their task effectively.

the Four Books, and testing the *isnāds* against the criteria of later *ḥadīth* critics. We know, for example, that al-Shaykh al-Ṭūsī used to cite a weak *isnād* with a report, when he knew of a sound *isnād* for the same report. He did this because the *isnād* we perceive as weak leads to a source book which he knew to be sound. The point is that al-Shaykh al-Ṭūsī did not record *isnāds* in order that they might be scrutinised by subsequent *ḥadīth* critics. Al-Ṭūsī cited them "simply to gain a blessing by completing the chains of verbal transmission and to defend the Shīʿa from the accusation of the Sunnis that their *ḥadīth* are without chains of transmission (*ghayr muʿanʿana*)."[8] The *isnāds* are, then, included within the early collections of the *ḥadīths* for decoration or for completeness. They are not supposed to demonstrate the authenticity of the accompanying report, and they are not supposed, therefore, to be scrutinised according to the usual tools of *isnād* criticism. Al-Ḥurr also produces some related consequentialist arguments against scrutinising the *isnāds* and declaring some reports unsound. If some of the *akhbār* were of dubious authenticity, then it would be prohibited to record them, since one would be recording an inaccurate (and religiously destructive) report.[9] If the *akhbār* are of dubious authenticity, then the Shīʿa have been in error for many centuries, even during the time of the Imams, since they have based their action upon some of these "inauthentic" *ḥadīths*. Consequentialist arguments such as these aim to persuade the reader by revealing the unpalatable (and possibly heretical) consequences of accepting the opponents' position. One finds them in Astarābādī's *al-Fawā'id*, and they are repeated by al-Ḥurr, with little new information. The arguments are also found in the work of later Akhbārīs, such as Yūsuf al-Baḥrānī who emphasises the authenticity of the Four Books of *akhbār*, and bases this claim on the history of their compilation (that is, that the authors collected their *ḥadīths* from the 400 *uṣūl*).[10]

In contrast to al-Ḥurr's replication of Astarābādī's argumentation is al-Karakī's presentation in his *Hidāyat al-abrār*. Al-Karakī's position is not at variance with that of Astarābādī and al-Ḥurr. The contrast lies in al-Karakī's argumentation for this position. For him, as for

---

[8] Ḥurr, *Wasā'il*, v. 30, p. 258. This passage follows closely the wording of Astarābādī's *al-Fawā'id al-Madaniyya* (see Astarābādī, *al-Fawā'id*, p. 377).

[9] Ḥurr, *Wasā'il*, v. 30, pp. 259–260.

[10] See Baḥrānī, *al-Ḥadā'iq*, v. 1, pp. 14–25. This is analysed in Gleave, *Inevitable Doubt*, pp. 31–46.

Astarābādī and al-Ḥurr, the *ḥadīths* in the Four Books are authentic The bulk of al-Karakī's arguments are found in a section entitled "The evidence which demonstrates the truth of what the three leaders [namely, al-Kulaynī, Ibn Bābūya and al-Shaykh al-Ṭūsī] transmitted and what they decreed to be sound".[11] He enumerates evidence (in the form of references to previous works) which establishes that the extant material (in the Four Books) was, in turn, selected from books written by the companions of the Imams. He demonstrates, then, not only that such books existed, but more importantly, that these books were used as sources for both belief and action by the companions (*yaʿmalūn bihā fī ʿaqāʾidihim wa-aʿmālihim*).[12] Indeed the Imams themselves encouraged their companions to commit their sayings and actions to writing.[13] The books were presented to the Imams and approved, and had thenceforth acted as sources of religious and practical guidance:

> It is known [from the references in extant books] that the companions of the Imams acted on the basis of these books even though it was possible for them to refer to [the Imams] directly, and gain information from [the Imams] directly concerning these [books]. This was only due to the certainty of [the companions] that what was in the books was sound.[14]

The assembled evidence also demonstrates that there were books in existence which were not relied upon because they contained untruths written by liars and deceivers (*al-kadhdhābūn wa'l-waḍḍāʿūn*). Such fabricated books existed, and were rejected: they did not form part of the material used by the three *akhbār* collectors. Admitting the existence of forgeries is undoubtedly a concession, yielded by al-Karakī, to Uṣūlī arguments. Astarābādī makes no reference to these "books of liars" (*kutub al-kadhdhābīn*), probably fearing that such an admission would endanger the persuasiveness of his argument. Al-Karakī, however, permits them a mention, but turns their presence into a strength: the *akhbār* collectors (namely, al-Kulaynī, Ibn Bābūya

---

[11] *al-qarāʾin al-dālla ʿalā ṣidq mā naqalahu al-aʾimma al-thalātha wa-ḥakamū bi-ṣiḥḥatihi* (Karakī, *Hidāyat*, p. 82).

[12] Karakī, *Hidāyat*, p. 83.

[13] Karakī, *Hidāyat*, p. 85. Amongst the *akhbār* cited at this point is Imam al-Sadiq's words, "Record in your books! Because you will need it." (Kulaynī, *al-Kāfī*, v. 1, p. 52, #10).

[14] Karakī, *Hidāyat*, p. 83.

and al-Shaykh al-Ṭūsī) knew of such works, but did not use them in their collections. We know this because they state that they only used *uṣūl* which had received community approval and had formed the basis of community action. The liars' books did not fall into this category. Furthermore, according to al-Karakī, the three collectors did not collect *ḥadīths* as such. Rather they collected books of *ḥadīths*, extracted the *ḥadīths* from these books and reordered them according to their own schema. When an *isnād* is cited in one of the Four Books, it is (most often) not an *isnād* of the oral transmission of a particular report:

> When an individual name is mentioned in a *sanad*, then the [author's] chain of transmission is merely his chain of transmission to [that person's] source (*aṣl*), from which he [cf the author] took the *ḥadīth*.[15]

The mistake made by Uṣūlīs is to treat the *isnāds* in the four collections as recording the oral transmission of individual reports. Inevitably this makes reports such as these *khabar al-wāḥid*. However, in most cases, the *isnād* actually records *ijāza* and transmission linkages pertaining to an individual work. Establishing an individual's *ijāza-isnād* to be reliable may formally require the fulfilment of the same criteria (such as the links being contemporary, collocated and reliable). However, it is not subject to the same need for multiple chains (that is, the *tawātur* test) in order to establish the right of the *ijāza* recipient to cite and employ the *ḥadīth* in question. Furthermore, the purpose of an *ijāza* is not entirely to prove the authenticity of the *ḥadīth* within a book, but to demonstrate the right of the individual author (through his religious training) to transmit these *ḥadīth*. In short, the *isnāds* in the Four Books represent *ijāza-isnāds* and not *ḥadīth-isnāds*, strengthening the claim for the authenticity of the quoted reports. A comparison of Astarābādī's argument with al-Karakī's here shows a developing sophistication. Al-Karakī's admission that there were unreliable collections of reports is an indulgence to the Uṣūlī position, but the community's ability to weed out rogue collections (either through testing them with the Imam or by their failure to gain communal recognition as a basis for law) is asserted subsequently. If an Uṣūlī were to criticise this view, he would, in effect, be undermining a fundamental pillar of Shīʿī salvation history. Turning an admission

[15] Karakī, *Hidāyat*, p. 88.

into a strength, as al-Karakī does here, most probably emerged in the cut and thrust of real Akhbārī-Uṣūlī debate after Astarābādī's death. Al-Karakī's arguments are also strengthened by the shift in the focus of the argument from the collection of the *akhbār*, back in time to their commitment to writing. Astarābādī's emphasis on the expert collection of the *akhbār* themselves left opponents a potential line of criticism: they could raise doubts, not about the probity of the collectors, but their method of presentation. There was a common Uṣūlī counter-argument that the three collectors merely presented all the *akhbār* they heard, and did not consider it their task to distinguish between the "reliable and unreliable". Al-Karakī, however, introduces the notion that the *isnāds* within the Four Books are records of the transmission of books, not reports. Since one does not require multiple sound *isnāds* in order to establish a pupil's right to recite a work (one *isnād* is normally sufficient), the critic is now forced to question not the methodology of the collectors, but the accuracy of the original authors of the *uṣūl*. This is considerably more difficult to do, since they were the nonesuches of early Shiʿism, and most often the Imams' favoured confidents. In short, al-Karakī's argument is that the collectors selected books of *ḥadīth*, not individual *ḥadīth*, and these books had either been approved by the Imam himself or tested for reliability by the community itself.

Once these arguments had reached their final form, they dominated subsequent Akhbārī discussions. Little is added by later Akhbārīs to the presentations of Astarabādī, al-Ḥurr and al-Karakī concerning this version of the history of the collection of Imāmī *akhbār*. Al-Baḥranī's version draws on the above accounts, and does not deviate from the established Akhbārī narrative.[16] Muḥammad al-Akhbārī also draws on this account, repeating the call to act on the contents of the "indubitable *uṣūl* of the companions" (*uṣūl al-aṣḥāb al-maqṭūʿa*) which are preserved in the early works of *akhbār*.[17] This element of Akhbārī polemic remained remarkably stable, from the beginnings of the school through to the early Qajar period.

[16] For a summary of al-Baḥrānī's argumentation see Gleave, *Inevitable Doubt*, pp. 40–46.
[17] Akhbārī, *Maʿāwil*, f.195 (p. 390).

Akhbārī isnād *Criticism*

The above arguments, whilst differing in detail and perhaps style, share an assertion that the *akhbār* in the Four Books, and (under certain conditions) those outside of the Four Books, are of assured authenticity. This was based on a general historical account of how the *akhbār* were committed to writing, and later collected. The historical account, it was argued, obviated the need for scrutiny and criticism of *isnāds*.[18] However, an alternative Akhbārī approach, which could be termed *isnād*-critical, is also evident in the literature. This approach attempts to delineate an alternative to (or modified form of) the established discipline of *isnād* criticism, and was outlined first by Muḥammad Taqī al-Majlisī (Majlisī I) in his introduction to his *Lawāmiʿ ṣāḥibqirānī*. This introduction comprises twelve *fāʾidas* on various aspects of *uṣūl al-fiqh*, including hermeneutic issues (such as the reconciliation of contradictory *akhbār*), the transmission of knowledge (the characteristics of the *ʿulamāʾ* and the *ijāza* system)[19] and the legitimacy (or rather, illegitimacy) of *ijtihād*. The eleventh *fāʾida* is a critical review of the technical terms (*iṣṭilāḥāt*) used for the classification of *ḥadīth*. Other Akhbārī authors merely rejected the utility of terms such as *mutawātir* and *khabar al-wāḥid* in any assessment of the reliability of reports. Majlisī I, however, seeks to redefine them and put them to use in the analysis of individual reports. His re-employment of the established technical terminology under new definitions can be found in the rest of his *Lawāmiʿ*, a commentary on the *akhbār* collection *Man lā Yaḥḍuruhu al-Faqīh* of Ibn Bābūya.[20] The aim is to demonstrate that when the correct method of *isnād* criticism is employed, all the *akhbār* of the Four

[18] Al-Karakī, for example, goes into great detail concerning the meaning of the various terms used in the science of *ḥadīth* criticism (i.e. *tawātur/khabar al-wāḥid*, the four categories, sub-categories of sound and weak *ḥadīth*s etc.). However, he ends his discussion with the statement, "All this is useless for us today because the ancient scholars (*al-qudamāʾ*) have provided sufficient information for us on this matter and have sorted through the *ḥadīth*s" (Karakī, *Hidāyat*, p. 130).

[19] This section has already been analysed above, pp. 145–148.

[20] See, for example, Majlisī I, *Lawāmiʿ*, v. 2, p. 214 (where *ḥadīths* are describes as "similar to *ṣaḥīḥ*" under the established terminology). One can also find this redefinition in his Arabic *ḥadīth* commentary on the same *akhbār* collection. See Majlisī I, *Rawḍat*, v. 1, p. 371 (where a *ḥadīth* in Ibn Bābūya's work is without an *isnād*, but is classified *ṣaḥīḥ* because of its additional appearances in the collections of al-Kulaynī and al-Ṭūsī).

Books can be tested individually and found to be authentic. Hence in the introduction to his commentary, an exposition of this "correct" method is necessary.

It is significant that Majlisī I is quite willing to employ at least some of the established Uṣūlī terminology in his scheme. He begins with the terms *mutawātir* and *khabar al-wāḥid*, providing his own definitions. As is well-known, *mutawātir* is used to describe a report which has been transmitted through a number of reliable chains such that the hearer cannot doubt that the information contained within the report is true. Majlisī I, however, reverses this definition:

> A *mutawātir* report is one which at least three persons have transmitted, and these reports give us knowledge (*az akhbār-i īshān ʿilm biham mīrasad*). Sometimes *ʿilm* is not obtained from 1000 individuals—such as the testimony of peasants concerning the ownership of land or water. Sometimes it is the case that the word of a single, pious man, who reports that he saw so-and-so, gives knowledge. The central point here is knowledge, not number.[21]

Majlisī I is, then, arguing that the term *mutawātir* describes a report which has three or more transmitters and the contents of which are true.[22] However, certain knowledge can be gained from the report of a single individual, and can be unobtainable from a report with 1000 transmitters. Describing a report as *mutawātir* does not, for Majlisī I, mean that the information contained within it necessarily holds a higher epistemological status than a non-*mutawātir* report. *Mutawātir* merely indicates that the report has two independent and unrelated qualities—(1) it is transmitted by three or more and (2) knowledge is obtained from it.[23] The second category is "a report which is surrounded by external evidence" (*khabar-i maḥfūf bi-qarīnah*).[24] This

---

[21] Majlisī I, *Lawāmiʿ*, v. 1, p. 99. A full description of Majlisī I's attributes towards the verification of *ḥadīth* can be found in Jadīdīnizhād, "Nigāhī".

[22] Three is, of course, an unusually low number of transmitters for a report to be classified as *mutawātir* (see Weiss, "Knowledge of the Past", pp. 94–95, where the standard numbers mentioned by al-Ghazālī are 40 and 70, though greater numbers are required; Hallaq, "The authenticity of Prophetic *ḥadīth*", p. 79 mentions 12, 20, 40, 70 and 313). The reason for the Akhbārīs' choice of three as the minimum is clear—a report found in different places within the Four Books (of which there are only *three* authors), can now be considered *mutawātir*.

[23] It is quite possible, then, for a report to have three or more transmitters and not provide us with *ʿilm*. These are termed *khabar-i mustafīd* by Majlisī I, which is a type of *khabar-i vāḥid*. See Majlisī I, *Lawāmiʿ*, v. 1, p. 100.

[24] Majlisī's idea is, of course, not new to Shīʿī *fiqh*. However, it is usually associated

category of report is accompanied by evidence which gives the hearer knowledge concerning the information within the report. Majlisī I gives examples of such "external evidence" (*qarāyin az khārij-i khabar*). A tells B a secret which no one else knows or could have known. B writes it down and A seals the scroll. C comes and opens the scroll, knowing the handwriting of B and seal of A. C now has knowledge (*ʿilm*) of the conversation between A and B. The external evidence (the handwriting, the seal etc.) confirm the high epistemological status of the information. The same can be said of the attribution of books to scholars of the past—the *ijāza-isnāds* bring information "that no one could doubt" (*hich kas shakk nimī-kunad*).[25] Finally, there is the category of "isolated reports" (*khabar-i vāḥid*), "a report from which knowledge is not obtained, whether the report is transmitted by a single individual or 1000 individuals." The number of transmitters is unimportant for a report to be classified as *khabar-i vāḥid*. The classification comes from the emergence (or in this case, the non-emergence) of *ʿilm* concerning the contents of the report.

These definitions differ from the established (Uṣūlī) definitions of these terms. In the established definitions, a *mutawātir* report gives *ʿilm* because of the number of its transmitters (that is, there is a causal connection between the number and the availability of *ʿilm*). Similarly, a *khabar al-wāḥid* does not bring *ʿilm* because of its low number of transmitters. Admittedly, the latter can bring *ʿilm* through external indicators (*qarāʾin*), and they then become a special sub-category of *khabar al-wāḥid*: "a *khabar* which can provide knowledge by its connection with external pieces of evidence" (*khabar qad yufīdu al-ʿilm bi-inḍimām al-qarāʾin ilayhi*).[26] However, Majlisī I's definitions depart from the usual Uṣūlī usage in that they make *ʿilm* the reason for classification, not the number of transmitters. His reformulation of these definitions centres around a clean break of the causal connection between the extent of a report's dissemination and its epistemological status. Instead all reports are to be tested by external pieces of evi-

---

with raising the status of "isolated reports" (*khabar al-wāḥid*) only. Isolated reports can bring *ʿilm* if they are "connected" (*inḍimām*) with external evidence (Ḥasan b. Shahīd II, *Maʿālim*, p. 261) or they are "encompassed" (*iḥtaffa*) by external reports (Tūnī, *Al-Wāfiya*, p. 81).

[25] Majlisī I, *Lawāmiʿ*, v. 1, p. 100.

[26] The classical expression of this categorisation scheme can be found in Ḥasan b. Shahīd II, *Maʿālim*, p. 261.

dence (*qarāʾin*) which may (or may not) provide the report's recipient with *ʿilm*. If these *qarāʾin* provide *ʿilm*, then the report is *maḥfūf bi-qarīnah*. If they do not, the report is *khabar-i vāḥid*. A report may be classified as *mutawātir* if it is both *maḥfūf* and transmitted by more than three transmitters. *Mutawātir* status does not, however, give a report any particular epistemological advantage over *maḥfūf* reports. This reformulation enables Majlisī I to maintain a connection with Shīʿī *ḥadīth* scholarship, by retaining the terminology of the established classification system. However, he is also able to pervert the purpose of that discipline through adjustment to the definitions of these categories. By allowing "external evidence" such a significant role, Majlisī I lowers the epistemological bar for the attainment of *ʿilm*. We have already seen how important the relaxation of the criteria for *ʿilm* was for the coherence of Astarābādī's theory.[27] Here Majlisī I applies this central Akhbārī doctrine to the classification of reports. By employing "external evidence", Majlisī I is able to present the *akhbār* from the Four Books, and from other sources, as providing unshakable bases for right action. For example:

> If each of the three [authors of the Four Books] transmit a report from Ḥusayn b. Saʿīd, and each of them agree in the transmission, then we have obtained knowledge (*ʿilm*) that they do not attribute lies to Ḥusayn b. Saʿīd. Following on from this, it is now possible that the presence of *mutawātir* reports in these Four Books will, praise be to God, enable other reports in other books to be confirmed (*muʾayyad*), such as the *Maḥāsin* of Barqī, the *Qurb al-isnād* of Ḥimyarī, the *Basāʾir al-Darajāt* of Ṣaffār and other books.[28]

It is, then, through the Four Books that other early *ḥadīth* collections can be confirmed as reliable and form the basis for action. A report is transmitted by the three authors of the Four Books, and this combines with other pieces of evidence (*qarāʾin*—that is, the authors' own statements that they "do not relate anything upon which they do not rely") to make the report *mutawātir*. The original work from which this report was taken is now considered reliable, and this reliability can then be used to authenticate other reports, found in works outside of the Four Books, but transmitted from the same original work. Majlisī I's position appears to be, then, that reports outside of the

[27] See above, pp. 70–73.

[28] Majlisī I, *Lawāmiʿ*, v. 1, p. 103.

Four Books can be sound providing there is evidence from within the Four Books which establishes their reliability.[29]

The effects of this reformulation of the *mutawātir/khabar al-wāḥid* classification system are mirrored by Majlisī I's version of the classification system based on *isnād*. The classification system (*ṣaḥīḥ, ḥasan, muwaththaq, ḍaʿīf*) favoured by the Uṣūlīs is not rejected, but a fifth category is added: "a report which contains one or more praised transmitters, but also one or more deviants (*bad madhhab dāshtah*)." The category has no name, but is known as *muwaththaq* by some Uṣūlīs and *ḥasan* by others.[30] Majlisī I considers it deserving of its own category, and including this additional category serves to undermine the neatness of the established scheme, and exposes its arbitrary nature. The category of "weak" (*ḍaʿīf*) is restricted by Majlisī I. Uṣūlīs considered reports with *isnāds* with an "unknown" link (normally termed *mursal*) as equivalent to *ḍaʿīf* in terms of reliability. For them, this was signalled in the *isnād* by locutions such as "from someone" (*ʿan rajul*), "from one of our companions" (*ʿan rajul min aṣḥābinā*) or "from a group of our companions" (*ʿan jamāʿa min aṣḥābinā*). For Majlisī I, however, there is a possibility that an unknown transmitter is reliable (*mumkin-ast kih majhūl al-ḥāl thiqa bāshad*), and this possibility raises the report above the "weak" category. *Ḍaʿīf*, as a category, is reserved for those reports which have a "known miscreant within their *isnāds*".[31] Once again, the categories are retained, but the means of categorising a report are adjusted to ensure that a minimal number of reports are rejected in the interpretive process.

Majlisī I makes the general point that the external evidence which demonstrates reports to be sound can be found in the testimony of the authors of the *akhbār* collections (that is, that they themselves claim to have only included reliable reports in their collections). This means that if a report is included in their collections, it has (by definition) a sound *isnād*. The Uṣūlīs may use works of *rijāl* (in which transmitters

[29] His position is, then, similar to that propounded by the later Akhbārī, Yūsuf al-Baḥrānī, who appears to argue that reports in the Four Books are immediately considered sound, but reports in other works are tested on a report by report basis. See Gleave, *Inevitable Doubt*, p. 46.

[30] Majlisī I, *Lawāmiʿ*, v. 1, p. 104.

[31] "*Hur chih dar ān fāsiqī bāshad, ḍaʿīf mī-shumuram*" (Majlisī I, *Lawāmiʿ*, v. 1, p. 104).

are assessed as to their reliability) in order to find weaknesses within *isnāds*. They do this primarily through condemning certain transmitters as unreliable on the basis of heresy. However, the testimony of the authors of the *akhbār* collections is more reliable (*bihtar yaqīn*[an]) than that of the *rijāl* authors. If a report is included in their collections, and is declared to be sound, then, *mutatis mutandis*, the *isnād* accompanying it must be sound. If such *isnāds* are sound, then the same *isnāds* found outside of the Four Books, but attached to other reports, are also sound. Once again, the volume of reliable reports is increased through establishing that the Four Books contain unchallengeable reports.

Majlisī I's technique of employing the terminology of established *ḥadīth* criticism in the defense of Akhbarism is also found in later Akhbārī works. Al-Jazāʾirī, the pupil of Muḥammad Bāqir al-Majlisī, portrayed himself in his difference list as a scholar who had adopted a "middle path".[32] As already noted, this self-definition should be treated with suspicion. Not only does he take the Akhbārī position in most areas of dispute between the two schools, but he is also portrayed as an Akhbārī in nearly all subsequent biographical literature. Most subsequent Akhbārīs claim him as a member of their own school, and no Uṣūlī (to my knowledge) defends him as an Uṣūlī. In his *Kashf al-asrār*, a commentary on al-Shaykh al-Ṭūsī's *al-Istibṣār*, he argues for an unusual Akhbārī position, which might be considered "moderate" in comparison with other Akhbārīs. What is particularly interesting here is the mode of argumentation which borrows from both established Akhbārī and Uṣūlī presentations. Al-Jazāʾirī begins in Akhbārī style, arguing that the Uṣūlī four-fold categorisation of reports is a modern convention (*iṣṭilāḥ tār*[in]) which was initiated by al-ʿAllāma or perhaps his teacher Ibn Ṭāwūs.[33] *Ṣaḥīḥ* had a quite different meaning for the earliest Shīʿī scholars: it did not mean that a report had an *isnād* which led back to the Imam, with "just" (*ʿadl*) transmitters at every stage. Rather, *saḥīḥ* described a report which should act as the basis for action. This sound basis could be established for a number of reasons:

(1) the report is known to come from a reliable source (such as the 400 *uṣūl*)

---

[32] See above, p. 201.

[33] Jazāʾirī, *Kashf*, v. 2, p. 39.

(2) it is repeated in one or more sources with different transmission chains

(3) it is found in a source, written by a person who, it is generally agreed, was reliable and trustworthy

(4) it is found in a work which was known to have been shown to one of the Imams and approved by him.[34]

Now, al-Jazāʾirī argues, Ibn Bābūya uses *ṣaḥīḥ* in this way when he says that the reports in his *Man lā Yaḥḍuruhu al-Faqīh* are "sound". However, a careful scholar cannot merely deduce from statements such as this that all the reports in Ibn Bābūya's works are sound—one must examine whether the reports in the early collections fulfil criteria such as those listed above, and only then can they act as secure bases for religious action. By allowing this line of argument, al-Jazāʾirī is departing from the view of his Akhbārī colleagues, who accepted the word of the early collectors of *akhbār* without question, and consequently declared all the *akhbār* to be sound. In order to carry out such an exercise, al-Jazāʾirī embarks on a study of the presentation techniques of the *akhbār* collectors, and their method of presenting *isnāds*.

The mistake of some modern scholars (by which he means his fellow Akhbārīs) is to presume that because the original sources have been lost (*indirās al-uṣūl*), we must simply rely on the word of the *akhbār* collectors that what they collected was sound. However, al-Jazāʾirī claims, the original sources (*al-uṣūl wa'l-kutub*) are not entirely lost. Al-Jazāʾirī has access to a number of them, and by comparing their content with that found in the Four Books of *akhbār*, some interesting conclusions can be made concerning the reports found in the Four Books. Firstly, one regularly finds that a number of *isnāds* are attached to the same report in the *uṣūl*, and the authors of the Four Books have simply cited one of these *isnāds*. Secondly, the *isnāds* in the *uṣūl* are of diverse quality: some are *ṣaḥīḥ*, but others are *ḥasan* or *muwaththaq*. Thirdly, when an author such as al-Kulaynī had access to these various *isnād* chains, he did not always choose the strongest *isnād*. "Things like this enable us to increase our faith in the content of the four sources [of *akhbār*], even when their *isnād* is not immaculate,"[35] says al-Jazāʾirī: even though some *isnāds* in the

[34] Jazāʾirī, *Kashf*, v. 2, p. 41.
[35] Jazāʾirī, *Kashf*, v. 2, p. 44.

Four Books are weak, there are known to have been other (uncited) *isnāds* which are sound. Why, then, did the authors of the Four Books cite weak *isnāds*? Al-Jazāʾirī says that the weak *isnād* differs from the others in that it is "higher" than the rest—that is, it goes back to the earliest Imam or is transmitted by the earliest companion to the Imam. The *akhbār* collector in choosing to cite this higher *isnād* is concerned with choosing, not the soundest *isnād*, but the oldest. For this reason, the weakness of the *isnāds* in the *akhbār* collections does not damage the authenticity of the *akhbār* there, because the authors' practice was to cite the highest *isnād*, and not mention the others. For every *isnād+khabar* in the Four Books, there are many others, un-cited by the collectors but nonetheless extant at the time. There are not only many more ("sounder") *isnāds*, but other versions of each report. They may appear *ḍaʿīf* and *khabar al-wāḥid*, but they are, in fact, *ṣaḥīḥ* and *mutawātir*

Furthermore, al-Jazāʾirī claims that the *akhbār* collectors present two sorts of *isnād* in their works, one leading back to a source, and the other leading back to the Imam himself. The former are the transmission chain from the *akhbār* collector to the author of the book (that is, the *ijāza-isnād*). The latter are augmented by the transmission chain from the *aṣl* author to the Imam. These extended *isnāds* are, he argues, each merely one amongst many possible *isnāds* which the *akhbār* collectors could have cited. The authors of the Four Books, when they cite a full *isnād* (that is, an *isnād* which goes back to the Imam and not merely to the *aṣl* author), are citing from the *aṣl*, but they are including in their citation only one of the many *isnāds* from the *aṣl* author to the Imam. The result is that al-Kulaynī, for example, cites the *isnāds* in *al-Kāfī* which are, in fact, a combination of two *isnāds*: one from the collector to the book's author, and the other from the author to the Imam himself. Al-Jazāʾirī agrees with many of the "modern *muḥaddithūn*" who claim that although all the reports in the Four Books are *ṣaḥīḥ*, only those that go back to the Imams themselves (and not merely the *aṣl* authors) are *mutawātir*.[36]

---

[36] The distinction between a report which is *mutawātir* and one which is merely *ṣaḥīḥ* is important for al-Jazāʾirī's method of deciding between contradictory reports. See below, p. 293. According to al-Jazāʾirī, this is why al-Sayyid al-Murtaḍā says all "our *ḥadīths*" are *mutawātir*. He does not mean that all the *akhbār* are *mutawātir*—he means by "our *akhbār*" the *akhbār* leading back to the Imams, since he himself was a descendent of Imam Mūsā al-Kāẓim. Jazāʾirī, *Kashf*, v. 2, p. 45.

Muḥammad al-Akhbārī, writing a century after al-Jazāʾirī, pinpoints another element of the Uṣūlī method of *ḥadīth* criticism which could be exploited for Akhbārī ends. Muḥammad al-Akhbārī draws on the notion presented by Majlisī I that certainty regarding the probative force of a report is not dependent upon the number of transmitters (namely the *tawātur* criterion). In his defence of Astarabādī, entitled *Maʿāwil al-ʿUqūl*, Muḥammad al-Akhbārī explores the epistemological problems of relying on the number of transmissions as the basis for classifying a *ḥadīth* as having probative force. His argument is that since all the Shīʿa admit the possibility of *taqiyya*, a large number of transmitters (both synchronically and diachronically) is not sufficient to guarantee probative force (*ḥujjiyya*). It may guarantee historical accuracy, but this is of little use in discerning God's law, if the report is irrelevant due to dissimulation. Hence, another technique is needed in order to ensure that a report is, not only historically accurate, but legally relevant. Whilst Akhbārīs had used this distinction before Muḥammad al-Akhbārī, it had been a hermeneutic issue.[37] Muḥammad al-Akhbārī uses it to destroy the basis of the Uṣūlī *tawātur/khabar al-waḥid* distinction, and instead erects a method which can measure both historical accuracy and legal relevance. If the first transmitter—that is, the person who heard the report from the Imam—is a Sunni, then there is a possibility that the report was issued under *taqiyya* since the Imam was aware of the Sunni's presence and may have been dissimulating. However, the Uṣūlīs use the presence of a Sunni later in the *isnād* as evidence of possible fabrication, and therefore unreliability, of a report. Muḥammad al-Akhbārī, true to his Akhbārī principles, does not view the presence of a Sunni in the *isnād* as evidence of possible fabrication. The report is certain in terms of its origin (*qaṭʿī al-wurūd, qaṭʿī al-ṣudūr*) whether the *isnād* contains Sunnis or not.[38] However, whilst an *isnād* with a Sunni transmitter does not invalidate a report's claim to historical accuracy, it could indicate that the report was issued under *taqiyya*. It could do this because the Imam was aware of the fact that whilst there may have been no Sunnis present at the time of him speaking, the report might be transmitted by a Sunni at some later time before being recorded in a collection of *akhbār*. If the Imam was aware of this at the time

---

[37] See below, p. 84.

[38] Akhbārī, *Maʿāwil*, f.54a.15 (p. 107.15).

of speaking, then he would need to guard against his true message (which may endanger the community at some future time) being known to the Sunnis through this non-Imāmī transmitter. The result is a method whereby a report might be considered *taqiyya* even though there was no Sunni present at the time of speaking. Muḥammad al-Akhbārī here is using *isnād* criticism to drive a wedge between historical accuracy and legal relevance, providing us with another example of how an Akhbārī writer was able to employ the techniques of established disciplines (such as *ḥadīth* criticism) in order to support his Akhbārī perspective.[39]

One can see from the argumentation of both al-Jazā'irī and al-Akhbārī here that they draw extensively on the existing discipline of *ḥadīth* criticism. Like Majlisī I, al-Jazā'irī employs existing terminology and intellectual structures whilst developing his argument concerning the authenticity of the *akhbār*. It is, clearly, a less radical approach than that of Astarābādī, al-Karakī and al-Ḥurr (for whom, the discipline of *isnād* criticism was rendered redundant due to the statements of the *akhbār* collectors). We have, then, (at least) two broad trends within Akhbārī argumentation concerning the authenticity (and hence probative force) of the *akhbār* within *ḥadīth* collections, particularly the Four Books. In the first (more radical) approach, *isnāds* were considered merely decorative, included for the sake of completeness and played no role in the assessment of the probative force of the *akhbār*. In the second (more conciliatory) approach, the techniques and terminology of *isnād* criticism were employed. It was hoped that the discipline, as it was conceived by the Uṣūlīs themselves, would support the Akhbārī conclusion that the *akhbār* were authentic and had insurmountable probative force in legal argumentation. Proponents of both approaches were contemporaneous and found within the ranks of the Akhbārī school. There does not appear to have been a time when one or other of the approaches dominated. Indeed individual Akhbārīs were, at times, willing to employ both approaches, as they cast around for arguments which established the authenticity and probative force of the *akhbār* collections. Furthermore, the conclusions of scholars using these two approaches are not radically different. On this doctrine there is near unanimity amongst the Akhbārīs,

---

[39] Akhbārī, *Maʿāwil*, f.53b–54b (pp. 106–108). He cites, with approval, Majlisī I's approach to the validation of *akhbār* in *Maʿāwil*, f.52b.

indicating that it is a central quality of admittance into the ranks of the Akhbāriyya. On this issue, variety within the Akhbārī school is expressed not in opinion, but in novel argumentation devised to defend almost unanimous doctrine.

Marginal Akhbārīs

As with the interpretation of the Qur'ān, distinctive Akhbārī views on the *akhbār*'s authenticity and probative force can be used to position scholars normally considered on the margins of the Akhbārī school. On the interpretation of legal material in the Qur'ān, al-Tūnī, al-Māzandarānī and Majlisī II held positions remarkably similar to those within the Akhbārī school, even though their Akhbārī identity is debated in later scholarship. Similar conclusions can be drawn from their positions on the *akhbār* as a legal source, perhaps in an exclusive manner.

Unlike Majlisī I, al-Tūnī accepts the established definitions of *mutawātir* and *khabar al-wāḥid*, and, in particular, the causal link between the categorisation of a report and the number of its transmissions. *Mutawātir* reports bring knowledge (*yufīdu al-ʿilm*) because the number of their transmissions is such that collusion is unthinkable. *Khabar al-wāḥid* reports are simply those that do not have a sufficient number of transmissions to produce *ʿilm*. Al-Tūnī also accepts that there are reports which are classified as *khabar al-wāḥid*, but can produce knowledge due to external evidence (*qarā'in*). The debate, then, is the standard debate within Uṣūlī discourse concerning the probative force of *khabar al-wāḥid* which are not accompanied by evidence (*khabar al-wāḥid al-ʿārī ʿan qarā'in al-qaṭʿ*). According to al-Tūnī, most jurists (*al-akthar min ʿulamā'inā al-bāḥithīn fī'l-uṣūl*) have said that such reports have no probative force because they are *ẓannī*. Al-Tūnī, however, considers *khabar al-wāḥid* to have probative force even though they are *ẓannī*. This is not because he accepts *ẓannī* evidence (as the Uṣūlīs do). Rather, he surveys the evidence and concludes that it is certain that the Imams have given us permission to act on *khabar al-wāḥid*. So, even though the report itself may be *ẓannī*, using it as a basis for action is not *ẓannī*, since we have certainty (*al-qaṭʿ*) that it is allowed:

> Following *ẓann* [occurs] when the evidence for an action is *ẓannī*... but this case [that is, the case of following *khabar al-wāḥid*] is not like that. The basis of the action here is the words of the Sinless Ones, the reports of the cradles of revelation itself... acting on the basis of *khabar al-wāḥid* is merely following a certain indicator (*al-dalīl al-qaṭʿī*) which indicates that *khabar al-wāḥid* has probative force. Hence it is following *yaqīn* [not *ẓann*].[40]

Al-Tūnī is arguing here that reports classified as *khabar al-wāḥid* have probative force. In this he concurs with the Uṣūlīs. However, he holds this opinion not because *ẓannī* evidence is acceptable. Rather, for al-Tūnī, *khabar al-wāḥid* have the same, certain probative force of *mutawātir* reports because there are *mutawātir* reports which permit the use of *khabar al-wāḥid* reports. In short, Uṣūlī epistemology has been used to support an Akhbārī conclusion.[41]

On the issue of the authenticity and probative force of *ḥadīth*, al-Māzandarānī's commentary on Shaykh Ḥasan's *Maʿālim al-Dīn* also has echoes of Akhbārīsm. He comments on Shaykh Ḥasan's statement that the "gate to certain knowledge" (*bāb al-ʿilm al-qaṭʿī*) is closed, because "the indicators found in the extant books of *ḥadīth* give nothing but *ẓann*, and *mutawātir sunna* has been lost."[42] Al-Māzandarānī takes Shaykh Ḥasan to mean that the existence of a report in the four canonical collections is insufficient to guarantee *mutawātir* status because three transmitters is too small a number. Therefore, the *akhbār* are *ẓannī*. Al-Māzandarānī takes exception to this:

> The opinion that the Four Books are based on three [authors], and therefore *tawātur* obtained for them, cannot be refuted. Number alone does not indicate *tawātur*. Rather the thing that indicates *tawātur* is knowledge that sound transmission (*ʿilm bi-ṣiḥḥat al-naql*) has taken place. It is clear to any just person that there may be difference in one of their reports, though there is agreement over the transmitters. There may also be disagreements over the *sanad*, whilst there is an

[40] Tūnī, *al-Wāfiya*, p. 165.

[41] It is important to note that not all *khabar al-wāḥid* reports acquire probative force for al-Tūnī. Only those which are to be found in "reliable books" (*al-kutub al-muʿtamida*) which the Shīʿa have agreed can act as the basis of action "such as *al-Kāfī*, *al-Faqīh*, *al-Tahdhīb* and the like." It does not matter whether the individual transmitters of these reports have moral probity or not, or what the state of the *isnād* might be. The Akhbārī tone of al-Tūnī's remarks here is clear (see Tūnī, *al-Wāfiya*, p. 166).

[42] Ḥasan, *Maʿālim*, p. 268.

> agreement over the text. [But in either case] there is knowledge that *tawātur*—either with respect to the words or with respect to the meaning of a report—has been achieved. This is especially the case when the three [authors of the Four Books] state explicitly that what they record is taken from some of the recorded [400] *Uṣūl*.[43]

Al-Māzandarānī goes on to say that the three collectors were pious, and therefore would not have recorded lies concerning God. Reports of the four canonical collections are, for al-Māzandarānī, all authentic, and this authenticity is secured by reference to the diligence and probity of the collectors.[44] The argument, then, replicates the points made by Astarābādī and al-Ḥurr, though for them, the argument's remit can be expanded to cover early *ḥadīth* collections beyond the Four Books. On this juristic issue, as with the issue of the probative force of the Qur'ān, al-Māzandarānī's position is not a mere replication of Astarābādī's views. He has his own personal take in restricting authenticity to the Four Books. However, both his argumentation and his eventual position fit within the range of Akhbārī opinion current at the time.

Majlisī II appears undiscriminating in his attitude towards the authenticity of the *akhbār*. His *Biḥār al-anwār* is the most extensive collection of Shīʿī *ḥadīth* from the pre-modern period. It is divided into topic-based chapters, each containing *ḥadīth* relating to a particular subject. There are chapters (and hence categorised *ḥadīth*) on an enormous range of topics. The *isnāds* of the *ḥadīths* are usually included, particularly when they have been lifted from previous *ḥadīth* collections. Comment from Majlisī II himself is kept to a minimum, and there is very little assessment of the authenticity or probative force of the *ḥadīth* in question. For much of the collection, Majlisī II sees it as his task to merely collect the *akhbār*, not judge them. Such an attitude might spring from a less stringent application of *isnād* criticism which, as indicated above, was a popular approach of some Akhbārī jurists in their attempt to maximise the available revelatory material. Furthermore, there is little argumentation as such. Rather Majlisī II prefers simply to present the revelatory material to his reader with minimal exegetical comment. The doctrinal conclusions

[43] Māzandarānī, *Sharḥ al-Maʿālim*, p. 230.

[44] Al-Māzandarānī also admits that though the reports are authentic, the legal ruling they indicate may be uncertain (i.e. they may be *ẓannī al-dalāla*).

drawn by the reader appear to be based on his or her reception of the material, and not on any engagement with the issues raised by it. In short, it mimics the *akhbār*-based commentary of al-Huwayzī and Hāshim al-Baḥrānī examined in the previous chapter. On the other hand, Majlisī II's *Mir'āt al-ʿuqūl*, an extensive commentary on al-Kulaynī's *al-Kāfī*, includes detailed discussions of the reliability or otherwise of the *isnāds* attached to *matns*. The categorisation of the *isnāds* does not prevent Majlisī II commenting at great length on the doctrinal significance of *ḥadīths* which he clearly considers very weak, and in most cases, simply inauthentic. On the rare occasions when he does enter into theoretical discussion over the implications of classifying a report according to the usual categorisation schema, he appears to fully accept the established definitions of *mutawātir* and *khabar al-wāḥid*.[45] However, what effect describing a report as *khabar al-wāḥid* has on the evaluation of its probative force is unclear. His view on the probative force of the various *isnād*-based categories of *ḥadīth* appears to include at least the categories of *ḥasan* and *muwaththaq*. For example, in a commentary on the famous "Maqbūla" *ḥadīth* of ʿUmar b. Hanẓala,[46] Majlisī II states that the acceptance by the Imam of probative force of well-known (*mashhūr*) reports related by "*thiqa*" transmitters is evidence that the Imam is permitting his community to rely on *muwaththaq* reports.[47] Majlisī II is clearly attempting to marry the pre-technical language of the report with the technical definition of *muwaththaq* in later Imāmī *ḥadīth* criticism. In this he is not only accepting the established definitions, but also defending them as rooted in the *akhbār*. However, as Majlisī notes, the Maqbūla text appears to accept *muwaththaq ḥadīths* but with one reservation. The stipulation (*qayd*) attached to them relates to the inclusion of the word "well-known" (*mashhūr*) when describing the *thiqa* related *ḥadīth*. For Majlisī II *muwaththaq* report can act as the basis for action, but it must also be "well-known" amongst the Muslims. The stipulation is included

---

[45] See, for example, the constant gloss of any mention of "*al-sunna*" in a *ḥadīth* as referring to "*al-sunna al-mutawātira*" (e.g. Majlisī II, *Mir'āt al-ʿuqūl*, v. 1, p. 227).

[46] A translation of this report can be found in Gleave, "Two Classical Shii theories of *qaḍā'*", pp. 119–120. Discussions of the report and its interpretation in Shīʿī jurisprudence can be found in Calder, "Judicial Authority". For a quite different interpretation, see Elias, "Misconceptions", pp. 13–25.

[47] Majlisī II, *Mir'āt al-ʿuqūl*, v. 1, p. 226.

in the report, and affirmed by the Imam, in order that reports from unbelievers, which may fulfil all the criteria of *muwaththaq*, are not accepted. An unbeliever may well be "just" (*ʿādil*) says Majlisī II, and in objective terms, therefore, act as a reliable transmitter. The added stipulation that a report be well-known amongst the Muslims means that *muwaththaq* reports must be disseminated by Muslims before they can be the basis for action. The significance of the discussion here for the Akhbārī-Uṣūlī debate is Majlisī II's implicit acceptance of central Akhbārī doctrine. Whilst the Uṣūlīs only accepted reports which were transmitted by just, reliable Imāmīs, Akhbārīs placed no such restriction on the transmitters. As al-Samāhījī puts it:

> The *mujtahids* define *ṣaḥīḥ* as meaning [a report] transmitted by just, reliable Imāmīs all the way back to the Imam . . . The Akhbārīs define *ṣaḥīḥ* to mean anything which can be truly attributed to the Imam, be it . . . *khabar al-wāḥid* . . . [or any other category], or is found in one of the sound *uṣūl*.[48]

Now Majlisī II's view is that even an unbeliever can be "just", and that a report need only be *muwaththaq* and "well-known" before it can act as a basis for (religiously) valid action. In short, he accepts the established categories of *isnād* classification, but is willing to loosen the Uṣūlī restrictions on what can and what cannot act as a basis for action, or have probative force in deriving legal rules. His position on the authenticity and religious significance of the *akhbār*, then, appears Akhbārī in motivation, though Uṣūlī in expression, and this makes him a particularly important, "marginal" Akhbārī.

Conclusions

The plethora of Akhbārī arguments presented in this chapter demonstrate the ingenuity of the Akhbārī writers. They collated and interpreted historical evidence concerning the collection of the *akhbār* in early Shiʿism. In doing so, they were attempting to demonstrate that the Four Books of *akhbār* which were later considered the primary material for religious (particularly legal) argumentation were sound and indubitable sources of knowledge. These arguments went through a number of versions before reaching a stable expression within the

[48] Newman, "Akhbārī-Uṣūlī Dispute, pt. 1", pp. 25–26.

school. Once stability in the argument had been achieved, the history of transmission was reiterated by Akhbārīs in the hope that it might become the dominant narrative against which the Uṣūlī school would be forced to kick. Their version of how reports were transmitted, from the Imams' words to the 400 *uṣūl* to the extant early collections of *akhbār*, was, for Akhbārīs, so obvious that to deny it prompted ridicule. Whilst this narrative was being established, some Akhbārī scholars (perhaps with greater classical training) also devised a series of arguments which aimed not simply to reject the Uṣulī sciences, but to subvert them. By scrutinising the premises upon which *isnād* criticism was based, Majlisī I, al-Jazā'irī and Muḥammad al-Akhbārī devised intricate and detailed arguments with which to defeat the Uṣūliyyūn. Their arguments show an attention to detail which rivals that of the *isnād* critics amongst the Uṣūlīs. As with the internal debates over the legal interpretation of the Qur'ān, Akhbārīs vacillated between rejecting and accepting the intellectual sciences which had dominated Shī'ī learning for many centuries. When they did incorporate the terminology (and even the conclusions) of these disciplines, they did so in order to win converts from within the Shī'ī scholarly elite.

CHAPTER NINE

AKHBĀRĪ HERMENEUTICS

Whatever their assessment of the potential of the Qur'ān to act as an independent legal source, all Akhbārīs considered the sayings and actions of the Prophet and Imams (*akhbār*) to be more plentiful sources of legal indicators (*al-adilla al-shar'iyya*). Indeed, the school owes its name to the doctrine that the *akhbār*, as scripture, provide sufficient legal guidance for the Shī'ī community. The need for additional legal indicators, from the Qur'ān or elsewhere, was championed by the Uṣūlīs, but considered obsolete by the Akhbārīs. Reason (*al-'aql*) was rejected as a source, along with consensus (*ijmā'*). Furthermore, for most Akhbārīs, the Qur'ān's meaning is only available through the *akhbār*. This rule applied to individual verses (exemplified by the Akhbārī method of *tafsīr*), and more generally—that is, the significance of God's revelation in the Qur'ān could only be understood through reference to the *akhbār* of the Imams. As we have seen, this method did not always produce an unequivocal reading of a Qur'anic verse, though it did produce a range of possible meanings, each sanctioned by the Imams themselves.[1] No other possible source of knowledge could aid the exegetical process. Some Akhbārīs allowed direct interpretation of the Qur'ān, but the restrictions on the occasions when direct interpretation was possible were such as to minimise its utility as a source of law.

Underlying these doctrines was a presupposition that the meaning of *akhbār* is readily available, whilst the meaning of the Qur'ān (either in part or whole) is not. In other words, the coherence of the Akhbārī doctrine concerning the dependency of Qur'anic meaning upon the exposition found in the *akhbār* itself required a theory (however rudimentary) of how the language of the *akhbār* conveys meaning in a simple and uncomplicated manner to the reader/listener. Uṣūlī writers were well aware of the general need for a theory of language in a text-based jurisprudence such as *uṣūl al-fiqh*. Invariably works of

[1] See above, pp. 242–243.

uṣūl al-fiqh in both the Shīʿī and other Muslim traditions begin with an examination of the history and operation of language. It is, then, a little surprising that discussions of linguistic issues are not better represented in Akhbārī works of legal theory. It is possible that most Akhbārīs considered language to be a simple and transparent vehicle through which meaning is conveyed to the reader/listener, and therefore no explanation of the workings of language was required. However, even if the *akhbār*, the main sources of knowledge of God's law, are easily understood, other hermeneutic problems emerge. The *akhbār* may appear to contradict each other, and so consequently a procedure must be devised to solve such apparent contradictions, resulting in a clear ruling which the community can obey. The *akhbār* may appear to be silent on a particular issue, and there is also a need for an appropriate procedure on these occasions. These second level problems—that is, issues which emerge once a framework for language's operation has been established—are extensively explored in Akhbārī hermeneutic discussions. The Uṣūlīs, of course, had laid out procedures for such eventualities, subsumed under the general rubric of *ijtihād*. The Akhbārīs considered *ijtihād* illegitimate, not only because *ijtihādīs* questioned the authenticity of the *akhbār* (through *isnād* criticism and the reduction of *khabar al-wāḥid* to *ẓannī* status), but also because they made inappropriate leaps from the texts to the putative rulings they derived from the texts. This chapter examines the Akhbārī replacement hermeneutic for *ijtihād* by examining these three issues—the operation of language, the correct procedure when the texts are silent and the proper means of reconciling contradictory *akhbār*. Since the authority of the scholarly class (*ʿulamāʾ*) flowed, in part, from the Uṣūlī theory of *ijtihād*, Akhbārī views on the authority of the interpreter of the law are summarised in the conclusions. Astarābādī's position on these matters has already been outlined,[2] and subsequent Akhbārī thought is both a development and modification of his views.

Hermeneutic Theory and Exegetic Practice

The process of discerning the intended meaning of utterances (recorded in texts) set out in works of *uṣūl al-fiqh* was described through a series

[2] See above, pp. 98–99.

of binary classifications. An utterance was either univocal (*naṣṣ*) or ambiguous (*mujmal*). It might be a general statement (*ʿāmm*), which other texts might particularise (*takhṣīṣ*). It may appear unconditional (or unrestricted, *muṭlaq*), but be subject to a restriction (*taqyīd*) by other texts. In works of *uṣūl al-fiqh*, these binary categories are explored (supplemented by others including *ẓāhir/bāṭin, ḥaqīqa/majāz, muḥkam/mutashābih*), and through the classification of a verse in relation to these binary categories, the intended meaning of a statement supposedly emerges. In relation to the interpretation of the Qurʾān, the partial Akhbārī acceptance of the terminology of the *uṣūl al-fiqh* discipline has already been demonstrated.[3] The variety of Akhbārī positions on the question of the independent availability of Qurʾanic meaning was expressed through this established terminology, and terms were often used in a manner consonant with their meanings in the mainstream of *uṣūl al-fiqh*. In short, though Akhbārī views were distinctive, the manner in which they were usually expressed was within established disciplinary parameters. Whilst many Akhbārīs may have regarded *uṣūl al-fiqh* as a futile discipline, they were, nonetheless, forced to adopt some of its descriptive tools. To Akhbārīs, these tools were a natural part of any juristic discussions which hoped for intellectual respectability.

This feature of Akhbārī exegesis can be further exemplified through an example of exegetical practice, randomly selected. In Muslim purity law, certain substances impurify any water with which they have contact, making the water unuseable for ritual ablutions. Amongst these impurifying substances is (it was thought) the animal cadaver. When asked whether standing water in which lies an animal cadaver could be used for ritual purification purposes, the Imam answered:

> Do your ritual ablution on the other side, and do not do your ritual ablutions by the side of the cadaver.[4]

Majlisī I, in his *Lawāmiʿ ṣāḥibqirānī*, quotes this report and adds his commentary, consisting of a translation of the verse into Persian and a comparison with a similar report in al-Kulaynī's *al-Kāfī*. In this other report, a more specific question is asked: is the cleansing of genitalia (*istinjāʾ*) with water which has come into contact with an

---

[3] See above, p. 244 and pp. 261–262.
[4] Ibn Bābūya, *Man Lā*, v. 1, p. 16; Ṭūsī, *al-Istibṣār*, v. 1, p. 21.

animal cadaver permitted? The Imam's answer is the same. From this, Majlisī I deduces that in the first report, the Imam's phrase "Do your ritual ablution..." (*tawaḍḍa'*) refers to ablution generally, not the specific ritual for the elimination of a minor purity infraction (namely, the *wuḍū'*). *Istinjā'* is not included in *wuḍū'*, but is part of the more extensive ritual washing called *ghusl*. Hence, despite the etymological relationship between *wuḍū'* and *tawaḍḍa'*, the latter should be taken as having a linguistic, unrestricted meaning, rather than a (legally) technical and specific one (*wuḍū-rā bi-maʿnā-yi lughawī iṭlāq nimūdah-and*).[5] In another commentary on this report, found in his Arabic *Rawḍat al-Muttaqīn*, Majlisī I quotes yet another report, this time taken from al-Shaykh al-Ṭūsī's *al-Tahdhīb*. Here a similar question is asked of the Imam, but instead of referring to "standing water" (*al-mā' al-sākin*), the word for "water" (*al-mā'*) is left unqualified. The Imam's answer is, once again, the same. Since the word "water" is of general reference (*ʿumūm*), the rule outlined in the first report applies to both running and standing water.

Now that "water" is discovered to be of general reference, the obvious meaning of the report (*ẓāhir al-khabar*) is that any amount of water (even a small amount) which comes into contact with a cadaver is not rendered impure (and hence unusable for ritual purification purposes). The Imam's answer allows ritual washing to take place, whether the water is standing or running, whether it is a small amount or a large amount (though he does add the caveat that one should not wash in the water right next to the cadaver). This, Majlisī I argues, is in accordance with the views of the Akhbārīs who say that unequivocal texts (such as this one) must form the basis of a believer's correct action.[6] Citing yet more reports with the same apparent meaning, Majlisī I states that the apparent meaning of the texts (*ẓāhir*) is that the impurifying effects of a cadaver do not apply to water, be it in small or large amounts, be it standing or running. To argue otherwise is to propose that the *akhbār* which indicate that a small amount of water can be impurified by contact with other impure substances (blood, urine etc.) should be conjoined (*jamʿ*) with those that indicate that the phrase "standing water" (*al-mā'*

[5] Majlisī I, *Lawāmiʿ*, v. 1, p. 248.

[6] [*ẓāhir*] *al-akhbāriyyīn fī'l-ʿamal bi'l-naṣṣ*. See Majlisī I, *Rawḍat*, v. 1, p. 77. See also, the editor's gloss (n. 3 on the same page).

al-sākin) refers to a volume of water sufficient to dilute the presence of an impurifying substance (*kurr*). Majlisī I's position appears to be that the reports unambiguously indicate that the mere presence of a cadaver in a body of water is not, in itself, sufficient to cause the water becoming impure because water can only be impurified by substances which mix with it, either because they are themselves liquid (blood or urine, for example) or because they break up and mix with it on contact (faeces, or a rotting or punctured cadaver). An intact cadaver falls into neither category, and therefore, the water remains pure. However, ablution (be it *istinjā'* or *wuḍū'*) should (that is, it is recommended that it) take place on the other side of the water from the cadaver. Exactly why one should use water distant from the cadaver is not clear (to me, at least). If the purity status of the water is unaffected by contact with the cadaver, then why does the Imam stipulate that one should wash on the other side of the body of water (or in some reports, on the other side of the stream)? Is this a case of an inscrutable ruling from the Imam, the reason for which cannot be discerned, and the believer's task is simply to obey it?

Al-Jazā'irī's discussion of this report takes up this problem, through a discussion of whether the order (*amr*) to wash on the other side of the water from the cadaver indicates an obligation (*wujūb*) or a mere recommendation (*istiḥbāb*).[7] He does not indicate which he considers to be correct, but his brief discussion does outline arguments for both positions. If washing on the other side of the river/pond is recommended, it is because of a general rule that the water for *wuḍū'* should be free of repugnant substances (*khāliṣ*[an] *min al-akhbatha*), even if those repugnant substances do not have a ritually impurifying effect. If washing on the other side of the river/pond is obligatory, then it must be because only the water with which the cadaver actually comes into contact (that is, the area around it) has changed.[8] Now the imperative form (*amr*) to wash on the other side

[7] Jazā'irī, *Kashf*, v. 2, pp. 194–195.

[8] It is of interest here that *isnād* criticism plays no part in the discussion. Whilst al-Jazā'irī introduces his commentary with a discussion of the *isnād* (pronouncing it "weak", *ḍa'īf*), this has no effect on which of the two possible interpretations (*istiḥbāb* or *wujūb*) is taken as authoritative. Uṣūlīs would, of course, be tempted to reduce the assessment of such an order (to recommendation only) on the basis of their doubts concerning the authenticity of the report. Unsurprisingly, this Uṣūlī line of reasoning does not appear in al-Jazā'irī's Akhbārī-style discussion (see Gleave, *Inevitable Doubt*, pp. 40–41).

of the river/pond from the cadaver is a grammatical form which was the subject of much debate within works of *uṣūl al-fiqh*. Most legal theorists agreed that an imperative, when used by God (or the Prophet, or for Shīʿīs, the Imams), could mean that the action ordered fell into one of three categories of legal assessment. These were obligatory (that is, unavoidable if one wished to fulfil the requirements of the law), recommended (that is, an optional action which could be performed out of piety) or permitted (an action which had no legal effect). Which of these was the intended meaning of the speaker when a particular imperative was uttered could only be discovered after a consideration of the contextual evidence (*qarāʾin*) external to the imperative utterance itself. The question which was debated most vigorously was the meaning of an imperative in the absence of contextual information (or rather, in the absence of external evidence sufficiently convincing to place the ordered action in one of the three categories). Al-Jazāʾirī lays out the potential external evidence in this case. The imperative could, depending on one's perspective, establish an obligation or a recommendation, though al-Jazāʾirī does not indicate which evidence is, in his opinion, the stronger. Al-Baḥrānī, in his *al-Ḥādāʾiq al-Nāḍira*, mentions this *ḥadīth* in the course of an extended argument about the manner in which water changes when it comes into contact with an impure substance.[9] The debate concerns contradictory reports, some of which indicate that water becomes immediately impure and others which indicate it does not. The details need not concern us here, but, al-Baḥrānī argues, firstly at least one set of contradictory *akhbār* must be interpreted as issued under *taqiyya*: that is, the reports may be historically accurate but they are not legally effective. Secondly, he argues that the requirement to wash on the other side comes out of caution. This, building on al-Baḥrānī's legal theory, is the view that when one is not certain of a ruling in a particular situation, then one should act cautiously. Hence, the argument would presumably progress as follows: one is not certain whether the cadaver is entirely intact. Consequently, one is not certain whether or not blood or some other substance has dissipated into the water from the cadaver. Hence the Imams' ruling to wash on the other side of the pond/stream is an act of caution in which contact with impure water is minimised by distancing oneself

[9] Baḥrānī, *al-Ḥādāʾiq*, v. 1, pp. 476–477.

from the cadaver. The water around the cadaver is, on this reasoning, most likely to have been impurified.[10] The arguments of Majlisī I, al-Jazā'irī and al-Baḥrānī all display Akhbārī hermeneutic concerns. For example, these Akhbārīs argue that a word of general reference in one *ḥadīth* ("water") can modify a ruling in another *ḥadīth* where the same word is used, but is qualified by an adjective ("standing water"). The use of "standing" in the question in the first of the above *ḥadīths*, for example, does not restrict the implications of the Imams' answer only to standing water. One cannot, the Akhbārīs argue, deduce anything about running water from the fact that the water described in the question is qualified with the adjective "standing". To do so, would be to employ the exegetical device known in the works of *uṣūl al-fiqh* as *mafhūm al-ṣifa* (an implication drawn from a quality).[11] Majlisī I, al-Jazā'irī and al-Baḥrānī are all resistant to the idea that the adjective "standing" places a restriction on the applicability of the ruling concerning the purity of running water which has come into contact with an animal cadaver. This resistance is an indication of their opposition (as Akhbārīs) to the uncontrolled use of the hermeneutic device known as *mafhūm al-ṣifa*. Other *akhbār* indicate that the rule concerning the animal cadaver applies equally to running water, immediately demonstrating that nothing about running water can be deduced from this report. In short, when the Imam gives a report, it is not permitted for the exegete to use presumptive hermeneutic devices and, thereby, extend the relevance of the report to unmentioned areas of the law. This indicates a careful application of Akhbārī hermeneutics to exegetical practice.

Another example of the encroachment of Akhbārī hermeneutic views on the interpretation of this particular report concerns whether washing on the other side of the pond/stream is obligatory or recommended. The debate displays a concern for the "given" meaning of linguistic

[10] Al-Baḥrānī does not spell out his reasoning here, but this would appear to be implied in his analysis of the contradictory *akhbār*, combined with his legal theory generally. See Baḥrānī, *al-Ḥādā'iq*, v. 1, pp. 476–479. The reasoning here is similar to that of the famous case of the mouse which falls in a vat of fat and dies. The fat around the mouse is removed, but the rest is useable. See Baḥrānī, *al-Ḥādā'iq*, v. 1, p. 293.

[11] For example, the inclusion of the word "grazing" in the Prophetic order "Pay *zakāt* on grazing goats" is taken to mean that one should not pay *zakāt* on non-grazing goats. See Zysow, "Economy of Certainty", pp. 164–174 and Weiss, *God's Law*, pp. 490–501. On other *mafhūmāt* in Akhbārī thought, see Akhbārī, *Ma'āwil*, f.163a (in particular his analysis of *mafhūm al-sharṭ*).

forms (in technical language, *ḥaqīqa ṣīghat al-amr*). Furthermore, the assumption that the cadaver is neither punctured nor rotting (and therefore not impurifying the water) is justified solely by the Imams' words, and not through any process of legal reasoning (the debate relates to the technical *uṣūl* concepts of *istiṣḥāb al-ḥāl* and *al-barā'a al-aṣliyya*). The command to wash on the other side of the water is interpreted as cautionary (*iḥtiyāṭ*)—not because one does not know the law, but because one is uncertain of the facts (has the cadaver been punctured or decayed in such a way that the water around it has been impurified?). Finally, the argument that some contradictory reports are *taqiyya* (that is, historically accurate but legally irrelevant) presented by al-Baḥrānī concerns the method whereby contradictory opinions are to be reconciled.

The opinions reached through Akhbārī legal reasoning may not be different from those reached by their Uṣūlī opponents: that is, the same legal ruling may be reached by following either Akhbārī hermeneutic principles or Uṣūlī reasoning. However, what is distinctive here is the manner in which the conclusions are demonstrated, and the manner in which a distinctive Akhbārī legal methodology is employed. These include the authenticity of the *akhbār*, the disavowal of speculatively combining reports, the suspicion with which linguistic implications (*mafāhīm*) are treated and the underlying conviction that the modes of expression found within the *akhbār* cannot be considered ambiguous. There is, then, a close link between hermeneutic theory and exegetical practice within Akhbārī texts, and exegetic practice can, at times, illuminate Akhbārī legal theory in areas not covered by the usual source of a scholar's jurisprudence (namely, works of *uṣūl al-fiqh*, or at least discussions of *uṣūl*-related issues). Whilst the terminology and general presentation of the legal problems in Akhbārī texts owes much to the established Uṣūlī tradition of scholarship, the methodology, and (at times) the legal conclusions are distinctively Akhbārī.

Akhbārī Conceptions of Language

The Akhbārīs, it seems, worked under similar linguistic presumptions to those employed by their Uṣūlī opponents. They did not present (or invent) a novel theory of how language conveys meaning. It is perhaps for this reason, that there are few Akhbārī discussions of how language works (that is, how it conveys the speaker's meaning to the listener in an effective manner).[12] Most Akhbārīs had little to add to existing paradigms of language's operation.

One exception to this is al-Karakī's *Hidāyat al-abrār*, confirming my earlier comment that this work represents the most thorough explication of Astarābādī's Akhbārī views to all areas of *uṣūl al-fiqh*.[13] In the seventh chapter (*bāb*) of the *Hidāyat* (entitled "speech according to the science of jurisprudence"), al-Karakī outlines his view of how language operates.[14] The discussion is wide-ranging and technical, covering the relationship between utterance (*lafẓ*) and meaning (*maʿnā*), homonymy, synonymy, etymology and implication. On the question of language's origins, al-Karakī explores a number of possible "inventers" of language (God, human practice, a mixture of God and human practice etcetera). His own view is that language, as it exists today, came about through a mixture of God associating meanings to sounds, and human beings supplementing this basic structure with additional associations of meaning with words. He describes this position as *madhhab al-tawzīgh* ("allotment"—that is, God and humanity are "allotted" certain roles in the development of language). He claims it to be the chosen opinion of most scholars (*akhtarahu al-akthar*).

---

[12] One exception to this is, possibly, al-Jazāʾirī's son, Nūr al-Dīn al-Jazāʾirī, who though not renowned as an Akhbārī, may well have been one. Both his father and son were Akhbārīs. He composed a work against al-Shaykh al-Bahāʾī's *al-Qūswiyya*, a work on riddles (entitled *al-Sayfiyya fīʾl-Laghz*), a commentary on Fayḍ's work of *furūʿ al-fiqh, al-Nukhba* and an interesting linguistic work, *Furūq al-lughāt fī tamyīz bayn mufād al-kalimāt*. This latter has survived and been published. The work is designed as a handbook to distinguish between "almost" synonyms, and has some specifically Shīʿī, though not necessarily Akhbārī, features. So, for example, the difference between *Islām* and *Īmān* is the difference between the general (all Muslims, Islām) and the Shīʿa (*al-muʾminīn*) (Nūr al-Dīn al-Jazāʾirī, *Furūq al-Lughāt*, p. 30). There is little evidence of any Akhbārī leaning in the work, though there is an emphasis on *shawāhid* from the *akhbār* of the Imams. The difference between *ẓann, shakk* and *wahm* is not used as an opportunity to vilify *ẓann*, for example (p. 152). There is a perhaps unusual concentration on the differences between *ʿilm, yaqīn* and *fahm* (p. 172, p. 175 and p. 176), but there is no mention of the standard Akhbārī category of *al-ʿilm al-ʿādī*. There is, however, a link with the question of food and clothing not used at the time of the Prophet or Imams, and whether it is permitted to wear them (see below, p. 200). They are, he claims, *mubāḥ* since *mubāḥ* means "that which has no *naṣṣ* indicating it to be forbidden, either in terms of a general or specific ruling" (pp. 111–112). This distinguishes *mubāḥ* from *ḥalāl*, which is described as "that which has a *naṣṣ* from the Lawgiver indicating it is permitted." Such a distinction might indicate an acceptance of *ibāḥāt al-aṣl* (though perhaps not *al-barāʾa al-aṣliyya*), which would run counter to general Akhbārī doctrine.

[13] See above, pp. 166–167.

[14] Karakī, *Hidāyat*, pp. 232–306.

However, the question of language's origins does not seem to affect al-Karakī's general theory of language. He accepts that words have meanings "placed" (*wuḍiʿa*) upon them, though the precise process of "placing" (was it by God or human beings?) is not a matter of great concern. Utterances (that is, sounds, and when written, words on a page) "indicate" (*dālla*) or point to meanings. The utterance can indicate a single meaning even though it may be used to describe a number of items. For example, the word "white" is used to describe a number of things with the general quality of whiteness. This is the case even though the "white" items may differ in terms of intensity (snow is whiter than ivory, but both are called white). The same quality is being ascribed to the various items through the same process of indication. It is not that the whiteness of snow is true whiteness, and that of ivory some sort of metaphorical or diverted use of the word (*majāz*). Rather, al-Karakī argues, when one describes both snow and ivory as white one is using an utterance which covers a range of non-identical but similar qualities.[15] This distinguishes these uses of "white" from the purely metaphorical (such as the use of white to indicate moral purity or virginity). On the other hand, the use of the term white is not without limits. A thing which can be truthfully described as white, whatever the intensity of its whiteness, cannot simultaneously be truthfully described as black. The use of the word white distinguishes (*tabāyun*) the described item from items which can be described as black. The point here is that sounds (that is, the sounds of words) have been assigned a relation to particular phenomena, and this assignment excludes a sound's use to describe other phenomena. For this reason the utterance (that is, the sound) and meaning are "unified" (*ittiḥād*). This is not to say that words cannot be employed in ways which deviate from this "unified" manner. However, when this happens, the word is not being used in its designated manner. The speaker is borrowing (*istiʿāra*) the word, and making it indicate something other than its meaning.

Al-Karakī's theory is complicated by the fact that certain meanings have two words attached to them. *Layth* and *asad* both mean lion, and al-Karakī treats this as a case of synonymy (*mutarādif*). Similarly, a single word may have more than one designated meaning, and there

[15] The items "come together in the quality" (*tawāfuq al-afrād fīhi*). Karakī, *Hidāyat*, p. 242.

is homonymy (*ishtirāk*). The existence of such elements in language means that in most cases a single utterance can be understood in a large number of ways. When the word "white" is used, it may mean the white of ivory or the white of snow. It may be that the word "white" is a homonym and some quality other than a colour is meant. Since this is not a borrowed or metaphorical meaning, it may be that no additional indication is deemed necessary by the speaker. Furthermore, it may be that white is meant in some borrowed manner. In short, the possibilities for alternative interpretations of an utterance are considerable, and this, al-Karakī states, is why there are very few occasions when a Qur'anic verse can be considered *muḥkam*—that is, allowing one and only one interpretation. All other occasions are *mutashābih*, and in need of explication (*bayān*), "and one refers to the guiding Imams for their explication."

The implication in all this is that the Imams' words do not require explication. They are never ambiguous, and this was certainly the position of some Akhbārīs.[16] However, this position is complicated by the Imams' own admittance that there are *mutashābih* phrases in the *akhbār*:

> In our *akhbār* there is *mutashābih*, like the *mutashābih* of the Qur'ān. So refer the [*akhbār*'s] *mutashābih* to the [*akhbār*'s] *muḥkam*. Do not follow the *mutashābih* without the *muḥkam*, lest you might err.[17]

The report itself is found only in Ibn Bābūya's *ʿUyūn akhbār al-Riḍā* (that is, not a canonical collection of *ḥadīth*), and receives little attention from Akhbārīs, presumably because it is considered awkward for the general thrust of Akhbārī theory.[18] The only explicit comment on the *ḥadīth* by an Akhbārī that I have found is by al-Ḥurr.[19] For

[16] See Astarābādī, *al-Fawā'id*, p. 179; al-Baḥrānī, *al-Ḥādā'iq*, v. 1, p. 157 (see also, Gleave, *Inevitable Doubt*, pp. 157–158).

[17] Ibn Bābūya, *ʿUyūn akhbār al-Riḍā*, v. 2, p. 271. Furthermore, Akhbārīs writing commentaries on the *akhbār* collections, clearly felt the need to explicate the meaning of reports for their readers. The report is also found in Ṭabarsī, *al-Iḥtijāj*, v. 2, p. 192 where it is presented as referring to the *ṣifāt Allāh* (God's attributes), and defectively in Majlisī II, *Biḥār*, v. 2, p. 185, #8 (#9 on the same page is an accurate quotation from the *ʿUyūn akhbār al-Riḍā*).

[18] It is cited by al-Ḥurr (*Wasā'il*, v. 27, p. 115, #33355), Muḥsin Fayḍ (*al-Ṣāfī*, v. 1, p. 321) and by al-Huwayzi (*Nūr al-Thaqalayn*, v. 1, p. 219), but with no comment.

[19] Ḥurr, *al-Fuṣūl al-Muhimma*, v. 1, p. 573. Al-Ḥurr's understanding of the report is indicated not by a direct statement, but by the chapter heading under which the report is listed: *wujūb radd al-mutashābih min al-aḥādīth ilā al-ḥukm* [probably

him, the report refers to the particularisation (*takhṣīṣ*) or restriction (*taqyīd*) of a ruling in one report by another when they appear to contradict. It is not that the Imams' words are in themselves ambiguous, merely that the inter-relationship between *akhbār* can bring about a modification of a report's unambiguous meaning. For example, the general ruling that blood, semen or a rotting cadaver impurify water is subject to an exception (*khāṣṣ*). When the volume of water reaches a particular amount (termed *kurr*), the impurifying effects of the *najis* substance are nullified. Furthermore, the Imam does not recommend referring the Qur'ān's *mutashābihāt* to the Qur'ān's own *muḥkamāt* in order to reach a direct understanding of the Qur'anic meaning. Only the Imams (*makhṣūṣ bi'l-a'imma*) could perform such a process and only because the Imams truly understand the Qur'ān. The Shīʿa are merely advised to refer the *mutashābihāt* of the *akhbār* to the *akhbār*'s own *muḥkamāt*, and this process is understood, not as the clarification of ambiguity, but rather the particularisation and restriction of one report by another. The rather inconvenient report which implies that on occasions the Imams' intentionally spoke in an ambiguous manner is thereby reinterpreted to refer not to ambiguity within the *akhbār* (as it is when the term *mutashābih* is used to describe Qur'anic verses). Instead, it refers to general (*ʿumūm*) and unrestricted (*muṭlaq*) statements in the *akhbār* being particularised (*takhṣīṣ*) and restricted (*taqyīd*) by other statements within the *akhbār*. In the case described above, the general application of the ruling in one report (which appeared restricted to standing water by the questioner's use of the phrase "standing water") was deduced through a comparison with other similar reports where the word is unqualified. Such a hermeneutic process exemplified how, for Akhbārīs, the *akhbār* corpus must be taken as a whole body of evidence, and individual reports cannot be isolated from the general law as it was revealed by the Imams.

For Akhbārīs then, it is quite possible for language to be used in such a way that ambiguity is avoided. This, it would seem, is what the Imams themselves did, and their "clear speech" is recorded in the *akhbār* themselves. The Qur'ān, though, remains irretrievably ambiguous (either in part or *in toto*) and its ambiguity is only to be

should be read as *al-muḥkam*] *bi-an yuḥmala al-ʿāmm ʿalā al-khāṣṣ wa'l-muṭlaq ʿalā al-muqayyad maʿa al-taʿāruḍ wa'l-tanāfī khāṣṣat*[an].

dispelled, and its meaning fully understood, through reading the *akhbār*. This dogmatic attachment to the straightforward use of language as a vehicle for conveying meaning may explain why so few Akhbārīs were interested in matters of linguistic and stylistic expression. The doctrine, however, seems not to fit with the plethora of Akhbārī commentaries on the early *ḥadīth* collections. Astarābādī himself authored commentaries (many of them consisting of slight asides or *hawāshī*) on the Four Books. Similarly Majlisī I, Khalīl al-Qazwīnī and al-Jazā'irī all wrote commentaries on the *akhbār* collections. The method of exegesis employed in these commentaries has already been analysed,[20] and it is clear that, ironically, the "unambiguous" expression of the Imams required extensive explication. Whilst theoretically the *akhbār* were clear to any who read them, in practice Akhbārīs recognised a need to retrieve and present the Imams' intended meaning to their readership.

Evidence for Diverted Meaning

As argued above, Akhbārīs appear committed to a similar theory of language's operation to that of their opponents. Utterances (be they individual words or grammatical constructions) have meanings which somehow inhere in them, such that when they are used by a speaker, the presumed intended meaning is this inherent meaning. The intended meaning may not accord exactly with the inherent meaning of a report, but when the former deviates from the latter, evidence (*qarā'in*) is required for the listener to make the shift. This understanding of language dominated medieval Muslim legal and literary theory, and there were few, if any, challenges to its basic assumptions. Scholars did, however, disagree over what counts as evidence of diversion (*majāz*). What was needed in order for the listener (or exegete) to believe that the "intended" meaning was other than the given meaning? It was not the paradigm of language use, but its application which was debated.

Al-Karakī examines the various opinions of past legal theorists concerning linguistic issues such as the given meaning of the imperative (*ṣīghat al-amr* or the negative imperative, *al-nahy*), and concludes

[20] See above, pp. 37–38, p. 145 and p. 165.

that their various arguments are pointless. He asks rhetorically, "In matters relating to the derivation of rulings, how can the reliance on someone other than the Imams be sound?"[21] The Imams have indicated that anyone with a natural, untainted understanding of language (*dhū al-tabʿ al-sālim*) should refer to his own instinct. He will find that the imperative form, when it is used without any evidence for a diversion, can only mean one thing: the thing ordered by the imperative is being demanded by the speaker such that non-performance is condemned, and in law this means that non-performance entails punishment (*istiḥqāq al-ʿiqāb*). These, al-Karakī states, are all signs of the obligatory nature of the commanded action. Any downgrade to recommendation (or indeed permission) requires evidence. Of interest here is al-Karakī's reliance on a natural understanding of language—or rather, his view that the Imams demand that the believer rely on his natural understanding of language. This, he argues, is sufficient to prove the point. "Whoever wishes to prove this point by debating and arguing (*biʾl-jadal waʾl-baḥth*) may do so" if he wishes, though al-Karakī does not feel the need to involve himself in such discussions. The impression given is that a substantial amount of evidence is required to shift al-Karakī from a presumption that an imperative indicates an obligation.

Al-Karakī's position here is much less circumspect than that expressed by Astarābādī in his *al-Fawāʾid*. Astarābādī had argued that it is possible for the imperative to mean either obligation or recommendation: that is, the evidence for one does not overrule the other. In such circumstances, he states, the believer acts as if the ordered action is obligatory out of caution. The believer is uncertain of the actual status of the assessment, and so he acts cautiously. Caution dictates that the believer do everything within his power to ensure fulfilment of the law of God. This means treating the ordered action as obligatory.[22] The difference between Astarābādī and al-Karakī here is subtle, but important. For Astarābādī, when an ordered action could be recommended or obligatory, caution dictates that we take it as the latter. For al-Karakī, when an ordered action could be recommended or obligatory, the action is taken to be obligatory because the grammatical form of an order has the given meaning of obligation.

---

[21] Karakī, *Hidāyat*, p. 275.
[22] See above, pp. 80–82.

"Given" meanings do not require evidence to be taken as the intended meaning of an utterance; diverted meanings do. When the evidence is less than conclusive, the given meaning always predominates. This, al-Karakī claims, is a hermeneutic principle sanction by the Imams and is not based on caution.

Two different Akhbārī positions emerge here. On the one hand, Astarābādī seems to think that the given meaning of words can be easily made redundant, and replaced with a diverted meaning. The *qarā'in* necessary to effect a diversion (or, more accurately, to cause suspicion of a diversion) need not be particularly strong. On the other hand, al-Karakī seems to approach potential ambiguity with the assumption that the given meaning will always be preferred in cases of ambiguity. For Astarābādī it is caution which dictates the preference for obligation. For al-Karakī, it is the primacy of the given meaning which demands it take preference over any other possible meaning, and in this case the given meaning of the imperative form is obligation. The difference between the two thinkers rests on different views as to the ease with which given meanings can be dislodged, which in turn rests on different conceptions of the fixity of language. The shortage of theoretical discussions of language within the Akhbārī corpus makes tracing the history of these alternative views of language through the school's history difficult. However, the clearest exposition of later Akhbārī jurisprudence (namely al-Baḥrānī's introduction to his *al-Ḥādā'iq al-Nāḍira*) indicates that al-Karakī's position was more influential. For al-Baḥrānī, as for al-Karakī, words and grammatical constructions have clear given meanings. These can only be dislodged by strong evidence (*qarā'in*) that diversion has occurred. For later Akhbārīs, it would seem, it is not through caution that the imperative form (*ṣīghat al-amr*) is taken to mean obligation. Rather, it is because obligation was its given meaning and any evidence to dislodge it needs to jump a high bar before casting doubt on its presumed intended meaning. Of course, the demand for a high bar of evidence is, itself, a reflection of a cautionary approach, but that is caution operating on quite a different level, and constitutes a higher order of legal enquiry.

Akhbārī Hermeneutic Maxims

Akhbārīs could not write legal theory without reflecting upon the nature of the Imāmī revelatory corpus. They had given the *akhbār* a central role in the exegesis of the law, and argued vehemently for their reliability, both as historical records and as indicators of legal injunctions. However, the *akhbār* do not provide specific rulings on all issues and the rulings that they do provide are at times contradictory and (apparently) irreconcilable. The Uṣūlīs were also cognisant of these problems, and presented solutions which involved: firstly, downgrading the authenticity (and hence probative force) of the *akhbār* (in order to facilitate preference between conflicting reports); secondly, having a more relaxed attitude towards the meaning of the Imams' words and grammatical constructions (such as reading the imperative form as meaning recommendation, permission or merely "a request for action"); thirdly, producing a set of maxims (normally derived from reason or rational reflection on the nature of the juristic effort) whereby a ruling might be deduced in cases where revelation is (allegedly) silent, or a procedure whereby one possible ruling or indicator can be "preferred" over another (*tarjīḥ*). The Akhbārīs, of course, rejected the first proposition, as they considered the history of the collection of reports to be a clear indication of the *akhbār's* accuracy as a historical record of the Imams' words. The sketchy Akhbārī theories of language's operation constitute their response to the second proposition. With regard to the third Uṣūlī position, the Akhbārīs presented their own alternative set of maxims, derived, not from reason, but from revelation. Hermeneutic maxims, termed *qawāʿid* or *uṣūl*, formed a part of *uṣūl* literature well before the emergence of the Akhbāriyya. Sunni writers had collated maxims which served two purposes. Firstly, they aimed to explain why the indicators (from the Qur'ān and the Sunna) took the form they did by an appeal to underlying principles upon which the Sharīʿa was supposed to be based. Secondly, they provided principles whereby the jurist provides putative rulings in areas of the law not covered by the texts.[23] Within the Shīʿī *uṣūl* tradition, works cataloguing these principles began to appear in the Eighth *Hijrī* Century with Ḥasan al-Ḥillī's (d. ca. 740/1340) *ʿIqd al-jawāhir*.[24] Akhbārī scholars drew on this scholarship, but objected to the rational manner in which some principles were justified. For Akhbārīs, principles must be justified by reference to the Imams. The Imams sanctioned the use of certain interpretive principles which the scholar can use when encountering

[23] Heinrichs, "*Qawāʿid* as a Genre".
[24] See Stewart, *Islamic Legal Orthodoxy*, pp. 16–17.

a hermeneutic problem (revelatory silence, obscurity or contradiction). The Imams, then, not only provided legal rulings. They also provided a jurisprudence whereby legal rulings might be determined in problematic cases, and made the use of these principles obligatory for the Shīʿa in cases of legal uncertainty. The scholar, when applying these principles to the novel case, is certain that he is obeying the Imams' command. This is not because he is certain that the ruling at which he arrives is identical with the Imams' opinion, but rather because he has followed the procedures laid down by the Imam. "God has a ruling for every situation" was a common Akhbārī slogan, and by this the Akhbārīs meant not only that there is a ruling in reality (*fīʾl-wāqiʿ*) for every situation (that is, the Imam had total knowledge of the law, and hence knew God's opinion on every possible human circumstance). They meant also that God had a ruling for those occasions when the community (excluding the Imams) is uncertain as to what God's ruling might be. His ruling in these cases is to follow the hermeneutic maxims revealed by the Imams. In a number of reports, the Imams state that they have supplied the *uṣūl* ("roots") and it is the Shīʿa's task to work out the *furūʿ* ("branches").[25] Uṣūlīs interpreted these as a general invitation to carry out *ijtihād* in their explication of the law, seeing the disjunction of *uṣūl/furūʿ* as referring to legal rules (*uṣūl*) and their application to specific circumstances (*furūʿ*). Akhbārīs, however, generally interpreted the term *uṣūl* in the reports as referring to these hermeneutic maxims, and the maxims form part of the Akhbārī replacement for *ijtihād*. A late Akhbārī attempt to delineate these hermeneutic principles (*qawāʿid*) can be found in al-Baḥrānī's *al-Ḥadāʾiq al-Nāḍira*, and as I have indicated elsewhere, they constitute a rather disjointed collection.[26]

One principle which the Imams appeared to recommend to the Shīʿa concerned the assumption that things which are not known to be prohibited can be considered permitted. How this principle was to be understood was a matter of some debate amongst Akhbārīs. A similar principle was widely used by Uṣūlī jurists in cases of revelatory silence, and was termed *al-barāʾa al-aṣliyya* (or *aṣālat*

[25] See, for example, Ḥurr, *Wasāʾil*, v. 27, p. 61, #51 and 52. Both reports are to be found, not in the Four Book, but in Ibn Idrīs's *al-Sarāʾir*, cited from the *aṣls* of Hishām b. Sālim and Aḥmad b. Muḥammad b. Abī Naṣr respectively. Neither *aṣl* is extant to my knowledge.

[26] See Gleave, *Inevitable Doubt*, pp. 227–237.

al-barā'a—"fundamental freedom"). This principle, found also in Sunni jurisprudence, decreed an assumption that cases which were unmentioned in revelation were assumed to have no assessment. Of course, determining whether an action is unassessed is not, in itself, a simple matter. The Sunni jurists, who had in their exegetical armoury devices such as *qiyās* and *istiḥsān*, could systematically reduce the number of areas which fell into the unassessed category. Shīʿī jurists, at least at the level of polemic, rejected *qiyās* and *istiḥsān*. Hence there was a greater potential for areas of the law being free of an explicit assessment in the sources.[27] This risk of legal lacunae was overcome by activating alternative exegetical devices, but the limited nature of the texts could not simply be discarded by these mechanisms. The principle of *al-barā'a al-aṣliyya*, however it might have been formulated, became particularly important for Shīʿī jurists. The lack of an assessment means, according to this principle, the lack of an obligation. Hence, Uṣūlī jurists had argued that in such cases, *mujtahids* are justified in declaring (be it in a *fatwā* or in a judge's ruling) that the action in question is without legal consequences (*mubāḥ*) or at least not falling into the prohibited (*ḥarām*) or obligatory (*wājib*) categories. This was extended to purity law, where the presumption was that any substance not subject to an assessment in the sources of the law was to be declared pure (a principle termed *aṣl al-ṭahāra*). The Uṣūlīs generally accepted that such rulings were presumptuous (*ẓannī*). However, since they permitted *fatwās* on the basis of *ẓann*, the principle could operate in all areas of the law. Some Uṣūlī jurists extended the application of *al-barā'a al-aṣliyya* further. They also used it in cases where the texts were unclear or ambiguous. In such circumstances, they argued, the jurists could presume the action described in the source text (but with an unclear assessment) to be permitted. Furthermore, when two sources indicate conflicting rulings (one indicating obligation and the other permission), *al-barā'a* decrees a ruling of permission be given. This reasoning, unsurprisingly, troubled Akhbārī jurists. On this latter extension of *al-barā'a al-aṣliyya* to cases of revelatory ambiguity, Akhbārī scholars were united in their rejection. For them, as has just been outlined, the language found in the *akhbār* operates in such a way that the Imams' meanings are conveyed, smoothly and without misunderstanding,

[27] See Gleave, "Qiyās".

to the reader. This was based on a rather simplistic view of the communicative role of language, which though uncomplicated must have had extensive popular appeal.

Concerning the employment of *al-barā'a al-aṣliyya* in areas of revelatory silence, Astarābādī had declared that *al-barā'a al-aṣliyya* was an illegitimate exegetical principle, precisely because it was based on *ẓann*:

> The noble *ḥadīths* declare that for every situation which the community might encounter until the day of resurrection, there is a certain statement by God. Nothing remains in an "original" state of permission. So, adhering to *al-barā'a al-aṣliyya* does not allow [one] to deny that there is an assessment.[28]

This was generally the position taken up by subsequent Akhbārī writers, though with some modifications.[29] Al-Karakī, for example, repeats the slogan, *lillāh fī kull wāqiʿa ḥukm*[an] *muʿayyan*[an] (God has a specific assessment for every situation).[30] He argues, therefore, that to assume permission (*ibāḥa*) is an unacceptable presumption (*ẓann*). The scholar cannot give a *fatwā* on the basis of *al-barā'a al-aṣliyya*. Instead, he must suspend judgement in such matters (*tawaqquf*) and only recommend what he considers to be the "cautious" course of action (*iḥtiyāṭ*). The replacement of the assumption that an unassessed act was permitted with the principles of *tawaqquf* and *iḥtiyāṭ* was considered a hermeneutic principle revealed by the Imams themselves. Al-Karakī does not explore in detail how this cautious course of action might be known, though the principle of *iḥtiyāṭ* was well-known in Shīʿī jurisprudence. The believer should act in such a way as to maximise the potential fulfilment of the law (*ishtighāl al-dhimma*). In some cases, this might involve avoiding the performance of an action which has no explicit assessment. In other cases it might involve performing the action. Meals and clothing not known at the time of the Imams are one such case. Unfortunately, al-Karakī does not lay down a means whereby a scholar might decide in a particular case what the most cautious course of action might be.[31]

---

28 Astarābādī, *al-Fawā'id*, p. 216.

29 In addition to the discussions outlined below, see Ḥurr, *al-Fawā'id al-Ṭūsiyya*, pp. 196–215 (written as an atomised response to a passage from "a modern scholar"—al-Ṭihrānī speculates that the title of the response was *al-Shihāb al-Thāqib*, though he names no author).

30 Karakī, *Hidāyat*, p. 266.

31 Al-Karakī does give 12 examples of how *iḥtiyāṭ* functions with respect to

Akhbārī jurists were, however, faced with a more fundamental problem. Whilst *al-barā'a al-aṣliyya* had been justified on purely rational grounds by the Uṣūlīs, they had also put into service some reports from the Imams which seemed to justify using the principle of *al-barā'a al-aṣliyya* and *aṣālat al-ṭahāra*. These included:

> All things are unrestricted (*muṭlaq*) until a prohibition is referred to them.[32]
> All things are permitted (*ḥalāl*) until you know them to be forbidden specifically (*ḥarām bi-ʿaynihi*).[33]
> All things are clean until you know them to be dirty.[34]

These reports undermined Akhbārī rejections of *al-barā'a al-aṣliyya*, and were clearly an element in the counter-arguments proposed by the Uṣūlīs. These reports allowed the Uṣūlīs to argue that if the Akhbārīs truly followed the *akhbār*, then they should accept *al-barā'a al-aṣliyya* since it was permitted by the Imams. That most Akhbārīs deal with these reports in their rejection of the Uṣūlī version of this principle is a further indication of the perceived threat posed to the Akhbārī jurisprudence found in these inconvenient reports. Al-Karakī's counter-argument consists of an assertion that these reports do not mean that actions or substances should be treated as unrestricted, permitted or pure in the absence of a revelatory indicator. Rather, he argues, they exonerate an ignorant believer from unwittingly acting in a manner which transgresses the law. The reports, then, refer to the moral assessment of those ignorant of the law, not to the manner in which scholars of the law should proceed in their search for rulings. Such reports are outnumbered by the many others in which *tawaqquf* and *iḥtiyāṭ* are declared to be obligatory in cases of revelatory silence, and which recommend that the individual believer, in cases of personal ignorance, ask the scholars (*su'āl min al-ʿulamā'*).

---

particular legal rulings (Karakī, *Hidāyat*, pp. 223–231), with a brief introduction outlining the various Imāmī views on *iḥtiyāṭ*, but this does not constitute a guide for those puzzled over the correct course of action in cases of revelatory silence.

[32] Ibn Bābūya, *Man Lā*, v. 1, p. 317, #937.

[33] Kulaynī, *al-Kāfī*, v. 5, p. 313, #40; Ṭūsī, *al-Tahdhīb*, v. 7, p. 226, #989/9.

[34] "*kull shay' naẓīf hattā taʿlima annahu qadhr*" (Ṭūsī, *al-Tahdhīb*, v. 1, pp. 284–285, #832/199—the phrase is found at the end of the *ḥadīth*). Al-Baḥrānī cites it incorrectly in his *al-Ḥadā'iq* (v. 1, p. 42 and elsewhere) as "*kull shay' ṭāhir* . . ." which has a greater connection with the legal categories of ritual purity. His incorrect citation is a combination of this and another report (namely, "All water is *pure—ṭāhir*—until you know it to have been made dirty", Ḥurr, *Wasā'il*, v. 1, p. 142, #351/2).

Al-Karakī's interpretation of these reports was not, however, the only Akhbārī defence. His contemporary, Muḥsin Fayḍ al-Kāshānī, argues that Astarābādī was wrong to reject *al-barāʾa al-aṣliyya* completely. In *al-Uṣūl al-Aṣīla*, he quotes Astarābādī's blanket rejection (cited above) and argues instead that these *akhbār* establish a principle upon which the believer can act. *Al-barāʾa al-aṣliyya* is forbidden for anyone who either knows the law in its entirety or has access to a source of knowledge of the law. The former category is populated by the Imams alone, but the second category includes those who were present with the Imams and could ask them questions as members of their coterie. God has a ruling for every occasion, and one who knows the law, or has access to one who has perfect knowledge of the law, has no need of *al-barāʾa al-aṣliyya* and is, therefore, forbidden from using it:

> What is related from al-Ṣādiq, "all things are unrestricted until a prohibition is referred to them" means: they are unrestricted to you, and made available to you until such time as a prohibition reaches you. It does not mean that God has said that, in reality, [all things] are unrestricted.[35]

After the occultation, we no longer have access to God's law through the Imams, but the Imams knew this would be the case, and hence revealed this principle. However, this does not permit us to give *fatwās* on the basis of *al-barāʾa al-aṣliyya*. The Imams revealed the principle of *al-barāʾa al-aṣliyya* as a principle of action, not knowledge:

> The truth is that adhering to *al-barāʾa al-aṣliyya* is only permitted in action (*fīʾl-ʿamaliyyāt*) alone—not in matters relating to knowledge. That is, it is not permitted for us to give *fatwās* and rulings on the basis of *al-barāʾa*, though it is permitted to say: "it is permitted for one not to adopt [a particular ruling] because it is not proven for us", or to say, "we are searching for [the answer to a question] such that it might become clear [to us]".[36]

Fayḍ here is picking up on Astarābādī's theme of there being two levels of the law: one for action and one for giving *fatwās*. The former, at times, operates according to different principles to the latter, and in this way he accommodates *al-barāʾa al-aṣliyya* in a modified form into Akhbārī jurisprudence. In this, he is followed by al-Jazāʾirī, who concludes that "*al-barāʾa al-aṣliyya* is an indicator in certain

---

[35] Fayḍ, *al-Uṣūl*, p. 20.
[36] Fayḍ, *al-Uṣūl*, pp. 19–20.

circumstances, but not the way the *mujtahids* present it."[37] Until one knows a thing to be forbidden, it remains permitted (*mustamarr ʿalā al-ḥalliyya*).

Al-Baḥrānī combines the approaches of al-Karakī and Muḥsin Fayḍ in what is probably the most comprehensive Akhbārī account of *al-barāʾa al-aṣliyya*. Like al-Karakī, he displays an emphasis on *tawaqquf* and *iḥtiyāṭ*. Like Muḥsin Fayḍ, however, he views *al-barāʾa* as a legitimate course of action in cases of revelatory silence (but not ambiguity, for which he recommends *tawaqquf* and *iḥtiyāṭ*). However, his theory represents a development on both approaches. For al-Baḥrānī, *al-barāʾa al-aṣliyya* is one of the hermeneutic maxims revealed by the Imams themselves, and should therefore form the basis of a scholar's explication of the law for the community. The *muftī* should, therefore, give *fatwās* on the basis of *al-barāʾa al-aṣliyya* because the Imams' have explicitly permitted him to do so. However, he can only declare a *fatwā* when he is certain (that is, has *al-ʿilm al-qaṭʿī*) that there is no revelatory indicator which might be relevant to the particular case in question. Al-Baḥrānī adopts the Akhbārī doctrine that God has a ruling for every situation, and hence if there is any indication that the texts contain such a ruling (even if the scholar cannot discern the ruling itself with certainty), then *al-barāʾa al-aṣliyya* becomes redundant. Once the scholar has an inkling that there is a relevant assessment in the sources, he may be unable to discern it, but he certainly can no longer assume that the assessment is permission. In such circumstances, the Imams have revealed an alternative set of hermeneutic principles which revolve around *tawaqquf* and *iḥtiyāṭ*. Unlike his predecessor Akhbārīs, al-Baḥrānī outlines a series of mechanisms whereby the most cautious (*al-aḥwaṭ*) course of action can be determined, all of which are supported by reports from the Imams.[38]

Revelatory Contradiction

The debate over whether (and on what occasions) *tawaqquf* and *iḥtiyāṭ* should operate in the delineation of the law can also be seen in the resolution of conflicting legal indicators. The Uṣūlīs had a number of

[37] Jazāʾirī, *Manbaʿ*, p. 64.
[38] I outline al-Baḥrānī's procedures in my *Inevitable Doubt*, pp. 105–112.

procedures whereby contradiction between legal indicators might be resolved. Of course, for Akhbārīs, some of these contradictions were unimportant. The various potential conflicts between legal sources are outlined by the late Akhbārī writer, Muḥammad al-Akhbārī, in his *Maʿāwil*.[39] The conflict between a rational indicator (*dalīl ʿaqlī*) and a transmitted (or revelatory) indicator (*dalīl naqlī*), extensively discussed by the Uṣūlī jurists, was irrelevant for Akhbārīs. The transmitted indicator always defeats the rational, since the rational has no basis in the revelatory texts. The conflict between the application of one of the hermeneutic maxims and an isolated report (*khabar al-wāḥid*) was also unimportant since the maxims were established for cases which had no revelatory evidence, and the category of *khabar al-wāḥid* is descriptive, not evaluative, for Akhbārīs.[40] The maxims are not underlying bases on which the law was promulgated by the Imams (they have no explanatory function); rather they are the exegetic ground rules for occasions when the sources are silent. Any clash involving a revelatory source and consensus (*ijmāʿ*) is ignored by Akhbārīs, since the latter does not constitute a source of law. Finally, any conflict between Qurʾanic verses, or between Qurʾanic verses and the *akhbār* is, for most Akhbārīs, unimportant since the Qurʾān cannot be understood independently from the *akhbār*. For the Akhbārīs, the only type of contradiction in the sources that requires resolution is a conflict between two reports from the Imams.

Uṣūlī solutions to *akhbār* contradictions primarily involved the scrutiny of the transmission chains (*isnāds*) of the reports in order to undermine the historical reliability of one of the reports relative to the other. The rejection (or extensive modification) of *isnād* criticism by Akhbārīs made such a solution impossible. Another Uṣūlī solution was to reduce the number of contradictions by combining apparently conflicting reports through viewing one report as a particularisation (*takhṣīṣ*) of another (more general) ruling, or seeing one report as having restricted legal implications (*taqyīd*) whilst the other was unrestricted (*muṭlaq*). Some Akhbārīs were receptive to the idea that although the grammatical form of one report implied general or unrestricted application, this did not necessarily mean that there

---

[39] Akhbārī, *Maʿāwil*, f.158a.

[40] That an isolated report has the same probative force as a well-attested report flows from the Akhbārī assertion that all the *akhbār* are historically reliable reports of the Imams' words. See above, pp. 252–257.

were not exceptions to such general rules. The linguistic form in one report did not necessarily represent the final formulation of a particular ruling, since the *akhbār* had to be taken as a revelatory corpus, with individual elements having equal probative force. Therefore the legal implications (though not the grammatical meaning) of some reports can be modified in the light of others.[41] As for all Shīʿī jurists, abrogation (*naskh*) of one report from the Imams by another was ruled out by Akhbārīs, since the Imams' task was to elucidate the law, not to change or introduce new laws.[42] In cases where *takhṣīṣ* and *taqyīd* were not possible (that is, when there was a real and irreconcilable contradiction between reports), the other resource offered by Shīʿī legal thought was to identify one of the reports as *taqiyya*—that is, issued under dissimulation by one of the Imams, for fear that revealing the true ruling would undermine the security and welfare of the oppressed Shīʿī community. It was this that was embraced by Akhbārīs as an explanation for contradictions between the *akhbār*, and not any questioning of the historical accuracy of the reports. As Majlisī I, states, "due to fear from the unjust sultans, there are *taqiyya akhbār* which have emerged from the Imams."[43] The issue for an exegete was, then, how to discern a *taqiyya*-generated report.

Within the *akhbār* corpus, the Imams had described procedures whereby a ruling might be arrived at, even though the relevant reports were contradictory. This was, in effect, a means whereby one report could be identified as *taqiyya* and the other as a reflection of the true law. The most widely used of these was the Maqbūla of Ibn Ḥanẓala.[44] In this report, the Imam recommends a series of tests, through which one report can be preferred over another. Omitting the first test (in which the Imam is simply asked which report to follow, since the occultation has rendered this impossible), the tests begin

---

[41] Al-Ḥurr's exposition of the legitimacy of this process was mentioned above, pp. 278–279, and the operations of *takhṣīṣ* and *taqyīd* are explained by al-Karakī in his *Hidāyat*, pp. 280–292.

[42] As we have seen, however, Akhbārīs did permit the *akhbār* to act as indicators that Qur'anic abrogation had occurred. A report could act as an indication that a Qur'anic verse, cited elsewhere in the *akhbār* as not abrogated, was in fact abrogated. In a sense, then, one report from the Imams could cancel out another report by acting as an indicator that the latter was issued under *taqiyya*. It is not clear whether such strictures applied to the *Sunna* also. See above, p. 84.

[43] Majlisī I, *Lawāmiʿ*, v. 1, p. 57.

[44] I translate the report in Gleave, "Two classical Shiʿite theories of *Qaḍā*'", pp. 119–120.

with the order to adopt the report of the most just, most learned, most truthful and most pious transmitter. If the transmitters are equal in these regards, the report upon which the community has agreed (and acted) should be adopted. If the community cannot agree (or have not yet done so), then the report which agrees with the Qur'ān (and if not that, then the *sunna* of the Prophet) is adopted. If these procedures do not produce a clear preference, then the ruling which disagrees with Sunni *fiqh* is preferred. If Sunni opinion is not united on an issue, then the Imam recommends suspension of judgement (*irjā'*). Akhbārī authors generally considered only the last two tests available—namely comparison with Sunni opinion and suspension of judgement. The reasons for the unavailability of the earlier tests are obvious: the early *ḥadīth* collectors had already approved the transmission of these reports, the community had not agreed on which reports to adopt (otherwise the dispute would not have emerged in the first place) and the Qur'ān (and possibly the Prophetic *sunna* also) were uninterpretable without the *akhbār*. One can see Astarābādī employing the test of comparison with Sunni doctrine in his *fatwā* (and the subsequent *risāla*) concerning the purity of wine, and this was a test of whether a report was *taqiyya* generated or not.[45] What is discounted by this and other similar reports are the various Uṣūlī tests, including determining which of the two rulings contained within the reports is in accordance with *al-barā'a al-aṣliyya* (that is, which report indicates that an action is permitted), and declaring that to be preferred (*rājiḥ*).

There is an almost unanimous Akhbārī position, from Astarābādī onwards, that the scholar should suspend judgement when the above enumerated tests fail to produce a clear preference. This position highlights the two-tiered nature of the law in Akhbārī jurisprudence generally. At times, the law can be known with certainty since the *akhbār* clearly indicate a particular ruling. At other times, the law itself may be unknown, but the legitimacy of a particular course of action can be assured by following revealed procedures encapsulated in hermeneutic maxims. In such cases, the important question is not whether the believer has acted in full compliance with God's particular ruling, but rather whether he has acted in a manner which can be justified by the legal sources. It is this distinction which Fayḍ was

---

[45] See Appendix 3 below, and Gleave, "The Purity of Wine".

exploiting when he argued that *al-barāʾa al-aṣliyya* can be used as a principle of action (*ʿamaliyyāt*), but not as a means of elucidating the law (that is, giving a *fatwā*).[46]

In contrast to the intra-Akhbārī agreement over suspending judgement in cases of revelatory contradiction, there is divergence on the procedure to be followed thereafter. Astarābādī, as we have seen, argued that in matters of personal devotion (*ʿibādāt*), the believer could follow either of the conflicting reports. His actions would be legally valid, since both reports form acceptable bases for action (*min bāb al-taslīm*).[47] In cases of public law (or more accurately, matters concerning human interrelationships—*muʿāmalāt*), one should act cautiously and in such a way that the duty to fulfil the law is most likely to be "discharged" (*ishtighāl al-dhimma*).[48] In this distinction between choice (*takhyīr* or *takhayyur*) and caution (*iḥtiyāṭ*) and its application to different areas of the law, Astarābādī was followed within the Akhbārī school by al-Jazāʾirī.[49]

Al-Karakī cites the Maqbūla of Ibn Ḥanẓala in full, and interprets it to mean that one must do *iḥtiyāṭ* "if possible" (*in amkana*), and if this is not possible, one can choose (*takhyīr*), as following either of the reports is acceptable.[50] This is a rare recognition in Akhbārī works that discerning the most cautious course of action (*al-ahwāṭ*) may not always be so simple, and if it is impossible to do so, then choice becomes the operating principle. For Muḥsin Fayḍ, the scholar should perform *iḥtiyāṭ* in terms of issuing a *fatwā* (that is, he should not issue a *fatwā*), but may exercise choice between the conflicting reports "because the Imams have permitted it".[51] For Fayḍ, then, *iḥtiyāṭ* does not mean choosing the most cautious course of action; it means

---

[46] See above, pp. 288–289.

[47] See Gleave, *Inevitable Doubt*, p. 113 and the useful corrective by Clarke (in "Review of R. Gleave, *Inevitable Doubt*").

[48] See above, p. 286.

[49] Jazāʾirī, *Kashf*, v. 2, p. 45. Al-Jazāʾirī is said to transmit Majlisī II's opinion that suspension is only recommended, and that choice is permitted (Baḥrānī, *al-Ḥādāʾiq*, v. 1, p. 102). This would mean that Majlisī II adopts the position that the scholar may suspend judgement if he wishes, and this is recommended, but that he is not obligated to do so (as the Akhbārīs would argue). Al-Baḥrānī, however, states that such a position is not evidenced in Majlisī II's *Biḥār*, and al-Jazāʾirī can only be relating what he heard Majlisī II say, not what the latter wrote (see Baḥrānī, *al-Ḥādāʾiq*, v. 1, p. 102, n. 1).

[50] Karakī, *Hidāyat*, p. 171.

[51] Fayḍ, *Safīnat al-Najāt*, f.4a.l6.

avoiding giving a *fatwā*, since this would give the impression that the *akhbār* are unambiguous on the legal point under examination. This would also appear to be Muḥammad al-Akhbārī's position.[52] Al-Baḥrānī argues that one does *tawaqquf* with regard to *fatwās*, but one does *iḥtiyāṭ* with respect to action, and he implies that one can give *fatwās* stating exactly what the most cautious course of action might be. One is not, of course, permitted to give the impression in one's *fatwā* that the *akhbār* are clear on this matter. Instead, one's *fatwā* must be clear concerning the cautionary nature of one's ruling.[53] Whilst the Akhbārīs reached almost total agreement on the need to suspend judgement in cases of contradictions between the *akhbār*,[54] they show a remarkable variation in how to proceed after this has been established. On *tawaqquf*, the Imams had laid out clear hermeneutic procedures; intra-Akhbārī dispute emerges over how to proceed once the necessity of *tawaqquf* has been established. The Imams seem, unfortunately, not to have provided clear guidance or revealed any procedural maxims for such cases. Ironically, the reports describing how to proceed when the *akhbār* contradict are, themselves, contradictory. Consequently, a single Akhbārī opinion was unachievable.

Conclusions

Akhbārī hermeneutic procedures exemplify both the unity and the diversity of the school as a whole. The unity of the school can be seen in a commitment to the doctrine that the Imams used language in an unambiguous manner. Though the Qur'ān may only be properly understood through the *akhbār*, the *akhbār* themselves pose no substantial linguistic or interpretive problems. This, at least, was the theory. In their exegetical practice, such as their commentaries on the *ḥadīth* collections, the Akhbārīs faced ambiguities in interpreting the Imams' words. The fact that they wrote commentaries indicates that

---

52 Akhbārī, *Maʿāwil*, f.194b.

53 For a full analysis of al-Baḥrānī's position, see Gleave, *Inevitable Doubt*, pp. 112–121.

54 The only dissenter is Majlisī I (*Lawāmiʿ*, v. 1, pp. 57–65), who seems, most unusually, to allow *isnād* evaluation to be employed (namely for determining whether the person who transmits from the Imam might have been the cause of the judgement being *taqiyya*) in order that the preferred report might be identified.

they considered the *akhbār* to be in need of some sort of explication, even if at times this consisted of a simple restatement of the perceived content of individual reports. In Persian *ḥadīth* commentaries, such as those by Majlisī I and Khalīl al-Qazwīnī, the translation of individual reports makes up a substantial part of the commentary, enabling the reader to both translate and interpret a report's meaning simultaneously. Akhbārī scholars also agreed that the maxims whereby the law might be extended to novel and unmentioned cases must be rooted in the *akhbār* themselves. Though they differed over the precise formulation of these maxims, Akhbārīs agreed on their necessity. The maxims were valid bases on which to carry out exegesis because the Imams had sanctioned them as such (they were activated "*bi-idhn al-aʾimma*"—with the Imams' permission). Hence the *akhbār* were seen as a source not only of substantive rulings on this or that aspect of the law. They also provided theoretical justification for interpretive practices. As we have seen, however, when the *akhbār* could not be mined for hermeneutic principles (such as the correct course of action after *tawaqquf*), Akhbārī scholars devised a range of different solutions.

The above analysis also demonstrates the centrality of scholarly authority in the Akhbārī juristic system. As I have said elsewhere in relation to al-Baḥrānī's legal system, scholarly authority is not diminished by the Akhbārī rejection of *ijtihād*.[55] The scholarly elite may be conceived of as less hierarchical (internally) than the elaborate system of later Uṣūlī jurists. For them, there were grades of *ijtihād*, and institutional levels of scholarly accomplishment became formalised in the Nineteenth Century CE, resulting in the establishment of the *marjaʿ* system. This hierarchy was in its infancy when the Akhbārī school lost its position of dominance in Iran and southern Iraq in the early Qajar period. Akhbārī scholarly authority was justified (and rationalised) on different foundations to that found in Uṣūlī theory. The Akhbārī distinction between those issues on which a scholar can give a *fatwā* and those on which he must suspend judgement implies a distinction between the elite and the general populous. The people must still "ask the *ʿulamā*" for knowledge, as al-Karakī puts

[55] Gleave, *Inevitable Doubt*, pp. 220–222. In particular, see Fayḍ, *Waṣf al-ʿUlamāʾ*, a *risāla* devoted to praising the *ʿulamāʾ*, which is blended with his own system of mystical authority.

it. In a society where most did not have access to the texts, and few of those who did had the inclination to study them in the casuistic detail of the seminary, the Akhbārīs did not see their jurisprudence as a threat to the status of scholars within Shīʿī society. If there was a perceived threat, then this potential was nullified by the elaboration of a complex hermeneutic which could only be mastered by a scholar. For example, al-Baḥrānī argues that a scholar may give a *fatwā* based on his assessment of the most cautious course of action; though the scholar must avoid giving the impression that this is the law "in reality" (*fī'l-wāqiʿ*). Hermeneutic issues not covered here, such as the Akhbārī doctrine that the *fatwās* of a dead scholar are still authoritative, extend the scholar's authority rather than diminish it. The Uṣūlī doctrine that the *fatwās* of a dead *mujtahid* were also dead was employed by them to maintain the current authority of the scholarly elite. The Akhbārī rejection of this doctrine was used to present the unchanging and unalterable nature of scholarly authority. The dispute between the two schools, then, revolved around technical aspects of the epistemology of scholarly opinion, not around the level of authority accorded to the scholarly elite. For Akhbārīs, the elaboration of hermeneutic maxims enabled them to argue that even though the law could not be known in its entirety, legitimate and legal action could be guaranteed. The *ʿulamāʾ* were indispensable both to the process of interpreting the law according to these maxims, and to its promulgation in the Shīʿī community. Akhbārī self-conceptions concerning the scholars' role were also reflected in history, as Akhbārī scholars involved themselves in positions of community authority in a manner indistinguishable from their Uṣūlī counterparts. In sum, the Akhbārī-Uṣūlī dispute focussed on juristic methodology and the means whereby the sources should be exploited to construct a publicly known legal code. In this respect, it was a dispute restricted to the *ʿulamāʾ*, and was primarily of scholarly interest. It only rarely intruded on either the operations of state or the organisation of Shīʿī society.[56]

[56] On this particular point, see Gleave, "The Qāḍī and the Muftī".

CONCLUSIONS

The analysis of Akhbārī writings carried out in the previous chapters demonstrates the variety of juristic opinion within the Akhbārī school. This variety did not, though, coalesce into well-defined, internal Akhbārī factions. A particular constellation of Akhbārī views on one juristic issue (such as the possibility of interpreting the Qur'ān directly) can be identified, but the various groups of opinions are not consistently transferred to other issues. A different constellation of views appears when examining another issue (such as the legitimacy of *al-barā'a al-aṣliyya*). It is possible, however, to talk of general trends within post-Astarābādī Akhbarism: they emerge when considering the more general question of how (and whether) Akhbārīs should use the intellectual resources of established Shī'ī *uṣūl al-fiqh* to describe their Akhbārī jurisprudence. Some Akhbārīs were more receptive to that tradition, incorporating key ideas which formed part of the *uṣūl* canon. Muḥsin Fayḍ, Majlisī I and al-Jazā'irī can be characterised in this manner. As we have seen, Muḥsin Fayḍ was willing to use established definitions of *muḥkam* and *mutashābih*, and Majlisī I and al-Jazā'irī attempted to reform *isnād* criticism to support Akhbārī doctrine. Other Akhbārīs were hostile to the study of *uṣūl al-fiqh*, considering it an importation from Sunni Islam, and hence a heretical and redundant science. The approaches of Astarābādī, al-Ḥurr and Mīrzā Muḥammad al-Akhbārī can be described in this way. Then there were those who composed works which were undoubtedly influenced by the standard format of works of *uṣūl al-fiqh*, but who, within that model, proposed radical solutions which, in effect, negated the science of *uṣūl al-fiqh*. This tactic—of using the structure of an established genre of writing to undermine its legitimacy, can be seen most clearly in the writings of al-Karakī and Yūsuf al-Baḥrānī. Even this rough and ready categorisation of Akhbārī opinion does not produce a perfect fit. Amongst the early writings of Muḥsin Fayḍ, for example, is his *Naqḍ al-uṣūl al-fiqhiyya*, an attack on the legitimacy of the *uṣūl* discipline; in his latter *al-Uṣūl al-Aṣīla* and *Safīnat al-Najāt*, he appears more willing to cherry pick the findings of past *uṣūl* writers in an exposition of Akhbārī jurisprudence. Similarly, all the Akhbārīs surveyed in this book work within the theory of language found within the *uṣūl* tradition. They offered unoriginal opinions on questions such as language's origin, the manner in which words signify meaning and the occasions on which given meaning might be diverted. Even Akhbārī scholars who rejected the findings of *uṣūl al-fiqh*, then, felt forced, on occasions, to act within its terms of reference.

Within the Akhbārī school, there were also those who saw themselves as part of the Shīʿī legal *ikhtilāf* (that is, the acceptable difference of opinion on matters of law). Majlisī I, for example, was asked about whether or not he agreed with Astarābādī's rejection of *ijtihād* and *taqlīd*. The questioner states:

> Mawlānā Muḥammad Bāqir [Majlisī II] has been heard to say, "The great Ākhund, my father does not believe in *ijtihād* and *taqlīd*, and can be considered a *muḥaddith*"[1]

Majlisī I's reply:

> Those *muḥaddiths* who say, with certainty (*yaqīn*^{an}), that the opinions of the *mujtahids* are invalid (*madhāhib-i mujtahidīn bāṭil-ast*) are totally ignorant—and *vice versa* as well. This question is, itself, a matter of *ijtihād* (*aṣl-i īn mas'ala ijtihādī-ast*), and no scholar can say; "I am certain (*yaqīn*) that my opinion is correct." If he says such a thing, he does not understand what certainty means.[2]

This leads Majlisī I to the paradoxical (perhaps contradictory) conclusion that choice between being an Akhbārī or a *mujtahid* (that is, choice as to whether or not *ijtihād* is a legitimate hermeneutic tool) is, itself, a matter of *ijtihād*. Hence, it is a matter on which the individual scholar makes his own personal investigation and reaches an opinion (*ẓann*) as to which *madhhab* to follow. In order to legitimise the Akhbārī school, Majlisī I has accepted the epistemological structures of established *uṣūl al-fiqh*, and made the choice between the Akhbāriyya and the Uṣūliyya a matter of legal preference (that is, one of the branches of the faith—*furūʿ al-dīn*). On these *furūʿ*, he says, there can be *ikhtilāf*. Such an approach can be well-contrasted with Majlisī I's contemporary al-Karakī, who argues that the choice between the Akhbārī and Uṣūlī methods, is a choice between following the Imams or ignoring them. Hence, at issue in the Akhbārī-Uṣūlī dispute is a fundamental issue of religion (*min ḍurūriyyāt*

[1] Majlisī I, *Kitāb al-Mas'ūlāt*, p. 691.
[2] Majlisī I, *Kitāb al-Mas'ūlāt*, p. 692.

al-dīn), not a matter of individual juristic preference. The task of the Akhbārīs is to lead the Uṣūlīs back to the right path, since scholars have a responsibility to lead the community, and currently the "modern scholars" (*muta'akhkhirīn*, that is, the Uṣūliyya) are leading the community astray.[3] These contrasting Akhbārī approaches towards Uṣūlīs were expressed both in varying degrees of enthusiasm for the structure and terminology of works of *uṣūl al-fiqh*, and the elastic extent to which the paradigms of the *uṣūl* genre were accommodated within Akhbārī jurisprudence.

The difference between these two approaches is partly tactical, and they represent two different methods whereby Uṣūlīs might be persuaded to adopt Akhbārī (or Akhbārī-style) argumentation. However, there is also a principled, internal Akhbārī dispute here, over whether the school constitutes one amongst many, or Akhbārī rectitude eclipses the ideas of any and all opponents. Goldziher asks, in his classic study of the Ẓāhiriyya, whether the Ẓāhirīs can be considered a *madhhab kalamī* or *madhhab fiqhī*.[4] The internal debates within the Akhbāriyya have a similar flavour. Stewart, in his careful and informed analysis, described the Akhbārīs as an "anti-*madhhab*" movement. By this, he means that the Akhbārīs, by refusing to accept the probative force of both consensus and plurality of legal opinions, were rejecting the whole *madhhab* system.[5] Now, my analysis in the preceding chapters, and my identification of broad trends within Akhbarism, indicate that there was a debate within Akhbarism between, in Stewart's terms, pro- and anti-*madhhab* groups. That is, there was a debate around whether Akhbarism was Twelver Shi'sm or merely part of Twelver Shi'ism. This debate was not resolved within Akhbarism. Until Akhbarism's demise, one sees important Akhbārīs of both opinions rising to prominence. Yūsuf al-Baḥrānī and Mīrzā Muḥammad al-Akhbārī, though teacher and pupil, had distinct approaches to this question. The former, at least in the form and tenor of his writings, attempted a conciliation with the Uṣūlīs, and hence could be termed "pro-*madhhab*". The latter was more clearly "anti-*madhhab*". It is interesting to note that the Uṣūlīs did not have a similar internal debate about whether to accept or reject Akhbarism. Uṣūlī polemic against Akhbarism was

[3] Karakī, *Hidāyat*, pp. 307–311.
[4] Goldziher, *Ẓāhirīs*, pp. 123–124.
[5] Stewart, *Islamic Legal Orthodoxy*, pp. 175–179.

consistently strident, with little or no evidence of any intellectual compromise (though there is much evidence of personal friendships between Uṣūlīs and Akhbārīs).[6]

In the early Nineteenth Century CE (the mid Thirteenth Century AH), Akhbārī approaches to juristic questions suffered a collapse in popularity amongst the Shīʿī *ʿulamāʾ*. After Mīrzā Muḥammad al-Akhbārī, who was killed in Kazimayn in 1232/1818, Akhbarism did not produce any scholars of note. His pupil Fatḥ ʿAlī Zand was the major Iranian proponent of Akhbarism after al-Akhbārī's death, but his surviving *oeuvre* is restricted to a single (though impressive) work, *al-Fawāʾid al-Shīrāziyya*. He appears to have received some sponsorship from the Qajar elite, as the work is dedicated to "ʿAbd al-Raḥīm Khān, the Khān of Yazd" who had specifically asked for a Persian language treatise on the difference between Akhbārīs and Uṣūlīs.[7] However, in the battle for royal patronage in early Qajar Iran, Akhbarism was generally unsuccessful. Mīrzā Muḥammad's pupils, children and grandchildren continued to promulgate Akhbarism, but no influential Akhbārī scholar subsequently emerged in either Iraq or Iran.[8] The Akhbārī school did not die with Mīrzā Muḥammad, but it certainly became a minority interest in the seminaries. Mīrzā Muḥammad's connections with India, and the export of his writings to Lucknaw and other Indian Shīʿī centres, enabled Akhbarism to survive in the subcontinent for a longer period of time, though the Indian Akhbārī school is outside of the scope of this book. There remains to this day a vibrant Akhbārī community in Haydarabad.[9] Whether these modern Indian Akhbārīs should be seen as a continuation of the school of the early Akhbārī Ḥusayn b. Shihāb al-Dīn al-Karakī, or connected with Mīrzā Muḥammad's influence or as an entirely different branch of the school, requires further research. Their belief in the altered nature of the Qurʾanic text, and their vehement polemic against Uṣūlīs, indicate (respectively) both a deviation from, and a debt to, the older Akhbārī school. Akhbārī scholars also continued to be influential amongst the Shīʿa of the Gulf, where today

[6] See Gleave, "Two Classical Theories", pp. 107–108.

[7] Zand, *al-Fawāʾid al-Shīrāziyya*, f.3a.10–12.

[8] See Appendix 1. At least 13 pupils or descendents of Mīrzā Muḥammad, all referred to as Akhbārīs, are mentioned in the biographical literature.

[9] Led by one Syed Waheed Uddin Hyder Jaffery Akhbārī, see www.akhbari.org (accessed 30.6.06).

the Akhbārīs of Bahrain continue to teach and promulgate Akhbārī *fatwās*. Once again, the exact relationship between these modern Shīʿī movements which claim the Akhbārī name and the Safavid and Qajar Akhbāriyya analysed in this book is yet to be researched. That the Akhbārīs, still today, serve as the object of Uṣūlī polemic is clear from the continued reference to (and refutation of) Akhbārī doctrine in Shīʿī *uṣūl al-fiqh*. Such refutations aim at promoting a history of Shīʿī jurisprudence in which the Uṣūlī victory lead by Muḥammad Bāqir al-Bihbihānī (d. 1206/1791–92) is portrayed as total.[10] They may also be directed towards the small surviving Akhbārī communities in India and the Gulf, though the ferocity of the polemic does not seem commensurate with the level of threat.

In terms of Akhbarism's legacy, the Shaykhiyya have been suggested as an influential Shīʿī movement which emerged out of the Akhbārī school.[11] To delineate any possible Shaykhī debt to Akhbarism would require a more thorough analysis of Shaykhī *uṣūl al-fiqh* (both by Shaykh Aḥmad al-Aḥsāʾī himself and other Shaykhī intellectuals). There were family connections between Iranian Akhbārīs and Shaykhīs,[12] and Cole has suggested that Shaykh Ahmad is best seen as plotting a middle course between Usulism and Akhbarism, with a preference for the former.[13] On the other hand, there is certainly some evidence that distinctive Akhbārī legal doctrines (such as the prohibition on being simultaneously married to two descendents of Fatima) were rejected by the Kermānī Shaykhīs,[14] and Sayyid Kāẓim Rashtī does not appear to express any sympathy for the Akhbārī approach.[15]

Akhbārī influence is also claimed in the emergence and popularity of the Maktab-i Tafkīk in Iran in the early Twentieth Century CE. This intellectual movement was promoted by Najaf-trained scholars such as Sayyid Mūsā Zarābādī (d. 1353/1934–35) who taught the separation

---

[10] I argue that this is, in part at least, a construction, in my "Akhbārī-Uṣūlī Dispute".

[11] Amanat, *Resurrection and Renewal*, p. 48.

[12] See Momen, "Usuli, Akhbari, Shaykhi, Babi".

[13] Cole, "Sources of Religious Authority", pp. 84–85. His refusal to give ground to Akhbārī ideas in his answers to the Akhbārīs Shaykh Ḥusayn al-ʿAṣfūr (analysed by Cole) and to Fatḥ ʿAlī Zand (Aḥsāʾī, *al-Risāla fī Jawāb Masāʾil Fatḥ ʿAlī Khān Zand*) is clear.

[14] For the Akhbārī position, see Gleave, "Marrying Fāṭimid Women". The *fatwā* of Karim Khan Kermani permitting this on the basis of past *ʿulamāʾ* approval of it can be found in Kirmānī, *Majmaʿ al-rasāʾil*, v. 2, pp. 103–105.

[15] Rashtī, "Bayān-i ḥaqqiyyat-i akhbāriyyīn", pp. 76–100.

of philosophy and religion, and rejected the intrusion of modern science in religious questions. The Tafkīkī separation of *ʿaql* (associated with *falsafa* and *ʿirfān*) and *naql* (revealed and transmitted truths) has led them to be considered a continuation of the Akhbāriyya.[16] They differ from the Akhbārīs in certain important doctrines, however. For example, the Tafkīkī valorising of Qurʾanic knowledge, and their assertion that all useful knowledge can be found in the Qurʾān is, as we have seen, at variance with majority Akhbārī doctrine.[17] Their analysis of the infiltration of foreign (Greek) philosophy into Islam has as much in common with modern Salafī analyses as it does with the doctrines of the historical Akhbāriyya.[18]

It is, perhaps, worth speculating as to why the Akhbārīs declined in the early Qajar period, and here the plausibility of possible explanations depends, to an extent, on the conception of Akhbarism one wishes to adopt. If one considers Akhbarism to be anti-hierarchical (and possibly anti-clerical also), then the growth of independent *ʿulamāʾ* power in the early Qajar period, coupled with the need for a theory to rationalise their authority, could be a possible explanation for Akhbarism's decline. The *ʿulamāʾ*, it has been argued, were gaining an independent power base, both in terms of finances and in terms of popular support.[19] Uṣūlī legal theory, with its emphasis on the exclusive nature of scholarly authority through the practice of *ijtihād*, inevitably held a number of advantages over Akhbarism's populist understanding of how the community can gain religious knowledge. However, as I have tried to show in the preceding chapters, Akhbārī writers were no less committed to the maintenance of scholarly authority than their Uṣūlī opponents—they merely justified that authority through different intellectual mechanisms. None of the Safavid and Qajar Akhbārīs, to my knowledge, considered the *ʿulamāʾ* redundant because the texts were available to the whole community. Scholars were needed, not only to pass on the knowledge found in the texts, but also to perform complex hermeneutic procedures when the texts

[16] That some Akhbārīs considered *ʿaql*, properly understood, as a valid (even, divinely supplied) source of knowledge is clear. See above, p. 114 (with respect to Muḥammad Amīn al-Astarābādī), Gleave, *Inevitable Doubt*, pp. 188–204 (with respect to Yūsuf al-Baḥrānī) and Gleave, "Scripturalist Sufism".

[17] See Hāshimī, "Naqḍ va barrasī".

[18] See Muʿīnī, "Tafkīk".

[19] Floor, "Economic role of the *Ulama* in Qajar Persia".

were silent, contradictory or ambiguous. These procedures differed from Uṣūlī hermeneutic procedures only in their provenance (that is, the Akhbārīs only permitted procedures which had been explicitly sanctioned by the Imams themselves). Speaking generally, both juristic systems required scholarly experts to act as *muftīs* and *qāḍīs*. The principal difference between the schools was not whether scholars had authority, but the means whereby that authority might be justified internally within the intellectual elite.

Another possible explanation for Akhbarism's demise could be traced to the rise of the Shaykhī school. Shaykh Ahmad's successful efforts in spreading this new challenge to Usulism, and his public condemnation by the Uṣūlī *'ulamā'*, made the Shaykhiyya the natural alternative to the Uṣūliyya. This was continued later in the Qajar period by the Bābiyya and Bahā'iyya. Shaykhism, therefore, robbed Akhbarism of its pre-eminence by presenting the Shaykhiyya as a more radical challenge to Usulism. The popularity of subsequent heterodox movements prevented Akhbarism from re-establishing itself in Iraq and Iran. Perhaps, Akhbarism became too institutionalised through the efforts of so-called "moderate" Akhbārīs such as Yūsuf al-Baḥrānī.[20] Usulism's rise (under al-Bihbihānī and his pupils), however, predates both the emergence of Shaykh Aḥmad's arrival in Iran and the ascendancy of his method amongst both the scholarly elite and the nobility. The question remains as to why Akhbarism was waning before the arrival of Shaykhism.

The demise of Akhbarism appears, rather, to have resulted from a confluence of historical accidents. Firstly, the plague which hit Iraq and the Akhbārī-dominated shrine cities of Najaf and Karbala in 1134–1135/1772–1773 claimed the lives of many learned scholars, and one can presume that a generation of promising Akhbārī scholars, who would have continued al-Baḥrānī's legacy, were lost.[21] With this setback, and whilst Akhbarism was attempting a recovery, al-Bihbihānī and his pupils were able to establish themselves in the shrine cities. Al-Bihbihānī had been teaching Usulism in secret for some time, and al-Baḥrānī's death enabled him to assume the position of scholarly pre-eminence in Najaf and Karbala. Secondly, the assumption of

---

[20] That al-Baḥrānī's self-portrayal as a moderate is more rhetorical than real is argued in Gleave, *Inevitable Doubt*, p. 250.

[21] 'Umarī, *Ghāyat al-Marām*, p. 321; Suwaydī, *Ta'rīkh*, pp. 41–43.

power of Fatḥ ʿAlī Shāh Qājār in 1212/1797 brought to the throne a monarch who was intensely interested in religious movements. This meant that the Shah's personal religious preference for Uṣūlī clerics (if not for Usulism itself), Sufism and even his flirtation with Shaykhism encouraged the growth of Akhbarism's competitors. This prevented scholars of undoubted intellectual weight (such as Mīrzā Muḥammad) from re-establishing Akhbarism within Iran following the onset of the plague and al-Baḥrānī's death. Mīrzā Muḥammad did, it seems, attempt to gain the Shah's favour through the famous Tsitsianov episode (in which he "magically" brought about the death of a Russian general), though the Shah seems to have reneged on their agreement.[22] Whilst Akhbarism did not die out, its influence was restricted to marginal areas (southern Iran and the southern Gulf littoral). The patronage of court nobles, to be found in the major towns and cities, was no longer available. Thirdly, as mentioned above, Akhbarism never managed to achieve internal coherence over its identity. By this, I am referring to the intra-Akhbārī debates about how to view their opponents (be they Uṣūlīs or other Shīʿī groupings). Whether to count Uṣūlīs as merely misguided fellow Shīʿa or heretical deviants from the true faith was never resolved, giving Akhbarism less coherence than Usulism. Because it had failed to resolve certain basic questions concerning its role within the Shīʿī scholarly elite, a disruption of Akhbarism's institutional structure (such as a reduction in the supply of high grade scholars following the 1772–1773 CE plague) would therefore dent the prospects of the school more severely than their more intellectually coherent opponents. Finally, whilst Akhbārī legal theory was not devoid of an emphasis on scholarly authority, Uṣūlī-*mujtahid* theory was, perhaps, a more attractive theory for the *ʿulamāʾ*. It left little room for doubt as to the identity of the keepers of religious knowledge. With its explicit division of the population into *mujtahid* and *muqallid*, and its pronouncement that the community need only follow the commands of a *mujtahid* without knowing the reasoning behind the command, Usulism was probably more appealing. This, combined with the already established position in Shīʿī *fiqh* concerning the distribution of community taxes and the leadership of public prayers, made Usulism a clear favourite. For the *ʿulamāʾ*, jockeying for community loyalty (and the power that it might bring)

[22] See Busse, *History of Persia*, pp. 111–114.

at the beginning of a new dynastic reign, Akhbārī legal theory may have seemed too much of a risk, too easily abused by those who wished to undermine their position as inheritors of the Imams' position within Shīʿī society.

APPENDIX ONE

AKHBĀRĪ SCHOLARS FROM ASTARĀBĀDĪ TO THE END OF THE THIRTEENTH CENTURY AH

The list below is by no means comprehensive, and contains only the most prominent Akhbārī scholars found in the bio-bibliographical literature. The nature of biographical compendia was such that in the first 150 years of the Akhbārī school, little reference was made to the Akhbārī (or any other) allegiance of the scholars.[1] I have restricted the list below to scholars who composed Akhbārī works and who were referenced as such in the Shīʿī biographical tradition. The list takes us to the end of the Thirteenth Century AH. The large number of Thirteenth Century Akhbārī scholars, many of whom had personal connections with Mīrzā Muḥammad al-Akhbārī and/or were based in India is an indication of the possibilities for future research on the post-Mīrzā Muḥammad and modern Akhbārī school. The increase in the number of Akhbārīs in this list after Mīrzā Muḥammad al-Akhbārī does not necessarily reflect increased Akhbārī activity. Rather, it probably reflects other factors: proximity of time (and hence greater records available to the principally modern biographical works cited here), a greater willingness of biographical writers to identify their fellow scholars as Akhbārīs in the Thirteenth Century (as compared to the Eleventh), and finally a minority loyal to Mīrzā Muḥammad himself.

Scholars not mentioned elsewhere in this book are provided with single reference (many more could be given). The scholars are ordered by death date when known. When death dates are not known I have inserted scholars in what I judge to be an appropriate position in the list. Scholars whose Akhbārī allegiance is doubtful or disputed are marked with an asterisk*.

---

[1] See above, pp. 40–60.

Eleventh Century AH

Mīr Yūsuf ʿAlī al-Ḥusaynī (d. early Eleventh Century). Described as an Akhbārī in Ṭihrānī, *Ṭabaqāt*, v. 5, p. 645, though this is a late source.
ʿAbd Allāh al-Tustarī (d. 1021)
Mājid al-Baḥrānī (d. 1028)
Muḥammad Amīn al-Astarābādī (d. 1036)
Fakhr al-Dīn Ḥaydar al-Lankarī (d. after 1031)
ʿĪsā al-Duzmārī (d. not known)
Ṣafī al-Dīn Muḥammad (d. after 1033)
ʿAbd al-Hādī al-Ḥusaynī al-Tustarī (d. not known)
Ḥusayn b. al-Ḥasan b. Yūnis al-Ẓahīrī (alive in 1051)
Zayn al-ʿĀbidīn b. Nūr al-Dīn al-Kāshānī (d. after 1040)
ʿAlī Naqī al-Shīrāzī (d. 1060)
Zayn al-Dīn b. Muḥammad b. al-Ḥasan al-ʿĀmilī (d. 1064)
Muḥammad Taqī al-Majlisī (Majlisī I, d. 1070)
ʿAbd al-ʿAẓīm b. ʿAbbās al-Astarābādī (d. late Eleventh Century). Described as an Akhbārī by al-Baḥrānī (*Luʾluʾa*, p. 66).
*ʿAbd Allāh al-Tūnī (d. 1071)
Ḥusayn b. Shihāb al-Dīn al-Karakī (d. 1076)
ʿAbd ʿAlī b. Jumʿa al-ʿArūsī al-Huwayzī (alive 1079)
Ibrāhīm b. ʿAbd Allāh al-Astarābādī known as al-Khaṭīb (d. after 1081)
*Muḥammad Ṣāliḥ al-Māzandarānī (d. 1081 or 1086)
Muḥammad Muʾmin b. Dawsat (or Dūst) al-Astarābādī (d. 1087)
Khalīl al-Qazwīnī (d.1089)
Muḥsin Fayḍ al-Kāshānī (d. 1091)
ʿAlī Aṣghar al-Qazwīnī (alive 1092)
Raḍī al-Dīn al-Qazwīnī (d. 1096)
Muḥammad Ṭāhir al-Qummī (d. 1098)

Twelfth Century

Aḥmad b. Muḥammad b. Yūsuf al-Māqābī al-Baḥrānī (d. 1102)
Muḥammad b. al-Ḥasan al-Ḥurr al-ʿĀmilī (d. 1104)
Hāshim b. Sulaymān al-Baḥrānī (d. 1107)
*Muḥammad Bāqir al-Majlisī (d. 1111)
Sayyid Niʿmat Allāh al-Jazāʾirī (d. 1112)

*ʿAbd Allāh b. al-Ḥusayn al-Yazdī (alive 1113), Al-Ṭihrānī suggests (*Ṭabaqāt*, v. 6, p. 447) that this is the same person as ʿAbd Allāh b. al-Ḥusayn al-Ujrī.

Nūr al-Dīn Muḥammad al-Kāshānī al-Akhbārī (d. 1115)

Muḥammad Ṣāliḥ al-Harawī/al-Harātī (alive in 1119). Pupil of al-Ḥurr, author of the Akhbārī work *Wasīlat al-maʿād fī dhamm al-ijtihād*: see (*Mashhad Fihrist*, v. 16, p. 603, MS#3030, where an *ijāza* to one Naṣr Allāh al-Mudarris is recorded, another possible other Akhbārī in Khorasan, see Ṭihrānī, *Ṭabaqāt*, v. 6, p. 376).

Muḥammad Raḥīm b. Muḥammad al-Harawī/al-Harātī (no death date known). Another pupil of al-Ḥurr with the same *nisba*, and author of the Akhbārī work of *fiqh, Anīs al-Mustawahishīn*, see *Mashhad Fihrist*, v. 16, pp. 74–75 MS#1600 and Ṭihrānī, *Ṭabaqāt*, v. 6, p. 261.

Sulaymān b. ʿAbd Allāh al-Baḥrānī (d. 1121)

ʿAbd Allāh b. Ṣāliḥ al-Samāhījī (d. 1135). He gave an *ijāza* to one Nāsir b. al-Jārūdī al-Qaṭīfī, another possible, Baḥraynī Akhbārī (see Schmidtke, "The *ijāza*").

ʿAbd Allāh b. Nūr al-Dīn al-Jazāʾirī (alive in 1151)

Ṣadr al-Dīn b. Nāṣir al-Dīn b. Mīrza Ṣāliḥ al-Ṭabaṭabāʾī al-Yazdī (d.1153). Akhbārī teacher in Yazd, his son Muḥammad b. Ṣadr al-Dīn al-Akhbārī, no death date known, was also an Akhbārī (see Ṭihrānī, *al-Dharīʿa*, v. 6, p. 42 and p. 198). The son was forced to defend his father's Akhbarism and *ʿirfānī* tendencies (see Ṭihrānī, *Ṭabaqāt*, v. 6, p. 377).

Nūr al-Dīn b. Niʿmat Allāh al-Jazāʾirī (d. 1158)

Unknown 12th Century Akhbārī, who wrote an important commentary on al-Ḥurr's *Hidāyat al-Umma*, see Ṭihrānī, *al-Dharīʿa*, v. 14, p. 172.

Muḥammad Jaʿfar b. Muḥammad Ṭāhir (born 1080, no death date known, though the recent editor of his work argues for 1175, making him 95 at the time of his death). An Akhbārī pupil of Majlisī II and al-Ḥurr, who wrote an appendix to Sāḥib al-Rijāl's biographical compendium entitled *Iklīl al-Manhaj*;[2] see Amīn, *Aʿyān al-Shīʿa*, v. 9, p. 204 and described as an Akhbārī at al-Ḥusaynī, *Tilāmidhat al-Majlisī*, p. 88.

---

[2] This has been published in 1425 AH. One would not be able to detect his Akhbārī allegiance from the work itself.

Yūsuf b. Aḥmad al-Baḥrānī (d. 1176)

Shaykh Ḥasan Akhbārī (d. 1181)—a poet and Akhbārī, with an *ijāza* for *ijtihād* from Karbala, but based in Bushehr and then Bahrayn (Amīn, *Aʿyān al-Shīʿa*, v. 5, pp. 260–261).

Thirteenth Century

Muḥammad b. ʿAbd al-Nabī al-Akhbārī (d. 1232)

Fath ʿAlī Zand al-Shīrāzī (d. after 1236)

Shaykh al-Ḥusayn b. Muḥammad b. ʿAlī b. ʿAythān (or Ghayshān) al-Baḥrānī (d. before 1240). Referred to as an Akhbārī by both Mīrzā Muḥammad al-Akhbārī and Fatḥ ʿAlī Zand al-Shīrāzī (see his *al-Fawāʾid al-Shīrāziyya*, f.107b.11). Probably the Akhbārī who posed questioned to Shaykh Aḥmad al-Aḥsāʾī (see Aḥsāʾī, *al-Risāla fī Jawāb Masāʾil Ḥusayn al-ʿUṣfūr*, pp. 42–46).

Ḥasan ʿAlī Khān al-Akhbārī (d.1240). Composed *al-Wazīriyya* on the difference between Akhbārīs and Uṣūlīs (Kintūrī, *Kashf al-Hajab*, p. 600 #3372) for one Wazīr al-Dīn al-Akhbārī.

Muḥammad al-Kirmānī (d. 1240?). An Akhbārī pupil of Fatḥ ʿAlī Zand (Ṭihrānī, *al-Dharīʿa*, v. 2, p. 67).

Muḥammad Bāqir al-Tabrīzī (died after 1242). Author of an Akhbārī work of *ʿirfān* (!) *Masālik al-Aṭwār* (Ṭihrānī, *al-Dharīʿa*, v. 20, p. 377).

Ḥabīb al-Dīn Muḥammad b. ʿAlī Aṣghar al-Jurfādaqānī (d. after 1244). Author of *al-Malhama biʾl-ṣawāb*, supposedly a work of *ʿirfānī akhbārī kalām* (Ṭihrānī, *al-Dharīʿa*, v. 22, p. 223).

Jawād Siyāhpūsh b. Muḥammad al-Zaynī b. Aḥmad b. Zayn al-Ḥasanī al-Ḥusaynī (d.1248). A pupil and *ijāza* recipient of Mīrza Muḥammad al-Akhbārī (Amīn, *Aʿyān al-Shīʿa*, v. 4, p. 280).

Muḥammad Saʿīd b. Mahdī al-Qummī (death date not known). A pupil of Mīrzā Muḥammad al-Akhbārī (Ṭihrānī, *al-Dharīʿa*, v. 22, p. 169).

Muḥammad Ibrāhīm b. Muḥammad ʿAlī al-Ṭabasī (death date not known). Another pupil of Mīrzā Muḥammad al-Akhbārī (Ṭihrānī, *al-Dharīʿa*, v. 21, p. 161 and Ṭihrānī, *al-Dharīʿa*, v. 24, p. 71).

Muḥammad Mahdī al-Astarābādī (d. 1259). An Iranian who moved to Lucknow.

Ḥasan b. Ḥusayn b. Muḥammad b. Aḥmad al-Darāzī al-Baḥrānī Al-ʿUsfūrī (d.1261) a pupil of Fatḥ ʿAlī Zand, who died in Bushehr, and is mentioned in *al-Fawāʾid al-Shīrāziyya* (f.123a.5). Composed various works refuting *ijtihād* and defending the Akhbāriyya.

ʿAlī b. Mīrzā Muḥammad al-Akhbārī (d. 1275). Youngest son of Mīrzā Muḥammad al-Akhbārī, wrote a refutation of *ijtihād* (Ṭihrānī, *al-Dharīʿa*, v. 24, p. 267).

ʿAbd al-Ṣāḥib b. Muḥammad Jaʿfar al-Dawwānī al-Khashtī al-Fārisī (d. circa 1274). A pupil of Mīrzā Muḥammad al-Akhbārī and author of the Akhbārī defence *Tuḥfat al-ḥabīb* (Amīn, *Aʿyān al-Shīʿa*, v. 8, p. 16).

Muḥammad Bāqir b. Muḥammad ʿAlī al-Dastī al-Lārī (date of death unknown). An Akhbārī pupil of Mīrzā Muḥammad who wrote a refutation of the Bāb (Amīn, *Aʿyān al-Shīʿa*, v. 8, p. 308).

Muṣṭafā b. Ismāʿīl al-Mūsawī (date of death unknown). A pupil of Mīrzā Muḥammad al-Akhbārī who wrote an Akhbārī work entitled *al-Lawāmiʿ al-Muḥammadiyya* for Muḥammad Shāh Qajar, a copy of which is to be found in the University of Tehran Central Library (*University Fihrist*, Elāhiyāt section MS#463D)

Ḥasan b. ʿAlī b Muḥammad al-Akhbārī (date of death unknown). Grandson of Mīrzā Muḥammad al-Akhbārī: see Ṭihrānī, *al-Dharīʿa,* v. 16, p. 286. Brother of Muḥammad b. ʿAlī b. Muḥammad al-Akhbārī (about whom nothing else is known).

Muḥammad Najaf al-Kirmānī al-Mashhadī al-Akhbārī (d. 1292). An *ʿārif* Akhbārī, who composed a number of works of an Akhbārī character (Iʿtimād al-Salṭanah, *Maʾathir waʾl-Āthār*, p. 173).

ʿAlī b. Ismāʿīl al-Akhbārī (d. after 1295). Wrote *Ḥamalāt al-Layth*, an attack on the Uṣūlīs (Ṭihrānī, *al-Dharīʿa*, v. 9.3, p. 740).

Yad ʿAlī b. Mumtāz (date of death unknown). A pupil of Mīrzā Muḥammad al-Akhbārī (*Mustadrak Aʿyān al-Shīʿa*, v. 2, p. 286).

Ināyat Allāh b. Ḥasan ʿAlī b. Mīrzā Muḥammad (date of death unknown). Grandson of Mīrzā Muḥammad al-Akhbārī, who wrote a work on *tawḥīd* (Ṭihrānī, *al-Dharīʿa*, v. 4, p. 481).

Muḥammad b. ʿAlī b. Muḥammad al-Akhbārī (date of death unknown). Grandson of Mīrzā Muḥammad al-Akhbārī who appears to have taken a moderate Akhbārī path, writing a work of inductive law (*fiqh istidlālī*, normally a hallmark of an Uṣūlī) (see Tihrānī, *Ṭabaqāt al-Aʿlām*, v. 6, p. 745).

Muḥammad b. ʿAlī Ḥusayn or al-Ḥusaynī (d. Thirteenth Century). Described as an Akhbārī (*Mustadarkat Aʿyān al-Shīʿa*, v. 2, p. 26).

APPENDIX TWO

"DIFFERENCE LISTS" DETAILING THE DIFFERENCES BETWEEN AKHBĀRĪS AND UṢŪLĪS

This is a preliminary list of works in which the differences between Akhbārīs and Uṣūlīs are listed and enumerated (in numerical or *abjad* forms). Such dedicated lists of differences appeared quite some time after Astarābādī. The first was produced at least a century after his death. As I have argued in Chapter 6 above,[1] the appearance of such works, written both by Uṣūlīs and Akhbārīs indicates the emergence of a stronger line of definition between the schools. Lists of differences help fix this line, but, more importantly, reflect the state of school self-depiction at the time of writing. School lines are relatively fixed some time before the composition of the first list, and the difference lists reflect this fixity. Hence the earliest dedicated work appears a few decades after the point at which I date the emergence of a definable Akhbārī *madhhab* (namely around the time of al-Ḥurr al-ʿĀmilī's death in 1104). School allegiances, when they can be ascertained from the available sources, are given in parenthesis. Excluded from this list are works which specifically refute the doctrines of either school (of which there are many more). The works included here have the principal aim of delineating the dispute, and only a secondary aim of proving one or other (or indeed neither) of the schools to be correct.

Undated (but probably 12th Century, Akhbārī) Sayyid Muḥammad Faraj Allāh al-Dizfūlī (or Muḥammad b. Faraj Allāh—see Amīn, *Aʿyān al-Shīʿa*, v. 10, p. 36), *Fārūq al-Ḥaqq fīʾl-Farq bayn al-Akhbārī waʾl-Uṣūlī*, detailing 83 differences (*Marʿashī Fihrist*, v. 8, p. 216, MS#3028/6)

[1] See above, pp. 179–180.

Undated (but probably 12th Century; manuscript dated 1195, said to be by a student of al-Samāhījī, Akhbārī—see Ṭihrānī, *al-Dharīʿa*, v. 16, p. 177, #548)—*al-Farq bayn al-Uṣūlī wa'l-Akhbārī* (*Marʿashī Fihrist,* v. 24, pp. 236–237, MS#9474/2)

Before 1206, Muḥammad Bāqir al-Bihbihānī (d.1206, Uṣūlī), *Al-Farq bayn al-Akhbārī wa'l-Uṣūlī* located in the library of ʿAlī Aṣghar Fuḥūl al-Qazwīnī (d. 1277, on whom see Ḥusaynī, *Tarājim*, v. 1, p. 216) in Qazwin, the descendent of the famous Akhbārī Raḍī al-Dīn al-Qazwīnī (d. 1092) (see al-Ḥusaynī, "Dalīl al-Makhṭūṭāt", p. 87)

Before 1212, Mīrzā Muḥammad b. ʿAbd al-Nabī al-Akhbārī (d. 1232, Akhbārī), *Ḥirz al-Ḥawāss ʿan wasawa al-khannās* (University of Tehran, MS#922 and *Marʿashī Fihrist*, v. 17, p. 193 MS#6623/4) detailing 39 differences, but this is a composite number, since the *Ḥirz* is actually a collection of different lists of Akhbārī-Uṣūlī differences (see also Ṭihrānī, *al-Dharīʿa*, v. 6, p. 393, #2441)

Before 1232, Mīrzā Muḥammad b. ʿAbd al-Nabī al-Akhbārī (d.1232, Akhbārī), *Al-Ṭuhr al-Fāṣil*, detailing 59 differences (see Ṭabaṭabā'ī, *Riyāḍ al-Masā'il*, v. 1, p. 105, editor's introduction; see also Ṭihrānī, *al-Dharīʿa*, v. 15, p. 194, #1295)

1232 Mīrzā Muḥammad b. ʿAbd al-Nabī al-Akhbārī (Akhbārī), *Al-Mīzān li-maʿrifat al-Furqān*, written in answer to questions by one ʿAbd Allāh al-Mubārak, listing 40 differences between Akhbārīs and Uṣūlīs (see Ṭihrānī, *al-Dharīʿa*, v. 23, p. 317, #9138)

1232 Mīrzā Muḥammad b. ʿAbd al-Nabī al-Akhbārī (Akhbārī), *Al-Muṭamar al Fāṣil bayn al-Ḥaqq wa'l-Bāṭil* detailing 59 particular characteristics of the Akhbāriyya which distinguish them from the Uṣūliyya (see Ṭihrānī, *al-Dharīʿa*, v. 21, p. 160, #4221)

1255 ʿAlī b. Muḥammad b. ʿAbd al-Nabī al-Akhbārī (Akhbārī), *Shabīkat al-Lajayn fi'l-farq bayn al-farīqayn* (see Kintūrī, *Kashf al-Ḥajab*, p. 307 and Ziriklī, *al-Aʿlām*, v. 5, p. 18)

1256 Riḍā b. Muḥammad ʿAlī al-Qazwīnī (school allegiance not known), *Al-Farq bayn al-Akhbāriyyīn wa'l-Uṣūliyyīn*, completed in 1256 (*Marʿashī Fihrist*, v. 10, pp. 119–120, MS#3720)

Before 1259, Sayyid Kāẓim Rashtī (Shaykhī), *Bayān-i ḥaqqiyat-i Akhbāriyyīn va-Mujtahidīn* (*Majmaʿ al-Rasā'il*, 16, pp. 76–100)

Undated (mid-12th Century) Ḥājj Zayn al-ʿĀbidīn Khān b. Karīm Khān al-Kirmānī (Shaykhī), *Sharḥ-i ikhtilāf-i Uṣūliyyīn va-Akhbāriyyīn* (*Majmaʿ al-Rasāʾil*, 44, pp. 362–393)

1267 Muḥammad b. ʿAlī b. Muḥammad al-Akhbārī (Akhbārī), *Risāla iṣlāḥ dhāt al-bayyin li-rafʿ nizāʿ al-khasamayn* (see Ṭihrānī, *al-Dharīʿa*, v. 11, p. 83, #519)

circa 1268 Muḥammad b. ʿAlī b. Muḥammad b. ʿAbd al-Nabī al-Akhbārī (Akhbārī), *al-Akhbāriyyūn waʾl-Uṣūliyyūn*, unmentioned number of differences, and written as a reply to ʿAlī b. ʿAbbās al-Baḥrānī al-Karazakānī, pupil of ʿAlī b. Muḥammad Bāqir al-Bihbihānī (See Ṭihrānī, *al-Dharīʿa*, v. 26, p. 33, #142)

1275 Mīrzā Muḥammad al-Akhbārī al-Hindī (Akhbārī, but not the above mentioned Mīrzā Muḥammad b. ʿAbd al-Nabī al-Akhbārī), a pupil of the famous Indian Uṣūlī Sayyid Dildār ʿAlī, *Muhyat al-Fuḥūl fī naqḍ al-uṣūl* (see Ṭihrānī, *al-Dharīʿa*, v. 22, p. 206, #8649)

1275 Ṣāliḥ al-Awālī and ʿAbbās al-Awālī (Uṣūlī?), *Al-Aṣīla al-Awāliyya*, answers sent to one ʿAbd ʿAlī b. Khalaf b. ʿAbd ʿAlī al-Baḥrānī (d. 1303), the *imām al-jumʿa* of Bushehr. Eight differences are mentioned, including the claim that the Akhbāriyya appeared in the 5th Century (Ṭihrānī, *al-Dharīʿa*, v. 2, p. 76, #303 and Ṭihrānī, *al-Dharīʿa*, v. 5, p. 214, #1006)

circa 1277 Maḥmūd al-Ḥasanī al-Khafrī (Shaykhī, a pupil of Karīmkhān al-Kirmānī), *Maqāla fīʾl-Farq bayn al-Uṣūlī waʾl-Akhbārī* (Ṭihrānī, *al-Dharīʿa*, v. 21, p. 403, #5691)

1285 Muḥammad Ṣāḥib al-Hindī (Akhbārī, possibly identical to Mīrzā Muḥammad al-Akhbārī al-Hindī), *Risāla fīʾl-furūq al-khamsa bayn al-Akhbārī waʾl-Uṣūlī*, mentioned as a pupil of Ḥasan b. Dildār ʿAlī (Ṭihrānī, *al-Dharīʿa*, v. 16, p. 186, #611)

1295 ʿAlī b. Ismāʿīl b. Zayn al-Dīn al-Ḥusaynī (Uṣūlī, known as Abū al-Faḍāʾil), *Ḥamalāt al-Layth*, detailing 22 differences (Ṭihrānī, *al-Dharīʿa,* v. 26, p. 281, #1405)

Before 1298 Muḥammad Bāqir b. Murtaḍā al-Ṭabaṭabāʾī al-Yazdī (d.1298, Uṣūlī), *Risāla fīʾl-farq bayn al-Akhbāriyyīn waʾl-Uṣūliyyīn* (*Marʿashī Fihrist*, v. 11, p. 95 MS#4082/2)

Before 1311 Luṭf Allāh al-Askī al-Arijānī al-Najafī (d. 1311, school allegiance not known), *Al-Fawāʾid al-ʿIlmiyya fī ḥall baʿḍ al-masāʾil al-uṣūliyya waʾl-kalāmiyya* (Ṭihrānī, *al-Dharīʿa,* v. 16, p. 349, #1623)

Before 1343 Mahdī b. ʿAlī al-Gharīfī al-Baḥrānī al-Najafī (d. 1343, school allegiance not known), *Jumanat al-Baḥrayn* in verse (Ṭihrānī, *al-Dharīʿa*, v. 5, p. 131, #541)

Before 1352 Muḥammad Bāqir b. Muḥammad Ḥasan b. Asad Allāh al-Qāʾinī al-Sharīf al-Bīrjandī (d. 1352, Uṣūlī?), *Īdāḥ al-Ṭarīq* (Ṭihrānī, *al-Dharīʿa*, v. 2, p. 496, #1947; Amīn, *Aʿyān al-Shīʿa*, v. 9, p. 181)

Before 1352 Muḥammad Bāqir b. Muḥammad Ḥasan al-Qāʾinī al-Sharīf al-Bīrjandī (Uṣūlī?), *Al-Muḥākama bayn al-Uṣūliyyīn waʾl-Akhbāriyyīn* (this work is perhaps identical with his *Īdāḥ al-Ṭarīq* mentioned above) (Ṭihrānī, *al-Dharīʿa*, v. 20, p. 135, #2271)

Before 1359 Ḥasan b. Aḥmad al-Ashkazārī (d.1359, Uṣūlī), *Hadiyat al-numla ilā marjaʿ al-milla* (See Ṭihrānī, *al-Dharīʿa*, v. 25, p. 216, #348)

Undated *Risāla fīʾl-furūq bayn al-Uṣūlī waʾl-Akhbārī* (allegedly Akhbārī)—found amongst the writings of Muhammad Jaʿfar al-Astarābādī in Fayzābād (Ṭihrānī, *al-Dharīʿa*, v. 16, p. 186, #610)

APPENDIX THREE

MUḤAMMAD AMĪN AL-ASTARĀBĀDĪ'S *FATWĀ* FOR SHĀH ʿABBĀS AL-ṢAFAWĪ ON THE PURITY OF WINE

This *fatwā*, preserved in a *majmūʿ* in the Kitābkhānah-yi Malik in Tehran,[1] is titled as a *fatwā* from Astarābādī to the Safavid Shāh ʿAbbās. Astarābādī's own *Risāla* on the same topic, entitled *al-Risāla fī ṭahārat al-khamr wa-najāsatihi*,[2] mentions that he provided a *fatwā* concerning wine to a Shah who had been drinking from an early age.

عریضه‌ای که مولانا محمد امین استرابادی بنواب اشرف وجنت مکان شاه عباس نوشته در جواب سؤال از طهارت شراب یا نجاست

آن داعی قدیمی به ذروه عرض می رساند که مذهب ابو حنیفه وشافعی ومالک واحمد حنبلی که شرب خمر حرام ونجس است ودر فتاوی ائمه معصومین صلوات الله وسلامه علیهم اجمعین نیز تصریحات واقع شده که خمر حرام است تا حدی که بر آن سفره که شخصی خمر خورده ودر جانب دیگر آن سفره طعام خوردن شخص دیگر حرام است لکن اکثر احادیث ائمه معصومین صریح است در پاک بودن خمر نماز می توان کرد در جامه‌ای که خمر در او ریخته باشد وبعضی از احادیث ائمه معصومین علیهم السلام مواقف مذهب آن چهار مجتهد است که مذکور شد وائمه معصومین قاعده شریفه وصع فرموده‌اند وآن قاعده در توقیعاتی که در زمان غیبت صغری بخط شریف امام زمان ناموس العصر والاوان صلوات الله وسلامه علیه وآله شده مذکور است وآن قاعده آن است که در هر مسئله که فتاوی مختلف از ما صادر باید فتاوی که موافق فتاوی اهل سنت است حمل آن بر تقیه کنید وحمل به آن نکنید وحمل کنید به آن فتاوی که مخالف فتاوی اهل سنت است پس به مقتضای قاعده شریفه ائمه معصومین علیهم السلام خمر پاک است ونماز در جامه‌ای که نقل فرموده‌اند ومذهب قدمای امامیه این است که خمر پاک است ولیکن خود رای شیخ مفید بر نجاست خمر بود وجمعی دیگر بعد از شیخ طوسی نور الله مراقده بوده‌اند تابع شیخ طوسی شده‌اند وعالم ربانی شهید ثانی رحمه الله در شرح درایه نقل فرموده‌اند که متاخرین در اکثر مسائل خلافیه تقلید شیخ طوسی کرده‌اند وشیخ طوسی تقلید شیخ مفید رحمهم الله تعالی واز عجائب آن که شیخ مفید رحمه الله حدیثی که دلالت بر طهارت خمر میکند حمل بر تقیه کرده‌اند وحال آن که حق آن است که احادیثی که دلالت بر نجاست خمر می کند موافق اهل سنت است واز جهت اطمینان

[1] MS#1563 (*Malik Fihrist*, v. 9, p. 203).

[2] University of Tehran Library, MS#1257 (see *University Fihrist*, v. 7, p. 2668 for a description and Gleave, "Purity of Wine" for an edition).

خواطر اشرف بعضي از عبارات اهل سنت وبعضى از عبارات ائمه طاهرين سلام الله عليهم اجمعين نقل مى شود پس در كتاب منهاج كه از جمله كتب اهل سنت است اين عبارت مذكور است:

أجمع الأئمة الأربعة على نجاسة الخمر

وچون در وقت تحرير عريضه كتاب علل وكتاب كافي وكتاب محاسن وكتاب بصائر الدرجات وكتاب قرب الاسناد وكتاب من لا يحضره الفقيه در طائف بود نقل احاديث از آن كتب مسير نشد پس اكتفاء به نقل حديث از كتاب كتاب الاحكام كه مؤلف آن رئيس الطائفة ميباشد[3] ودر كتاب عدة كه بعد از كتب اربعة حديث مصنف شد ورئيس الطائفة تصريح به آن فرموده‌اند كه از معصوم حكم به صحت آن يعنى به ورود آن از اصحاب عصمت واقع شده واجماع قدماى فرقه ناجيه بر آن شده پس در كتاب تهذيب مذكور است:

قلت للصادق عليه السلام\ إن أصاب ثوبي شيء من الخمر أصلى فيه قبل أن أغسله؟ قال عليه السلام لا بأس أن الثوب لا يسكر

ونيز مذكور است:

سأل رجل أبا عبد الله وأنا عنده عن المسكر والنبيذ يصيب الثوب قال لا بأس

ونيز مذكور است:

سألت رجل أبا عبد الله عليه السلام عن الرجل يشرب الخمر ثم يمجه من فيه فيصيب ثوبي فقال عليه السلام ليس بشيء[4]

ونيز مذكور است از علي بن مهزيار كه از معظم قدماى اماميه است كه:

قرأت في كتاب عبد الله إبن محمد إلى إبي الحسين[5] عليه السلام جعلت فداك روى زرارة عن أبي جعفر وأبي عبد الله عليهما السلام في الخمر يصيب ثوب الرجل إنهما قالا لا بأس به بأن[6] يصلي فيه إنما حرم مشربها[7] وروى غير زرارة عن أبي عبد الله عليه السلام أنه قال إذا أصاب ثوبك خمر[8] فاغسله إن عرفت موضعه وإن لم تعرف موضعه فاغسله كله وإن صليت فيه فأعد صلوتك فأعلمني ما أخذ به فوقع بخطه عليه السلام[9] خذ بقول أبي عبد الله عليه السلام

ونيز در أحاديث ائمة هدى عليهم السلام وارد شده كه در هر مسئله كه دو حديث مختلف از ما به شما برسد عمل كنيد به حديثى كه متأخراست و ممكن است كه خذ بقول أبي عبد الله عليه السلام اشاره به آن باشد وممكن اسن كه آن بر تقيه باشد والله أعلم

بحقائق الأمور حرره شيخ محمد باقر گلپايگانى

---

[3] The manuscript is damaged here and ميباشد is my own suggestion.

[4] في الاصل "لا بأس"

[5] في الاصل "الحسن" وهو صحيح

[6] في الاصل" لا بأس أن"

[7] في الاصل "شربها"

[8] في الاصل "خمر أو نبيذ"

[9] في الاصل "السلام وقرأته"

Translation

A petition which Mawlānā Muḥammad Amīn Astarābādī wrote to Shah ʿAbbās in reply to a question concerning whether wine was pure or impure. That venerable scholar presented to the Pinnacle [that is, the Shah] that:

The opinion of Abū Ḥanīfa and Shāfiʿī and Mālik and Aḥmad-i Ḥanbalī is that drinking wine was forbidden and [wine was] impure. In the *fatwās* of the Sinless Imams—may the blessings and peace of God be upon them all—there are also explicit statements that wine is forbidden to such an extent that when someone is drinking wine on a ground rug, it is forbidden for another to eat food on a rug next to him. However, the *ḥadīths* of the Sinless Imams are [also] explicit concerning the purity of wine in that one can pray wearing a piece of clothing upon which wine has been spilled. Some of the *ḥadīth* of the Sinless Imams, upon them be peace, [then] agree with the aforementioned opinion of the four *mujtahids* [and others do not]. The Sinless Imams have revealed a noble principle. This principle is one of those revealed during the time of the Lesser Occultation, and was recorded in the hand of the Imam of the age, the lawgiver of the age and all times—may the blessings and peace of God be upon him and his progeny. The principle is this:

> When contradictory *fatwās* come from us, those *fatwās* which agree with the *fatwās* of the Sunnis should be understood as [issued under] dissimulation, and hence not interpreted at all. Interpret [only] those *fatwās* which contradict the *fatwās* of the Sunnis.

So, in accordance with this noble principle of the Sinless Imams—upon them be peace—wine is pure, and prayer in clothing such as that which we have mentioned [is permitted]. The opinion of the early Imāmī scholars was that wine is pure. However, the opinion of Shaykh Mufīd was that wine is impure.[10] Another group [of scholars] followed Shaykh Ṭūsī—may God illuminate his tomb. The great scholar Shahīd II—may God have mercy on him—in his commentary on the *Dirāya*—recorded that the moderns, in most of the issues where there is a difference of opinion, followed Shaykh Ṭūsī, and Shaykh Ṭūsī followed Shaykh Mufīd—may God the Most High have mercy on

[10] See Mufīd, *al-Muqniʿa*, p. 73.

them. One of the most remarkable things is that Shaykh Mufīd—may God have mercy on him—interprets the *ḥadīth* which indicates wine to be pure as being *taqiyya*. However, the truth is that the *ḥadīths* which indicate that wine is impure are in agreement with the Sunnis. So in order to calm the most noble minds, both some of the views of the Sunnis and some of the views of the pure Imams—may the peace of God be upon them all—will be recorded [here].

In the book *Minhāj*,[11] which is one of the books of the Sunnis, the following passage is recorded:

> The four Imams agree that wine is impure[12]

However, since at the time of the request [from the Shah] [my copy of] the books *ʿIlal* [*al-Sharāʾiʿ* of Ibn Bābūya], *al-Kāfī* [*fī ʿilm al-Dīn* of al-Kulaynī], *al-Maḥāsin* [of al-Barqī], *Baṣāʾir al-Darajāt* [of al-Ṣaffār al-Qummī], *Qurb al-Isnād* [of al-Ḥimyarī], *Man lā Yaḥḍuruhu al-Faqīh* [of Ibn Bābūya] are all in Ṭāʾif, recording the *ḥadīths* from those books is not easy. So it must suffice to record *ḥadīths* from the *Kitāb [Tahdhīb] al-Ahkam* by the leader of the sect [that is, Shaykh Ṭūsī].[13] In the *Kitāb al-ʿUdda*, which was written after the Four Books of *ḥadīth*, the leader of the sect states explicitly that there is a ruling from the Sinless One that [the *ḥadīths* found in *Tahdhīb al-Aḥkām*] are reliable. By which he means that they have [assuredly] come from the Possessors of Sinlessness. The early scholars of the saved sect all agreed this. So, in the book *al-Tahdhīb* it is recorded:

> I said to al-Ṣādiq—upon him be peace—if some wine comes into contact with my clothing, do I pray in it before I wash it? He—peace be upon him—said, "There is no problem, for clothing cannot intoxicate.[14]

It is also mentioned that:

> A man asked Abū ʿAbd Allāh [al-Ṣādiq], and I was there at the time, concerning intoxicating substances or date-wine which comes into contact with clothing. He said, "There is no problem."[15]

---

[11] Probably a reference to the *Minhāj al-Ṭalibīn* of Muhy al-Dīn al-Nawawī.

[12] This phrase is not found in al-Nawawī's *Minhaj*, though the view expressed here can be found at Nawawī, *al-Majmūʿ*, v. 2, p. 563.

[13] The manuscript here is unreadable, so I read here *bāshad* or *mī-bāshad* for *shudd*.

[14] Ṭūsī, *Tahdhīb*, v. 1, p. 280, #822/109.

[15] Ṭūsī, *Tahdhīb*, v. 1, p. 280, #823/110.

It is also mentioned that:

> A man asked Abū Abd Allāh [al-Ṣādiq] about a man who drunk wine, and then he spat some of it out and some dropped on my clothing. He—peace be upon him—said, "It is not important."[16]

It is also recorded, from ʿAlī b. Mahrziyār, who is one of the great scholars of the early Imāmīs, that:

I saw in the letter of ʿAbd Allāh b. Muḥammad, to Abū al-Ḥusayn[17]—upon him be peace—[the following]:

> Make me your sacrifice! Zurāra relates from Abū Jaʿfar [al-Bāqir] and Abū ʿAbd Allāh [al-Ṣādiq]—upon them both be peace—concerning wine which comes into contact with a man's clothing that they both said there was no problem with it, such that one could pray in it, [and that] it is only forbidden to drink [wine]. It is related from Zurāra, from Abū ʿAbd Allāh [al-Ṣādiq]—upon him be peace—that he said, "When your clothing comes into contact with wine—then wash it if you know where [it came into contact]. If you do not know the place, then wash all of it. If you have prayed in it, then repeat your prayer.
> So tell me, which [rule] should I adopt?"
> Then it is written in his own hand—upon him be peace: "Adopt the opinion of Abū ʿAbd Allāh." [18]

Also in the *ḥadīth* of the Imams of guidance—upon them be peace—it is related that when there are two contradictory *ḥadīths* from us on an issue, you should act on the more recent *ḥadīth*. It is possible that "take the opinion of Abū ʿAbd Allāh—upon him be peace" is an indication of this [rule]. It is possible also that this was [issued] out of *taqiyya*. God knows best.

[To record] the truth of the matter, this was written by Muḥammad Bāqir Gulpaygānī.

---

[16] Ṭūsī, *Tahdhīb*, v. 1, p. 280, #825/112.
[17] Abu al-Ḥasan in the original text. See above, n. 5.
[18] Ṭūsī, *Tahdhīb*, v. 1, p. 281, #826/113.

REFERENCES

Unpublished References

Akhbārī, *Ḥirz al-Ḥawāss*—Mīrzā Muḥammad al-Akhbārī, *Ḥirz al-Ḥawāss ʿan waswasat al-khannās*, University of Tehran Central Library, MS#3/261.

—, *Fatḥ al-Bāb*—Mīrzā Muḥammad al-Akhbārī, *Fatḥ al-Bāb ilā ṭarīq al-ṣawāb*, Āyat Allāh Marʿashī Library, Qum, MS#3682/1.

—, *Maʿāwil*—Mīrzā Muḥammad al-Akhbārī, *Maʿāwil al-ʿUqūl li-qaṭʿ Asās al-Uṣūl*, Punjab University Library, Lahore, MS#2164.

Astarābādī, *Badāʾ*—Muḥammad Amīn al-Astarābādī, *al-Risāla fīʾl-Badāʾ*, Āstān-i Quds-i Raḍawī Library, Mashhad, MS#6543.

—, *Dānishnāmah*—Muḥammad Amīn al-Astarābādī, *Dānishnāmah-yi Shāhī*, Majlis Library, Tehran, MS#3071/1.

—, *Fatwā*—Muḥammad Amīn al-Astarābādī, *Fatwā Muḥammad Amīn fī Ṭahārat al-Khamr wa-najāsatihi*, Mālik Library, Tehran, MS#1563.

—, *al-Fawāʾid*—Muḥammad Amīn al-Astarābādī, *al-Fawāʾid al-Madaniyya*, Majlis Library, Tehran, MS#2706.

—, *al-Fawāʾid al-Iʿtiqādiyya*—Muḥammad Amīn al-Astarābādī, *al-Fawāʾid al-Iʿtiqādiyya*, Āyat Allāh Marʿashī Library, Qum, MS#488/1.

—, *al-Fawāʾid al-Makkiyya*—Muḥammad Amīn al-Astarābādī, *al-Fawāʾid al-Makkiyya*, Āstān-i Quds-i Raḍawī Library, Mashhad, MS#213.

—, *al-Ḥāshiya ʿalā Man lā Yaḥḍuruhu al-Faqīh*—Muḥammad Amīn al-Astarābādī, *al-Ḥāshiya ʿalā Man lā Yaḥḍuruhu al-Faqīh*, Markaz Iḥyāʾ al-Turāth al-Islāmī, Qum, MS#2750.

—, *al-Ḥāshiya ʿalā al-Istibṣār*—Muḥammad Amīn al-Astarābādī, *al-Ḥāshiya ʿalā al-Istibṣār* (also known as *al-Fawāʾīd al-Makkiyya*), Markaz Iḥyāʾ al-Turāth al-Islāmī, Qum MS#2750.

—, *Mabāḥith*—Muḥammad Amīn al-Astarābādī, *al-Mabāḥith al-Thalāth*, Āstān-i Quds-i Raḍawī Library, Mashhad, MS#132.

—, *Jawāb*—Muḥammad Amīn al-Astarābādī, *Jawāb Masāʾil Shaykhinā Ḥusayn al-Ẓahīrī al-ʿĀmilī*, University of Tehran Central Library, MS1067/4.

—, *Risāla*—Muḥammad Amīn al-Astarābādī, *al-Risāla fī Ṭahārat al-Khamr waʾl-najāsatihu*, University of Tehran Central Library, MS1257/12.

—, *Sharḥ al-Tahdhīb*—Astarābādī, *Sharḥ Tahdhīb al-Aḥkām*, Āyat Allāh Marʿashī Library, Qum, MS#3789.

Calder, "Structure of Authority"—Norman Calder, "The Structure of Authority in Imāmī Shīʿī Jurisprudence". Unpublished PhD thesis, School of Oriental and African Studies, University of London, 1980.

Dizfūlī, *Fārūq al-Ḥaqq*—Muḥammad b. Faraj Allāh al-Dizfūlī, *Fārūq al-Ḥaqq*, Āyat Allāh Marʿashī Library, Qum, MS#3028.

Zand, *al-Fawāʾid al-Shīrāziyya*—Fatḥ ʿAlī Zand, *al-Fawāʾid al-Shīrāziyya fī bayān madhāhib al-ijtihādiyya waʾl-akhbāriyya*, Majlis Library, Tehran, MS#2047.

Fayḍ, *Safīnat al-Najāt*—Muḥsin Fayḍ al-Kāshānī, *Safīnat al-Najāt*, University of Tehran Central Library, MS#1799/1.

Newman, "Development and Political Significance"—Andrew J. Newman III, "The Development and Political Significance of the Rationalist (Uṣūlī) and Traditionalist (Akhbārī) Schools in Imāmī Shīʿī History from the third/Ninth to the Tenth/Sixteenth Century," Unpublished Ph.D. thesis, University of California Los Angeles, 1986.

Rashtī, *Bayān-i ḥaqqiyat*—Sayyid Kāẓim Rashtī, *Bayān-i ḥaqqiyat-i akhbāriyyīn va-mujtahidīn, Majma' al-Rasā'il* in the Kirmānī Shaykhī Microfilm Collection, University of Michigan, Ann Arbor, reel 3, Bā' 5, pp. 76–100, Fihrist #237, p. 320.
Samāhījī, *Munyat al-Mumārisīn*—'Abd Allāh b. Ṣāliḥ al-Samāhījī, *Munyat al-Mumārisīn*, University of Tehran Central Library, MS#1160.
Zahīrī, *Masā'il*—Ḥasan Ẓahīr al-Dīn b. al-Hussām al-Ẓahīrī, *al-Masā'il al-Ẓahīriyya*, Majlis Library, Tehran, MS#1836.
Zayn al-'Ābidīn Khān b. Karīm Khān al-Kirmānī, *Sharḥ-i ikhtilāf Uṣūliyyīn va-Akhbāriyyīn* (*Majma' al-Rasā'il* in the Kirmānī Shaykhī Microfilm Collection, University of Michigan, Ann Arbor, reel 19, Hā' 10, pp. 392–393, Fihrist #874, p. 605.
Zysow, "The Economy of Certainty"—Aron Zysow, "The Economy of Certainty: An Introduction to the Typology of Islamic Legal Theory," Unpublished Ph.D. Thesis, Harvard University, 1984.

Published References

Abisaab, *Converting Persia*—Rula Jurdi Abisaab, *Converting Persia: Religion and Power in the Safavid Empire* (London: I.B. Tauris, 2004).
Abṭaḥī, "Naqsh-i Sunnat"—'Abd al-Ḥamīd Abṭaḥī, "Naqsh-i Sunnat dar ḥujjiyyat-i fahm-i ẓawāhir-i kitāb," *Qabasāt*, 29 (1392Sh), pp. 143–162.
Afandī, *Riyāḍ*—'Abd Allāh Afandī al-Iṣbahānī, *Riyāḍ al-'ulamā' wa-ḥiyāḍ al-fuḍalā'* (Qum: Maktabat Āyat Allāh al-Mar'ashī, 1403).
—, *Ta'līqāt Amal al-Āmil*—'Abd Allāh Afandī al-Iṣbahānī, *Ta'līqāt Amal al-Āmil* (Qum: Maṭba'at al-Khayyām, 1410).
Aḥsā'ī, *al-Risāla fī Jawāb Masā'il Fatḥ 'Alī Khān Zand*—Shaykh Aḥmad al-Aḥsā'ī, *al-Risāla fī Jawāb Masā'il Fatḥ 'Alī Khān Zand* in *Jawāmi' al-Kalim* (Tabriz: Muḥammad Taqī Nakhjavānī, 1273–1276), vol. 2, pp. 92–100.
—, *al-Risāla fī Jawāb Masā'il Ḥusayn al-'Uṣfūr*—Shaykh Aḥmad al-Aḥsā'ī, *al-Risāla fī Jawāb Masā'il Ḥusayn al-'Uṣfūr* in *Jawāmi' al-Kalim* (Tabriz: Muḥammad Taqī Nakhjavānī, 1273–1276), vol. 2, pp. 42–46.
'Allāma, *Mabādi'*—al-Ḥasan b. Yūsuf b. al-Muṭahhar al-'Allāma al-Ḥillī, *Mabādi' al-wuṣūl ilā 'ilm al-uṣūl* (Najaf: Maṭba'at al-Ādāb, 1970).
—, *Mukhtalaf*—al-Ḥasan b. Yūsuf b. al-Muṭahhar al-'Allāma al-Ḥillī, *Mukhtalaf al-Shī'a fī aḥkām al-Sharī'a* (Qum: Mu'assasat al-Nashr al-Islāmī, 1412–1420).
"Alta Discussion"—"Alta Discussion" in B. Weiss (ed), *Studies in Islamic Legal Theory* (Leiden: Brill, 2003), pp. 385–428.
Āl 'Uṣfūr, "'Aqlgirī"—Sayyid Muḥsin Āl 'Uṣfūr, "'Aqlgirī va-naqlgirī dar fiqh-i siyāsī," *'Ulūm-i Siyāsī*, 14 (1380Sh), pp. 75–89.
Amanat, *Resurrection and Renewal*—Abbas Amanat, *Resurrection and renewal: the making of the Babi movement in Iran, 1844–1850* (London: Cornell University Press, 1989).
al-Qāḍī, *al-Kaysāniyya*—Wadād al-Qāḍī, *al-Kaysāniyya fī'l-ta'rīkh wa'l-adab* (Dār al-Thaqāfa: Beirut, 1974).
—, "Biographical Dictionaries"—Wadad al-Qadi, "Biographical Dictionaries: Inner Structure and Cultural Significance" in George N. Atiyer (ed.), *The Book in the Islamic World: The Written Word and Communication in the Middle East* (Albany: State University of New York Press, 1995), pp. 93–122.
Amiji, "The Bohras of East Africa"—Hatim Amiji, "The Bohras of East Africa," *Journal of Religion in Africa* 7.1 (1975), pp. 27–61.
'Amilī, *al-Shawāhid al-Makkiyya*—see Astarābādī, *al-Fawā'id*.
—, *Muntahā al-Maqāl*—Abū 'Alī Muḥammad b. Ismā'īl al-'Āmilī, *Muntahā al-Maqāl fī aḥwāl al-rijāl* (Beirut: al-Mu'assasa, 1998).

Arjomand, *Shadow of God*—S.A. Arjomand, *The Shadow of God and the Hidden Imam: religion, political order, and societal change in Shi'ite Iran from the beginning to 1890* (Chicago: University of Chicago Press, 1984).

As'adī, "Fahm-i Qur'ān"—'Alī As'adī, "Rīshah-yābī rūykard-i ẓāhirgiriyānah dar fahm-i Qur'ān," *Ma'rifat*, 73 (1382Sh), pp. 63–92.

—, "Qur'ān-i Mubīn"—Muḥammad As'adī, "Qur'ān-i Mubīn," *Ma'rifat*, 83 (1383Sh), pp. 5–27.

Ash'arī, *Maqālāt*—'Alī b. Ismā'īl al-Ash'arī, *Kitāb Maqālāt al-islāmiyīn wa-ikhtilāf al-muṣallīn: Die dogmatischen Lehren der Anhänger des Islam* (Wiesbaden: Franz Steiner, 1980).

Astarābādī, *al-Fawā'id*—Muḥammad Amīn al-Astarbābādī, *al-Fawā'id al-Madaniyya* (Qum: Mu'assasat al-Nashr al-Islāmī, 1424) with Nūr al-Dīn al-Mūsawī al-'Āmilī's *al-Shawāhid al-Makkiyya* in the footnotes.

—, *al-Ḥāshiya 'alā Uṣūl al-Kāfī*—Muḥammad Amīn al-Astarābādī, "al-Ḥāshiya 'alā Uṣūl al-Kāfī" edited by 'Alī Fāḍilī, *Mīrāth Ḥadīth al-Shī'ā*, 8 (1380Sh), pp. 229–410.

—, *al-Ḥāshiya 'alā Man lā Yaḥḍuruhu al-Faqīh*—Muḥammad Amīn al-Astarābādī, "al-Ḥāshiya 'alā Man lā Yaḥḍuruhu al-Faqīh" edited by 'Alī Fāḍilī, *Mīrāth Ḥadīth al-Shī'ā*, 10 (1382Sh), pp. 449–516.

Athar Ali, "Dabistān-e Mazāhib"—M. Athar Ali "Pursuing an elusive seeker of universal truth—the identity and environment of the author of the Dabistān-i Mazāhib," *Journal of the Royal Asiatic Society*, 9.3 (1999), pp. 365–374.

Ayoub, "Divine Preordination and Human Hope"—Mahmoud Ayoub, "Divine Preordination and Human Hope: A Study of the Concept of *Badā'* in Imāmī Shī'ism," *Journal of the American Oriental Society*, 106.4 (1986), pp. 623–32.

Babayan, *Mystics, Monarchs and Messiahs*—Kathryn Babayan, *Mystics, Monarchs and Messiahs: Cultural Landscapes in early Modern Iran* (Cambridge Mass: Harvard University Press, 2003).

Baḥrānī, *al-Ḥadā'iq*—Yūsuf b. Aḥmad al-Baḥrānī, *al-Ḥadā'iq al-Nādira fī aḥkām al-'itra al-ṭāhira* (Qum: Dār al-Kutub al-'Ilmiyya, 1363).

—, *Lu'lu'a*—Yūsuf b. Aḥmad al-Baḥrānī, *Lu'lu'at al-Baḥrayn* (Najaf: Maṭba'at al-Nu'mān, 1966).

—, *Anwār al-Badrayn*—al-Shaykh 'Alī al-Baḥrānī, *Anwār al-Badrayn fī tarājim 'ulamā' al-Qaṭīf wa'l-Aḥsā' wa'l-Baḥrayn* (Najaf: Maṭba'at al-Nū'mān, 1377).

Bayat, *Mysticism and Dissent*—Mangol Bayat, *Mysticism and dissent: socioreligious thought in Qajar Iran* (Syracuse: Syracuse University Press, 1982).

Bayhom-Daou, "The Imām's knowledge"—Tamima Bayhom-Daou, "The Imām's knowledge and the Qur'ān according to al-Faḍl b. Shadhān al-Nīsabūrī (d. 260 A.H./ 874 A.D.)," *Bulletin of the School of Oriental and African Studies* 64 (2001), pp. 188–207.

Browne, *Literary History of Persia*—E.G. Browne, *Literary History of Persia* (Cambridge: Cambridge University Press, 1902–1930).

Brockett, "The Value of the Ḥafṣ and Warsh Transmissions"—Adrian Brockett, "The Value of the Ḥafs and Warsh transmissions for the Textual History of the Qur'an," in A. Rippin (ed.), *Approaches to the History of the Interpretation of the Qur'an*. (Oxford: Clarendon Press, 1988), pp. 33–45.

Bulliet, "A Quantitative Approach"—Richard Bulliet, "A Quantitative Approach to Medieval Muslim Biographical Dictionaries," *Journal of the Economic and Social History of the Orient* 13 (1970), pp. 195–211.

—, Nishapur—Richard Bulliet, *The Patricians of Nishapur: A Study in Medieval Islamic Social History* (Cambridge, Mass: Harvard University Press, 1972).

Busse, *History of Persia*—Heribert Busse, *History of Persia Under Qajar Rule* (New York: Columbia University Press, 1972).

Calder, "Doubt and Prerogative"—Norman Calder, "Doubt and Prerogative: The

Emergence of an Imāmī Shīʿī Theory of Ijtihād," *Studia Islamica* 70 (1989), pp. 57–78.

Calder, "Judicial Authority"—Norman Calder, "Judicial Authority in Imami Shiʿi Jurisprudence," *Journal of the British Society for Middle Eastern Studies*, 6.2 (1979), pp. 104–108.

—, "al-Nawawī"—Norman Calder, "Al-Nawawī's typology of *Muftīs* and its significance for a general theory of Islamic Law," *Islamic Law and society* 3.2 (1996), pp. 137–164.

Calmard, "Le Chiisme Imamite en Iran"—Jean Calmard, "Le Chiisme Imamite en Iran à l'époque seldjoukide, d'après le *Kitāb al-Naqḍ*," *Le Monde iranien et l'Islam* (Geneva: Droz, and Paris: Minard, 1971), pp. 43–67.

Carlyle, "On History"—Thomas Carlyle, "On History," in Thomas Carlyle *Critical and Miscellaneous Essays* (Boston: Houghton, Mifflin and Company, 1881), v. 1.ii, pp. 230–31.

Clarke, "Review of Gleave, *Inevitable Doubt*"—Lynda Clarke, "Review of Gleave, *Inevitable Doubt*," *Journal of Islamic Studies*, 12.2 (2001), pp. 198–203.

Cole, "Akhbari-Usuli Dispute"—Juan R.I. Cole, "Shi'i Clerics in Iraq and Iran 1722–1780: The Akhbari-Usuli Controversy Reconsidered," *Iranian Studies* 18.1 (1985), pp. 3–34.

—, *Roots of North Indian Shi'ism*—Juan R.I. Cole, *Roots of North Indian Shi'ism in Iran and Iraq: Religion and State in Awadh, 1722–1859* (Berkeley: University of California Press, 1988).

Cooperson, *Classical Arabic Biography*—Michael Cooperson, *Classical Arabic Biography: The Heirs of the Prophet in the Age of al-Maʾmun* (Cambridge University Press, 2000).

Crone and Hinds, *God's Caliph*—P. Crone and M. Hinds, *God's Caliph: Religious Authority in the First Centuries of Islam* (Cambridge: Cambridge University Press, 1987).

Davānī, *Vaḥīd-i Bihbihānī*—ʿAlī Davānī, *Āqā Muḥammad Bāqir al-Bihbihānī maʿruf bih Vaḥīd-i Bihbihānī* (Tehran: Amīr Kabīr, 1362Sh).

Davidson, *Medieval Islamic Philosophy*—Herbert A. Davidson, *Proofs for eternity, creation, and the existence of God in medieval Islamic and Jewish philosophy* (Oxford: Oxford University Press, 1987).

Dawwānī, *Sharḥ al-ʿAqāʾid*—Muḥammad b. Asʿad al-Dawwānī, "Sharḥ al-ʿAqāʾid al-Aḍudiyya" in Sulaymān Dunyā, *al-Shaykh Muḥammad ʿAbduh bayna al-falāsifah wa'l-kalāmiyyīn* (Cairo: ʿIsā al-B ābī al-Ḥalabī, 1958).

Dhākirī, "Akhbārīgirī"—ʿAlī Akbar Dhākirī, "Akhbārīgirī: Paydayish va-piyāmad-hā," *Hawzah* 89–90 (1377Sh), pp. 315–369.

Dutton, *The Origins*—Yaseen Dutton, *The Origins of Islamic Law: The Qurʾan, the Muwaṭṭaʾ and Medinan ʿAmal* (Richmond: Curzon, 1999).

Eliash, "Misconceptions"—Joseph Eliash, 'Misconceptions Regarding the Juridical Status of the Iranian ʿUlamaʾ,' *International Journal of Middle East Studies*, 10 (1979), pp. 9–25.

Falaturi, "Die Zwölfer-Schia"—Abdoldjawad Falaturi, "Die Zwölfer-Schia aus der Sicht eines Schiiten: Probleme ihrer Untersuchung" in E. Graf (ed.) *Festschrift Werner Caskel: zum siebzigsten Geburtstag 5. Marz 1966 gewidmet von Freunden und Schülern* (Leiden: Brill, 1968), pp. 62–95.

Fāḍilī, *Ijāzāt*—ʿAlī Fāḍilī, "Ijāzāṭ Muḥammad Amīn al-Astarābādī," *Mīrāth Ḥadīth al-Shīʿa*, 6 (1378Sh), pp. 519–528.

Fani (attributed), *Dabistān-i Madhāhib*—Muḥsin Fānī (attributed), *Dabistān-i Madhāhib* (Bombay: n.p., 1292).

Fayḍ, *al-Ḥaqq al-Mubīn*—Muḥsin Fayḍ al-Kāshānī, *al-Ḥaqq al-Mubīn fī taḥqīq kayfiyyat al-tafaqquh fi'l-dīn* (Tehran: Sāzamān-i Chāp-i Dānishgāh, n.d.).

——, *Mafātīḥ al-Sharā'iʿ*—Muḥsin Fayḍ al-Kāshānī, *Mafātīḥ al-Sharā'iʿ fī fiqh al-Imāmiyya* (Qum: Maṭbaʿat al-Khayyām, 1401).
——, *al-Ṣāfī*—Muḥsin Fayḍ al-Kāshānī, *Tafsīr al-Ṣāfī* (Qum: Mu'assasat al-Hādī, 1416).
——, *al-Uṣūl*—Muḥsin Fayḍ al-Kāshānī, *al-Uṣūl al-Aṣīla* (Tehran: Sāzamān-i Chāp-i Dānishgāh, 1390).
——, *al-Wāfī*—Muḥsin Fayḍ al-Kāshānī, *Kitāb al-Wāfī* (Iṣfahān: Maṭbaʿat Amīr al-Mu'minīn, 1406).
——, *Waṣf al-ʿUlamā'*—Muḥsin Fayḍ al-Kāshānī, "Waṣf al-ʿUlamā'," *Mīrāth-i Islāmī-i Īrān*, (1373Sh), v. 2, pp. 585–612.
Floor, "Economic Role of the *Ulama* in Qajar Persia"—W. Floor, "Economic Role of the *Ulama* in Qajar Persia" in Linda S. Walbridge (ed.), *The Institution of the Marjaʿ Taqlid* (Oxford: Oxford University Press, 2001), pp. 52–81.
Gharāwī, *Maṣādir al-Istinbāṭ*—Muḥammad ʿAbd al-Ḥasan Muḥsin al-Gharāwī, *Maṣādir al-Istinbāṭ bayna al-uṣūliyyīn wa'l-akhbāriyyīn* (Qum: Markaz al-Nashr, Maktab al-Iʿlām al-Islāmī, 1413).
Ghazālī, *Tahāfut al-Falāsifa*—Muḥammad Abū Ḥāmid al-Ghazālī, *Tahāfut al-Falāsifa* (Cairo: Dār al-Maʿārif, 1972).
Gleave, "Akhbārī-Usūlī Dispute"—Robert Gleave, "The Akhbari-Usuli Dispute in *Tabaqat* Literature: an analysis of the biographies of Yūsuf al-Bahrānī and Muhammad Bāqir al-Bihbihānī," *Jusur*, 10 (1994), pp. 79–109.
——, "Between *Ḥadīth* and *Fiqh*"—Robert Gleave, "Between Ḥadīth and Fiqh: Early Imāmī Collections of *Akhbār*", *Islamic Law and Society*, 8.3 (2001), pp. 350–382.
——, "Biography and hagiography"—Robert Gleave, "Biography and Hagiography in Tunukabuni's *Qisas al-Ulama*" in Charles Melville (ed), *Proceedings of the third European Conference of Iranian Studies held in Cambridge, 11th–15th September 1995. Part 2: Mediaeval and modern Persian Studies* (Wiesbaden: Reichart, 1999), pp. 237–255.
——, *Inevitable Doubt*—Robert Gleave, *Inevitable Doubt: Two Theories of Shīʿī Jurisprudence* (Leiden: Brill, 2000).
——, "Intra-madhhab *Ikhtilāf*"—Robert Gleave, "Intra-Madhhab Ikhtilaf and the Late Classical Imami Shiite Conception of the Madhhab" in P.Bearman, R.Peters and F. Vogel (eds), *The Islamic School of Law: Evolution, Devolution, Progress* (Cambridge, Mass: Harvard University Press, 2006), pp. 126–146.
——, "Marrying Fatimid Women"—Robert Gleave, "Marrying Fatimid Women: legal theory and substantive law in Shīʿī jurisprudence," *Islamic Law and Society*, 6.i (1999), pp. 38–68.
——, "Qiyās"—Robert Gleave, "'Imāmī Shīʿī Refutations of *Qiyās*', in B. Weiss (ed), *Studies in Islamic Legal Theory* (Leiden: Brill, 2003), pp. 267–293.
——, "Questions and Answers"—Robert Gleave, "Questions and Answers on Akhbārī jurisprudence" (forthcoming).
——, "Scripturalist Sufism"—Robert Gleave, "Scripturalist Sufism and Scripturalist anti-Sufism: Theology and Mysticism amongst the Shīʿī Akhbāriyya" in A. Shihadeh (ed.) *Sufism and Theology* (forthcoming).
——, "The *Ijāza*"—Robert Gleave, "The *ijāza* from Yusuf al-Bahrani (d. 1186/1772) to Muhammad Mahdi "Bahr al-Ulum" (d. 1212/1797–8)," *Iran*, 32 (1994), pp. 115–123.
——, "The Purity of Wine"—Robert Gleave, "Akhbārī Shīʿī jurisprudence in a *Risāla* on the Purity of Wine by Muḥammad Amīn al-Astarābādī" (forthcoming).
——, "The Qāḍī and the Muftī"—Robert Gleave, "The Qāḍī and the Muftī in Akhbārī Jurisprudence" in B. Weiss and P.Bearman (eds), *Islamic Legal Studies* (forthcoming).

——, "Two Classical Shīʿite Theories of *qaḍā*"—Gleave, "Two Classical Shīʿite Theories of *qaḍā*" in J. Mojaddedi, A. Samely and G. Hawting (eds), *Studies in Islamic and Middle Eastern Texts and Traditions in Memory of Norman Calder* (Oxford: Oxford University Press, 2001), pp. 105–121.

Goldziher, *Ẓāhirīs*—Ignaz Goldziher, *The Ẓāhirīs: their doctrine and their history: a contribution to the history of Islamic theology* (Leiden: E.J. Brill, 1971).

Goody, *Savage Mind*—Jack Goody, *The Domestication of the Savage Mind* (Cambridge: Cambridge University Press, 1977).

Gowharshad Fihrist—Fihrist-i Nukhahhā-yi Khaṭṭī-i Kitābkhānah-yi Jāmiʿ-i Gawharshād-i Mashhad (Mashhad: Mu'assasah-yi Intishārāt-i Āstān-i Quds-i Raḍawī, 1371Sh).

Hafsi, "Recherches sur le genre ṭabaqāt"—Ibrahim Hafsi, "Recherches sur le genre *'tabaqāt'* dans la littérature arabe" *Arabica*, 23 (1976), pp. 227–265; 14 (1977), pp. 1–41 and pp. 150–186.

Hallaq, "The authenticity of Prophetic *ḥadīth*"—Wael Hallaq, "The authenticity of Prophetic *Ḥadīth*: A Pseudo-problem," *Studia Islamica* 99 (1999), pp. 75–90.

——, "Murder in Cordoba"—Wael Hallaq, "Murder in Cordoba: *Ijtihād, Iftā'* and the Evolution of Substantive Law in Medieval Islam," *Acta Orientalia* 55 (1994), pp. 55–83.

——, "*Uṣūl al-fiqh*"—Wael Hallaq, "*Usul al-Fiqh*: Beyond Tradition," *Journal of Islamic Studies*, 3.2 (1992), pp. 172–202.

Halkin, "Ḥashwiyya"—A.S. Halkin: "The Ḥashwiyya" *Journal of the American Oriental Society*, 54 (1934), pp. 1–28.

Halm, *Shīʿīsm*—Heinz Halm, *Shīʿīsm* (Edinburgh: Edinburgh University Press, 2004)

——, *Shiʿa Islam*—Heinz Halm, *Shiʿa Islam: from religion to revolution* (Princeton: Markus Wiener, 1997).

Ḥasan, *Maʿālim*—Shaykh Ḥasan b. al-Shahīd al-Thānī, *Maʿālim al-Dīn wa-Malādhdh al-Mujtahidīn* (Qum: Mu'assasat al-Nashr al-Islāmī, n.d.).

Ḥasan al-Amīn, *Mustadrakāt*—Ḥasan al-Amīn, *Mustadrakāt Aʿyān al-Shīʿa* (Beirut: Dār al-Taʿāruf li-l-Maṭbūʿāt, 1418).

Hāshim al-Baḥrānī, *al-Burhān*—Hāshim b. Sulaymān al-Baḥrānī, *Kitāb al-Burhān fī tafsīr al-Qur'ān* (Tehran: Chāpkhānah-yi Āftāb, 1375Sh).

Hāshimī, "Naqd va Barrasī"—Muḥammad Ismāʿīl Hāshimī, "Naqd va Barrasī-i Naẓariyyah-yi Maʿrifat dar Maktab-i Tafkīk," *Qabasāt*, v. 28 (1382), pp. 45–63.

Hawting, "The significance of the slogan *Lā Ḥukm illā lillāh*"—Gerard Hawting, "The significance of the slogan *Lā Ḥukm illā lillāh* and the References to the *Ḥudūd* in the Traditions about the Fitna and the Murder of ʿUthmān," *Bulletin of the School of Oriental and African Studies*, 41 (1978), pp. 453–463.

Heffenung, "Ṭaba'āt" in *EI1* (supplement), pp.

Heinrichs, "*Qawāʿid* as a Genre"—Wolfhart Heinrichs, "*Qawāʿid* as a Genre of Legal Literature" in B. Weiss (ed), *Studies in Islamic Legal Theory* (Leiden: Brill, 2003), pp. 265–284.

Ḥibshī, *Maṣādir al-Fikr*—ʿAbd Allāh Muḥammad al-Ḥibshī, *Maṣādir al-Fikr al-ʿArabī al-Islāmī fī'l-Yaman* (Sanaʿ: Markaz al-Dirāsāt al-Yamāniyya, n.d.).

Ḥillī, *Nihāyat al-wuṣūl*—al-Ḥasan b. Yūsuf al-ʿAllāma al-Ḥillī, *Nihāyat al-wuṣūl ilā ʿilm al-uṣūl* (Qum: Mu'assasat al-Imām al-Sādiq, 1425).

Ḥurr, *Amal*—Muḥammad b. al-Ḥasan al-Ḥurr al-ʿĀmilī, *Amal al-ʿĀmil* (Beirut: Mu'assasat al-Wafā', 1983).

——, *al-Fuṣūl al-muhimma*—Muḥammad b. al-Ḥasan al-Ḥurr al-ʿĀmilī, *al-Fuṣūl al-muhimma fī uṣūl al-a'imma* (Qum: Mu'assasat Maʿārif Islāmī Imām Riḍā, 1418).

——, *Ithbāt al-Hudā*—Muḥammad b. al-Ḥasan al-Ḥurr al-ʿĀmilī, *Ithbāt al-Hudā bi-nuṣūṣ wa'l-muʿjiza* (Tehran: Dār al-Kutub al-Islāmiyya, 1364Sh).

——, *al-Fawā'id al-Ṭūsiyya*—Muḥammad b. al-Ḥasan al-Ḥurr al-ʿĀmilī, *al-Fawā'id al-Ṭūsiyya* (Qum: al-Maṭbaʿa al-ʿIlmiyya, 1403).
——, *Riṣāla fī bayān ḥukm shurb al-Tutun*—Muḥammad b. al-Ḥasan al-Ḥurr al-ʿĀmilī, "Riṣāla fī bayān ḥukm shurb al-Tūtūn wa'l-Qahwa bih kūshish-i Rasūl Jaʿfariyān," *Mīrāth-i Islamī-i Īrān*, 7 (1419), pp. 81–92.
——, *Wasā'il*—Muḥammad b. al-Ḥasan al-Ḥurr al-ʿĀmilī, *Wasā'il al-Shīʿa* (Qum: Mihr, 1414).
Ḥusaynī, "Dalīl al-Makhṭūṭāt"—Sayyid Aḥmad al-Ḥusaynī, "Dalīl al-Makhṭūṭāt," *Turāthunā* 2 (1406), pp. 41–64.
——, *Tarājim*—Sayyid Aḥmad al-Ḥusaynī, *Tarājim al-Rijāl* (Qum: Maktabat Āyat Allāh al-Marʿashī, 1414).
——, *Tilāmidhat al-Majlisī*—al-Sayyid ʿAlī al-Ḥusaynī, *Tilāmidhat al-ʿAllāma al-Majlisī* (Qumm: Maktabat Āyat Allāh al-Marʿashī, 1310).
Huwayzī, *Nūr al-Thaqlayn*—ʿAbd ʿĀlī b. Jumʿa al-Huwayzī, *Nūr al-Thaqlayn* (Qum: al-Maṭbaʿat al-ʿIlmiyya, n.d.).
Ibn ʿĀdī, *Al-Kāmil fī'l-ḍuʿafā'*—ʿAbd Allāh Ibn ʿĀdī, *Al-Kāmil fī'l-ḍuʿafā'* (Beirut: Dār al-Fikr, 1984).
Ibn Bābūya, *Man lā*—al-Shaykh al-Ṣadūq Ibn Bābūya, *Man lā Yaḥḍuruhu al-Faqīh* (Tehran: Mu'assasat al-Nashr al-Islāmiyya, n.d.).
——, *ʿUyūn akhbār al-Riḍā*—al-Shaykh al-Ṣadūq Ibn Bābūya, *ʿUyūn akhbār al-Riḍā* (Beirut: Mu'assasat al-Aʿlamī, 1404).
Ibn Idrīs, *al-Sarā'ir*—Muḥammad Ibn Idrīs al-ʿIjlī, *al-Sarā'ir al-Ḥāwī li-taḥrīr al-Fatāwī* (Qum: Mu'assasat al-Nashr al-Islāmī, 1410).
Ibn Nadīm, *Fihrist*—Muḥammad b. Isḥāq Ibn al-Nadīm, *al-Fihrist* (Beirut: Dār al-Kutub al-ʿIlmiyya, 1996).
Ibn Salāḥ, *Muqaddima*—ʿUthmān b. ʿAbd al-Raḥmān Ibn al-Salāḥ, *Muqaddima* (Medina: al-Maktaba al-ʿIlmiyya, 1972).
Iskandar-Bayg, *Ta'rīkh-i ʿAlam-Ārā*—Iskandar-Bayg Munshī, *Ta'rīkh-i ʿAlam-Ārā-i ʿAbbāsī* (Tehran: Dunyā-yi Kitāb, 1377Sh).
Iskandarjū'ī, "al-Muqārana"—Marḍiyyah Iskandarjū'ī, "al-Muqārana bayn Ārā' al-Uṣūliyyīn wa'l-Akhbāriyyīn," *Faṣlnāmah-yi Nidā-yi Ṣādiq*, 5 (1376Sh), pp. 102–115.
Iʿtimād al-Salṭanah, *al-Maʿāthir wa'l-Āthār*—Muḥammad Ḥasan Khān Iʿtimād al-Salṭanah, *al-Maʿāthir wa'l-Āthār* (Tehran: Kitābkhānah-yi Sanā'ī, n.d.).
Jābirī, *al-Fikr al-Salafī*—ʿAlī Ḥusayn al-Jābirī, *al-Fikr al-Salafī ʿinda al-Shīʿa al-Ithnāʿashariyya ʿalā wajh al-khuṣūṣ min al-manṭiq wa'l-falsafat al-yūnān* (Beirut: Manshūrāt ʿUwaydāt, 1977).
Jacques, *Authority, Conflict*—Kevin Jacques, *Authority, Conflict and the Transmission of Diversity in Medieval Islamic Law* (Leiden: Brill, 2006).
Jadīdīnizhād, "Nigāhī"—Muḥammad Riḍā Jadīdīnizhād, "Nigāhī bih mabānī-i Majlisī-i avval dar ḥadīth va rijāl," *Ulūm-i Ḥadīth*, 28 (1382), pp. 114–135.
Jaʿfariyān, *Ṣafaviyyah*—Rasūl Jaʿfariyān, *Ṣafaviyyah dar ʿArṣah-yi Dīn, Farhang va Siyasat* (Qum: Pizhuhishkadah-yi Ḥawzah va Dānishkadah, 1389).
Jarfadāqānī, *Az Kulaynī tā Khumaynī*—Muḥammad Jarfadāqānī, *ʿUlamā-yi shīʿa Az Kulaynī tā Khumaynī* (Qum: n.p., 1364Sh).
Jawdat, "al-Ḥaraka al-Akhbāriyya"—Jawdat al-Qazwīnī, "al-Ḥaraka al-Akhbāriyya wa-ḥaqīqat al-sirāʿ al-Usulī," *al-Fikr al-Jadīd*, 1 (1412), pp. 83–92.
Jazā'irī, *Kashf*—al-Sayyid Niʿmat Allāh al-Jazā'irī, *Kashf al-Asrār fī Sharḥ al-Istibṣār* (Qum: Mu'assast Dār al-Kutub, 1408).
——, *Manbaʿ*—al-Sayyid Niʿmat Allāh al-Jazā'irī, *Manbaʿ al-Ḥayāt* in Muḥsin Fayd al-Kashānī, *al-Shihāb al-Thāqib* (Qum: Mu'assast al-ʿIlmī lil-Maṭbūʿāt, n.d.), pp. 5–84.
——, *Nūr al-Barāhīn*—al-Sayyid Niʿmat Allāh al-Jazā'irī, *Nūr al-Barāhīn aw Anīs al-Waḥīd fī Sharḥ al-Tawḥīd* (Qum: Muʿassasat al-Nashr al-Islāmī, 1417).

——, *Zahr al-Rabīʿ*—al-Sayyid Niʿmat Allāh al-Jazāʾirī, *Zahr al-Rabīʿ* (Beirut: Muʾassast al-Balāgh, 1999).
Jurjānī, *Sharḥ al-Mawāqif*—al-Sayyid al-Sharīf ʿAlī b. Muḥammad al-Jurjānī, *Sharḥ al-Mawāqif lil-Qāḍī ʿAḍud al-Dīn ʿAbd al-Raḥmān al-ʿĪjī* (Beirut: Dār al-Kutub al-ʿIlmiyya, 1998).
Karakī, *Hidāyat*—al-Ḥusayn b. Shihāb al-Dīn al-Karakī, *Hidāyat al-Abrār ilā ṭarīq al-Aʾimma al-aṭhār* (Baghdad: n.p., n.d.).
Kashmīrī, *Nujūm al-samāʾ*—Muḥammad ʿAlī Kashmīrī, *Nujūm al-Samāʾ fī tarājim al-ʿUlamāʾ* (Qum: Maṭbaʿat al-Baṣīratī, n.d.).
Kerr, *Islamic Reform*—Malcom Kerr, *Islamic Reform: The Political and Legal Theories of Muḥammad ʿAbduh and Rashīd Riḍā* (Berkley: University of California Press, 1966).
Khalidi, *Arabic Historical Thought*—Tarif Khalidi, *Arabic Historical Thought in the Classical Period* (Cambridge: Cambridge University Press, 1994).
Khʷānsārī, *Rawḍāt*—Muḥammad Bāqir al-Khʷānsārī, *Rawḍāt al-Jannāt fī aḥwāl al-ʿūlamāʾ waʾl-sādāt* (Beirut: al-Dār al-Islāmiyya, 1411).
Kintūrī, *Kashf al-Ḥajab*—Iʿjāz Ḥusayn al-Kintūrī, *Kashf al-Ḥajab waʾl-astār ʿan al-Kutub waʾl-asfār* (Qumm: Maktabat Āyat allāh al-Marʿashī, 1409).
Kohlberg, "al-Uṣūl al-Arbāʿumiʾa"—Etan Kohlberg, "al-Uṣūl al-Arbāʿumiʾa," *Jerusalem Studies in Arabic and Islam*, 10 (1987), pp. 128–66.
——, "Aḳbariya"—Etan Kohlberg, "Aḳbariya," *Encyclopedia Iranica*, v. 1, pp. 716–718.
——, "Akhbari"—Etan Kohlberg, Aspects of Akhbari Thought in the Seventeenth and Eighteenth Centuries," in N. Levtzion and J.O. Voll (eds), *Eighteenth-Century Renewal and Reform in Islam* (Syracuse: Syracuse University Press, 1987), pp. 133–155.
——, "Muḥaddath"—Etan Kohlberg, "The Term *Muḥaddath* in Twelver Shīʿism," *Studia. Orientalia Memoriae DH Baneth Dedicata* (Jerusalem: Magnes Press, 1979), pp. 34–47.
——, "Some Imāmī Shīʿī views"—Etan Kohlberg, "Some Imāmī Shīʿī views on the Ṣaḥāba," *Jerusalem Studies in Arabic and Islam*, 5 (1984), pp. 143–175.
Kulaynī, *al-Kāfī*—Muḥammad b. Yaʿqūb al-Kulaynī, *al-Kāfī fī ʿilm al-Dīn* (Tehran: Dār al-Kutub al-Islāmiyya, 1363Sh).
Lawson, "Akhbārī *Tafsīr*"—Todd Lawson, "Akhbārī Shīʿī Approaches to *tafsīr*," in G.R. Hawting & Abdul-Kader A. Shareef (eds), *Approaches to the Qur'an* (New York & London: Routledge, 1993), pp. 173–210.
——, "Note"—Todd Lawson, "Note for the Study of a 'Shi'i Qurʾān'," *Journal of Semitic Studies*, 36 (1991), pp. 279–295.
Leaman, *Averroes*—Oliver Leaman, *Averroes and his Philosophy* (Oxford: Clarendon Press, 1988).
Loth, "Die Ursprung"—Otto Loth, "Die Ursprung und Bedeutung der Tabaqat," *Zeitschrift der deutschen morgenlandischen Gesellschaft* 23 (1869), pp. 593–614.
Madanī, *Sulāfat*—ʿAlī Ibn Maʿṣūm al-Madanī, *Sulāfat al-ʿAṣr* (Qatar: Maṭbaʿa ʿAlī Ibn ʿAlī, n.d.).
——, *Riyāḍ al-Sālikīn*—ʿAlī Ibn Maʿṣūm al-Madanī, *Riyāḍ al-Sālikīn: sharḥ al-Ṣaḥīfa al-Sajjādiyya* (Qum: Muʾassasat Āl al-Bayt, n.d.).
Madelung, "Akhbāriyya"—Wilferd Madelung, "Akhbāriyya," *Encyclopaedia of Islam Second Edition*, supplement v. 1–2, pp. 56–57.
——, "Authority in Twelver Shi'ism"—Wilferd Madelung, "Authority in Twelver Shi'ism in the Absence of the Imam" in G. Makdisi et al. (eds), *La notion d'autorité au Moyen-Age* (Paris, 1982), pp. 163–174.
——, "Imamism and Mu'tazilite Theology"—Wilferd Madelung, "Imamism and Mu'tazilite Theology" in T. Fahd (ed.), *Le Shîʿisme imâmite: colloque de Strasbourg (6–9 mai, 1968)* (Paris: Presses Universitaires de France, 1970), pp. 13–30.

Majlis Fihrist—*Fihrist-i Kitābkhānah-yi Majlis-i Shūrā-yi Millī-i Irān* (Tehran: Majlisi 1933–).
Majlisī I, *Lawāmiʿ*—Muḥammad Taqī al-Majlisī, *Lawāmiʿ ṣāḥibqirānī al-Mushtahar bi-sharḥ al-Faqīh* (Tehran: Maktabat-i Ismāʿīliyyīn, 1414).
—, *Kitāb al-Masʿūlāt*—Muḥammad Taqī al-Majlisī, "Kitāb al-Masʿūlāt" in *Mīrāth-i Islāmī-i Īrān* 8 (1380Sh), pp. 685–705.
—, *Rawḍat*—Muḥammad Taqī al-Majlisī, *Rawḍat al-Muttaqīn fī sharḥ Man lā Yaḥḍuruhu al-Faqīh* (Qum: Bunyad-i Farhang-i Kūshānpūr, 1393).
Majlisī II, *Biḥār*—Muḥammad Bāqir b. Muḥammad Taqī al-Majlisī, *Biḥār al-anwār al-jāmiʿa li-durar akhbār al-aʾimma al-aṭhār* (Beirut: Muʿassasat al-Wafāʾ, 1403).
—, *Ajwibah*—Muḥammad Bāqir b. Muḥammad Taqī al-Majlisī, *Ajwibah-yi Mullā Muḥammad Bāqir Majlisī bih Mullā Khalīl maʿrūf bih mubāhith-i thulāth* (N.p.: n.pub., lithograph 1311 A.H.).
—, *Ijāzāt al-Ḥadīth*—Muḥammad Bāqir b. Muḥammad Taqī al-Majlisī, *Ijāzāt al-Ḥadīth* (Qum: Maṭbaʿat Āyat Allāh al-Marʿashī, 1410).
—, *Mirʾāt al-ʿUqūl*—Muḥammad Bāqir b. Muḥammad Taqī al-Majlisī, *Mirʾāt al-ʿUqūl fī sharḥ akhbār Āl al-Rasūl* (Tehran: Dār al-Kutub al-Islāmiyya, 1401).
Makdisi, "Ṭabaqāt-*Biography*"—George Makdisi "*Ṭabaqāt*-Biography: Law and Orthodoxy in Classical Islam," *Islamic Studies* 32 (1993) pp. 371–396.
—, *The Rise*—George Makdisi, *The Rise of Colleges: Institutions of Learning in Islam and the West* (Edinburgh: Edinburgh University Press, 1981).
Malik Fihrist—*Fihrist-i Kitābhā-yi Khaṭṭī-i Kitābkhānah-yi Millī-i Mālik* (Tehran: Kitābkhānah-yi Millī-i Malik, 1342Sh).
Malikī, "ʿAllāma Majlisī"—ʿAlī Malikī, "ʿAllāma Majlisī—Akhbārī yā Uṣūlī?" in M. Mahrīzī and Hādī Rabbānī (eds), *Yādnamah-yi ʿAllāmah-yi Majlisī* (Tehran: Intishārāt-i Vazīrat-i Irshād, 1379Sh), pp. 260–312.
Marʿashī Fihrist—*Fihrist-i Nuskhah-hā-yi Khaṭṭī-i Kitābkhānah-yi ʿUmūmī Haḍarat Āyat Allāh al-ʿUẓmā al-Marʿashī* (Qum: Kitābkhānah-yi ʿUmūmī Haḍarat Āyat Allāh al-ʿUẓmā al-Marʿashī, 1354Sh-).
Mashāyikh, "Akhbāriyyah"—Muḥammad Ḥusayn Mashāyikh Farīdānī, "Akhbāriyyah," *Dāʾirat al-maʿārif-i tashayyuʿ*, v. 2, pp. 7–13.
Mashhad Fihrist—*Fihrist-i Kutub-i Khaṭṭī-i Kitābkhānah-yi Markazī-i Āstān-i Quds-i Raḍawī* (Mashhad: Kitābkhānah-yi Markazī-i Āstān-i Quds-i Raḍawī, 1365Sh).
Māzandarānī, *Sharḥ*—Muḥammad Ṣāliḥ al-Māzandarānī, *Sharḥ Uṣūl al-Kāfī* (Beirut: Dār Iḥyāʾ al-Turāth al-ʿArabī, 1421).
—, *Sharḥ al-Maʿālim*—Muḥammad Ṣāliḥ al-Māzandarānī, *Sharḥ Maʿālim al-Dīn* (Tehran: n.p., 1278).
Melchert, "The Imāmīs"—Christopher Melchert, "The Imāmīs between Rationalism and Traditionalism" in L. Clarke (ed.), *Shīʿite heritage Essays on Classical and Modern Traditions* (Binghampton: Global; Publications, 2001), pp. 273–284.
—, *The Formation*—Christopher Melchert, *The Formation of the Sunni Schools of Law, 9th–10th Centuries C.E.* (Leiden: Brill, 1997).
Melli Fihrist—*Fihrist-i Nusakh-i Khaṭṭī-i Kitābkhānah-yi Millī-i Irān* (Tehran: Kitābkhānah-yi Millī-i Irān, 1365Sh).
Modarressi, *An Introduction to Shīʿī Law*—Modarressi, *An Introduction to Shīʿī Law: a Bibliographical Study* (London: Ithaca Press, 1984).
—, "Early Debates"—Hossein Modarressi, "Early Debates on the Integrity of the Qurʾān: a brief survey," *Studia Islamica*, 77 (1993), pp. 5–39.
—, "Rationalism and Traditionalism"—Hossein Modarressi, "Rationalism and Traditionalism in Shīʿī Jurisprudence: A Preliminary Survey," *Studia Islamica* 59 (1984), pp. 141–158.
Mojaddedi, *Biographical Tradition*—Jawid Mojadeddi, *The Biographical Tradition in Sufism: The Tabaqat Genre from al-Sulami to Jami* (London: Curzon, 2001).
Momen, "Uṣūlī, Akhbārī, Shaykhī, Bābī"—Momen, "Uṣūlī, Akhbārī, Shaykhī,

Bābī: The Tribulations of a Qazvin Family" *Iranian Studies*, 36.3 (2003), pp. 317–337.

—, *An Introduction to Shīʿī Islam*—Moojan Momen, *An Introduction to Shīʿī Islam: the history and doctrines of Twelver Shīʿism* (New Haven: Yale University Press, 1985).

Moore, "Four Forms of Skepticism"—"Four Forms of Scepticism," in G.E. Moore, *Philosophical Papers* (London: Allen and Unwin, 1959), pp. 196–225.

Moussavi, *Religious Authority*—Ahmad Kazemi Moussavi, *Religious authority in Shiʿite Islam: from the office of mufti to the institution of marjaʿ* (Kuala Lumpur: International Institute of Islamic Thought and Civilization, 1996).

Mufīd, *al-Muqniʿa*—Muḥammad b. Muḥammad b. al-Nuʿmān al-Shaykh al-Mufīd, *al-Muqniʿa* (Qum: Muʿassasat al-Nashr al-Islāmī, 1410).

Muḥammad Mu'min, *al-Rajʿa*—Muḥammad Mu'min b. Muḥammad Dūst al-Astarābādī, *Kitāb al-Rajʿa* (Qum: Dār al-Iʿtiṣām, 1415).

Muḥaqqiq, *Maʿārij*—Jaʿfar b. al-Ḥasan al-Muḥaqqiq al-Ḥillī, *Maʿārij al-uṣūl* (Qum: Mu'assasat Āl al-Bayt, 1403).

Muʿīnī, "Tafkīk"—Muḥsin Muʿīnī, "Tafkīk," *Dā'irat al-Maʿārif-i Buzurg-i Islāmī*, v. 5, pp. 7–10.

Muntajab al-Dīn, *Fihrist*—ʿAlī b. Bābūya al-Rāzī Muntajab al-Dīn, *Fihrist Muntajab al-Dīn* (Qum: Maktabat Āyat Allāh al-Marʿashī, 1377Sh).

Naṣīrī, "ʿAllāmah-yi Majlisī"—Naṣīrī, "ʿAllāmah-yi Majlisī va-naqd-i dīdgāh-hā-yi Falsafah," *ʿUlūm-i Ḥadīth*, 26 (1381), pp. 46–53.

Nawas and Bernards, "Geographic Distribution"—John Nawas and Monique Bernards, "The Geographic Distribution of Muslim Jurists During the First Four Centuries A.H.," *Islamic Law and Society*, 10.2 (2003), pp. 168–181.

—, "Netherlands Ulama Project"—J. Nawas and M. Bernards, "A preliminary report of the Netherlands Ulama Project: The evolution of the class of ulama In Islam with special emphasis on the non-Arab converts (*mawālī*) from the first through fourth century A.H." in U. Vermeulen and J.M.F. van Reeth (eds) *Law Christianity and Modernism in Islamic Society* (Leuven: Peters, 1998), pp. 97–107.

Nawawī, *al-Majmūʿ*—Yayḥā b. Sharaf al-Dīn al-Nawawī, *al-Majmūʿ fī sharḥ al-Muhadhdhab* (n.p.: Dār al-Fikr, n.d.).

—, *Taqrīb*—Yayḥā b. Sharaf al-Dīn al-Nawawī, *al-Taqrīb wa'l-taysīr li-maʿrifat sunan al-bashīr al-nadhīr* (n.p.: Dār al-Mallāh, n.d.).

Nawbakhtī, *Firaq al-Shīʿa*—Ḥasan b. Mūsā al-Nawbakhtī, *Kitāb Firaq al-Shīʿa* (Cairo: Dār al-Rashād, 1992).

Newman, "Akhbārī/Uṣūlī Dispute pt. 1"—Andrew Newman, "The Nature of the Akhbārī-Uṣūlī Dispute in Late Safawid Iran. Part 1: ʿAbdallāh al-Samāhījī's Munyat al-Mumārisīn," *Bulletin of the School of Oriental and African Studies*, 15 (1992), pp. 22–51.

—, "Akhbārī/Uṣūlī Dispute pt. 2"—Andrew Newman, "The Nature of the Akhbārī-Uṣūlī Dispute in Late Safawid Iran. Part 2: The Conflict Re-Assessed," *Bulletin of the School of Oriental and African Studies*, 15 (1992), pp. 250–261.

—, "Anti-Akhābrī Sentiments"—Andrew Newman, "Anti-Akhbārī Sentiments among the Qajar ʿUlamā': The Case of Muḥammad Bāqir al-Khwānsārī (d. 1313/1895)" in R. Gleave (ed.) *Religion and Society in Qajar Iran* (London: Routledge, 2004), pp. 155–173.

Nūr al-Dīn, *al-Muʿīn*—Nūr al-Dīn Muḥammad al-Kāshānī, *al-Tafsīr al-Muʿīn* (Qum: Maktabat Āyat Allāh al-Marʿashī, n.d.).

Nūr al-Dīn al-Jazā'irī, *Furūq al-Lughāt*—Nūr al-Dīn b. Niʿmat Allāh al-Jazā'irī, *Furūq al-Lughāt fī tamyīz bayn mufād al-kalimāt* (Damascus: al-Mustashariyya al-Thaqāfiyya lil-Jumhūriyya al-Islāmiyya al-Īrāniyya, 1987).

Qayṣarī, "Akhbāriyān"—Iḥsān Qayṣarī, "Akhbāriyān," *Dā'irat al-Ma'ārif-i Buzurg-i Islāmī*, v. 7, pp. 160–163.
Qazwīnī, *Naqḍ*—'Abd al-Jalīl b. Abī al-Ḥusayn al-Qazwīnī, *Kitāb al-Naqḍ ma'rūf bih Ba'ḍ mathālib al-nawāṣīb fī naqḍ Ba'ḍ Faḍā'iḥ al-Rawāfiḍ* (Tehran: Anjuman-i Athār-i Millī, 1980).
Qummī, *Favāyid*—'Abbās b. Muḥammad Riḍā al-Qummī, *Favāyid al-Raḍawiyya fī aḥwāl 'ulamā' madhhab al-ja'fariyya* (Tehran: Kitābkhanah-yi Markazi, 1327Sh).
Rāzī, *al-Maḥṣūl*—Fakhr al-Dīn al-Rāzī, *al-Maḥṣūl fī uṣūl al-fiqh* (Beirut: Mu'assasat al-Risāla, 1412).
Reinhart, *Before Revelation*—Kevin A. Reinhart, *Before Revelation: The Boundaries of Muslim Moral Thought* (Albany: State University of New York Press, 1995).
Riḍā, *Uṣūl al-fiqh*—Muḥammad Riḍā Muẓaffar, *Uṣūl al-fiqh* (Qum: Mu'assasat Intishārāt Dār al-'Ilm, 1378).
Rippin, *Muslims*—Andrew Rippin, *Muslims: their religious beliefs and practices* (London: Routledge, 1990).
Rizvi, *Mulla Muhammad Shirazi*—Sajjad Rizvi, *Mulla Mulla Sadra Shirazi: His Life, Works and Sources for Safavid Philosophy* (Oxford: Oxford University Press, 2007).
Robson, "The Study and Transmission of Ḥadīth"—J. Robson, "Ḥadīth: IV The Study and Transmission of Ḥadīth," *EI2*, v. 3, p. 24.
Schmidtke, "The *ijāza*"—Sabine Schmidtke, "The ijāza from 'Abdallāh b. Ṣāliḥ al-Samāhījī to Nāṣir al-Jarūdī al-Qaṭīfī: A source for Twelver Shī'ī scholarly tradition of Baḥrayn" in F. Daftary and J.W. Meri (eds), *Culture and Memory in Medieval Islam: Essays in Honour of Wilferd Madelung* (London: I.B. Tauris, 2003), pp. 64–85.
Ṣadr, *Takmila*—Sayyid Ḥasan al-Ṣadr, *Takmila Amal al-Āmil* (Qum: Maktabat Āyat Allāh Mar'ashī, 1403).
Salati, "La *Lu'lu'a al-Baḥrayn*"—Marco Salati, "La *Lu'lu'a al-Baḥrayn fī l-ijāza li-qurratay al-'ayn* di Shaykh Yūsuf b. Aḥmad al-Bahrānī (1107–1186/1695–1772): per lo studio della shī'a di Bahrayn," *Annali Cà Foscari*, 28.3 (1989), pp. 111–145.
Sander, *Charisma und Ratio*—Paul Sander, *Zwischen Charisma und Ratio Entwicklungen in der frühen imamitischen Theologie* (Berlin: Klaus-Schwarz-Vlg, 1994).
Scarcia, "Intorno alle controversie"—Gianroberto Scarcia, "Intorno alle controversie tra Aḫbārī e Uṣūlī presso gli imamiti di Persia," *Rivista degli Studi Orientali* 33 (1958), pp. 211–250.
Schoeler, "schriftlichen oder mündlichen"—Gregor Schoeler, "Die Frage der schriftlichen oder mündlichen Überlieferung der Wissenschaften im frühen Islam," *Der Islam*, 62/2 (1985), pp. 201–230.
Sefatgol, *Sākhtār-i Nahād*—Manṣūr Ṣifatgul (Mansour Sefatgol), *Sakhtār-i Nahād va andīshah-yi dīnī dar Īrān-i 'aṣr-i Ṣafavī: tārīkh-i taḥavvulāt-i dīnī-i Īrān dar sadah-hā-yi dahum tā davāzdahum-i Hijrī-i qamarī* (Tehran: Rasā, 1381Sh).
—, "Safavid Administration"—Mansur Sifatgol, "Safavid Administration of *Awqāf*: Structure, Changes and Functions, 1077–1135/1666–1722" in Andrew Newman (ed.), *Society and Culture in the Early Modern Middle East: Studies on Iran in the Safavid Period* (Leiden: Brill, 2003), pp. 397–408.
Sepāhsalār Fihrist—*Fihrist-i Kitābkhānah-yi Madrasah-yi 'Alī Sipāhsālār* (Tehran: Chāp-i Dānishgāh, 1936–).
Shahīdī, "Madrasah-yi Falsafī-i Qazvīn dar 'Aṣr-i Ṣafavī"—'Abd al-Ḥusayn Ṣāliḥī Shahīdī, "Madrasah-yi Falsafi-i Qazvīn dar 'Aṣr-i Ṣafavī: Junbish-hā-yi fikrī va-Uṣūlī va-Akhbārī," *Ḥawzah*, 58 (1366), pp. 33–45.
—, "Sahm-i Ḥawzah-yi 'Ilmiyyah-yi Qazvīn"—'Abd al-Ḥusayn Ṣāliḥī Shahīdī, "Sahm-i Ḥawzah-yi 'Ilmiyyah-yi Qazvīn dar Nahḍat-i Fiqh-i Ja'farī," *Ḥawzah*, 66 (1373), pp. 135–150.

Sharastānī, *al-Milal*—Muḥammad b. ʿAbd al-Karīm al-Sharhrastānī, *Al-Milal wa'l-Naḥl* (Beirut: Dār al-Maʿrifa, 1990).
—, *Kitāb al-Musāraʿa*—Toby Mayer and Wilferd Madelung, *Struggling with the Philosopher: a Refutation of Avicenna's Metaphysics. A new Arabic edition and English translation of al-Shahrastanī's Kitāb al-Musāraʿa* (New York, London: I.B. Taurus, 2001).
Sharīfī, "ʿAql az Dīdgāh-i Akhbāriyān"—Sharīfī, "ʿAql az Dīdgāh-i Akhbāriyān-i Shīʿah," *Shīʿahshināsī*, 6 (1383), pp. 23–57.
Shāṭibī, *al-Muwāfaqāt*—Ibrāhīm b. Mūsā al-Shāṭibī, *al-Muwāfaqāt fī uṣūl al-fiqh* (Beirut: Mu'assasat al-Risālā, 1999).
Shaykh Ḥasan, *Maʿālim*—al-Ḥasan b. Zayn al-Dīn al-Shahīd al-Thānī al-ʿAmīlī, *Maʿālīm al-Dīn wa-malādhdh al-mujtahidīn* (Qum: Dār a-Fikr, 1374).
Shihada, *Teleological Ethics*—Ayman Shihadeh, *The teleological ethics of Fakhr al-Dīn al-Rāzī* (Leiden: Brill, 2006).
Sibā'ī, *Ta'rīkh Makka*—Aḥmad al-Sibā'ī, *Ta'rīkh Makka* (Riyad: Dār Quraysh, 1385).
Stewart, "Capital, Accumulation"—Devin Stewart, "Capital, Accumulation and the Islamic Academic Biography" *Edebiyat* 7 (1997), pp. 139–56.
—, *Islamic Legal Orthodoxy*—Devin Stewart, *Islamic legal orthodoxy: twelver Shiite responses to the Sunni legal system* (Salt Lake City: University of Utah Press, 1998).
—, "The Doctorate"—Devin Stewart, "The Doctorate of Islamic Law in Mamluk Egypt and Syria" in Joseph Lowry, Devin Stewart and Shawkat Toorawa (eds), *Law and Education in Medieval Islam* (Cambridge: E.J.W. Gibb Memorial Trust, 2004), pp. 45–90.
—, "The Genesis"—Devin Stewart, "The Genesis of the Akhbari Revival" in Michel Mazzaoui (ed.), *Safavid Iran and her Neighbours* (Salt Lake City: University of Utah Press, 2003), pp. 169–193.
Suwaydī, *Ta'rīkh*—ʿAbd al-Raḥmān b. ʿAbd Allāh, *Tārīkh Ḥawādith Baghdād wa'l-Baṣra, min 1186 ilā 1192 H/1772–1778 M* (Baghdad: Wizārat al-Thaqāfah wa'l-Funūn, 1978).
Ṭabarsī, *al-Iḥtijāj*—Aḥmad b. ʿAlī al-Ṭabarsī, *al-Iḥtijāj* (Najaf: Dār al-Nuʿmān, 1386).
Ṭabāṭabā'ī, *Riyāḍ al-Masā'il*—al-Sayyid ʿAlī al-Ṭabāṭabā'ī, *Riyāḍ al-Masā'il* (Qum: Mu'assasat al-Nashr al-Islāmī, 1412).
Tabrīzī, *Rayḥānat al-Adab*—Muḥammad ʿAlī Mudarris al-Tabrīzī, *Rayḥānat al-Adab dar tarājim-i aḥvāl* (Tehran: Chāpkhānah-yi Shirkat-i Sihāmī, n.d.).
Tanukābunī, *Qiṣaṣ*—Muḥammad b. Sulaymān al-Tanukābunī, *Qiṣaṣ al-ʿUlamā'* (Tehran: Intishārāt-i ʿIlmī va Farangī, 1383).
—, *Tadhkirat al-ʿUlamā'*—Muḥammad b. Sulaymān al-Tanukābunī, *Tadhkirat al-ʿUlamā'* (Mashhad: Āstān-i Quds-i Raḍawī, 1999).
Ṭihrānī, *al-Dharīʿa*—Muḥammad Muḥsin Āghā Buzurg al-Ṭihrānī, *al-Dharīʿa ilā tasānif al-Shīʿa* (Beirut: Dār al-Aḍwā, 1983–1988).
—, *Ṭabaqāt*—Muḥammad Muḥsin Āghā Buzurg al-Ṭihrānī, *Ṭabaqāt al-Aʿlām* (Qum: Mu'assasat Ismāʿīliyān, nd.).
Tūnī, *al-Wāfiya*—ʿAbd Allāh b. Muḥammad al-Tūnī, *al-Wāfiya fī uṣūl al-fiqh* (Qum: Bāqirī, 1415).
Turner, *Islam without Allah*—Colin Turner, *Islam Without Allah? The Rise of Religious Externalism in Safavid Iran* (Richmond, Curzon Press, 2000).
Ṭūsī, *al-Istibṣār*—Abū Jaʿfar Muḥammad b. Ḥasan al-Ṭūsī, *al-Istibṣār fī mā ukhtalifa al-Akhbār* (Tehran: Dār al-Kutub al-Islāmiyya, 1363Sh).
—, *Khilāf*—Abū Jaʿfar Muḥammad b. Ḥasan al-Ṭūsī, *Kitāb al-Khilāf* (Qum: Mu'assasat al-Nashr al-Islāmī, 1407).

——, *al-Tahdhīb*—Abū Jaʿfar Muḥammad b. Ḥasan al-Ṭūsī, *Tahdhīb al-Aḥkām fī Sharḥ al-Muqniʿa* (Tehran: Dār al-Kutub al-Islāmiyya, 1364Sh).

ʿUmarī, *Ghāyat al-Marām*—Yāsīn b. Khayr Allāh al-ʿUmarī, *Ghāyat al-Marām fī ta'rīkh maḥāsin Baghdād dār al-salām* (Baghdad: Maṭbaʿat Dār al-Baṣrī, 1968).

University Fihrist—*Fihrist-i Nuskhah-hā-yi Khaṭṭī-i Kitābkhānah-yi Markazī-i Dānishgāh-i Tihrān* (Tehran: Kitābkhānah-yi Markazī-i Dānishgāh-i Tihrān, 1367–).

van Ess, *Die Erkenntnislehre des ʿAḍudaddīn al-Īcī*—Josef van Ess, *Die Erkenntnislehre des ʿAḍudaddīn al-Īcī: Übersetzung und Kommentar des ersten Buches seiner Mawāqif* (Wiesbaden: Steiner, 1966).

——, *Theologie und Gesellschaft*—Josef van Ess, *Theologie und Gesellschaft im 2. und 3. Jahrhundert Hidschra: eine Geschichte des religiösen Denkens im frühen Islam* (Berlin: Walter de Gruyter, 1991–97).

Wansbrough, *Quranic Studies*—John Wansbrough, *Qur'ānic Studies: sources and methods of scriptural interpretation* (Oxford: Oxford University Press, 1977).

Wardī, *Lamaḥāt*—ʿAlī al-Wardī, *Lamaḥāt Ijtimāʿiyya min ta'rīkh al-ʿIrāq al-Ḥadīth* (Baghdad: Maṭbaʿat al-Irshād, 1969–1976).

Weiss, *God's Law*—Bernard Weiss, *The Search for God's Law: Islamic jurisprudence in the writings of Sayf al-Dīn al-Āmidī* (Salt Lake City: University of Utah Press, 1992).

——, "Knowledge of the Past"—Bernard Weiss, "Knowledge of the Past: The Theory of *Tawātur* According to Ghazali," *Studia Islamica* 61 (1985), pp. 105, 451.

Woolf, "The New Biography"—Virginia Woolf, "The New Biography" in Virginia Woolf, *Collected Essays* (New York: Harcourt, 1967), v. 4, pp. 229–35.

Zayn al-ʿĀbidīn al-Kāshānī, *Mufarraḥat al-anām*—Zayn al-ʿĀbidīn al-Kāshānī, "Mufarraḥat al-anām fī ta'sīs bayt Allāh al-Ḥarām bih Kūshish-i Rasūl Jaʿfariyān," *Mīrāth-i Islāmī-i Īrān*, 1 (1373Sh), pp. 367–392.

Zayn al-Dīn, *al-Durr al-Manthūr*—Zayn al-Dīn al-ʿĀmilī, *al-Durr al-Manthūr min al-ma'thūr wa-ghayr al-ma'thūr* (Qum: Maktabat Āyat Allāh al-Marʿashī, 1398).

Ziriklī, *Aʿlām*—Khayr al-Dīn al-Ziriklī, *al-Aʿlām* (Beirut: Dār al-ʿIlm, 1980).

INDEX

ISLAMIC PHILOSOPHY, THEOLOGY AND SCIENCE

TEXTS AND STUDIES

ISSN 0169-8729

8. Fakhry, M. *Ethical Theories in Islam.* Second expanded edition. 1994. ISBN 90 04 09300 1
9. Kemal, S. *The Poetics of Alfarabi and Avicenna.* 1991. ISBN 90 04 09371 0
10. Alon, I. *Socrates in Medieval Arabic Literature.* 1991. ISBN 90 04 09349 4
11. Bos, G. *Qusṭā ibn Lūqā's Medical Regime for the Pilgrims to Mecca.* The Risāla fī tadbīr safar al-ḥajj. 1992. ISBN 90 04 09541 1
12. Kohlberg, E. *A Medieval Muslim Scholar at Work.* Ibn Tāwūs and his Library. 1992. ISBN 90 04 09549 7
13. Daiber, H. *Naturwissenschaft bei den Arabern im 10. Jahrhundert n. Chr.* Briefe des Abū l-Faḥl Ibn al-ʿAmīd (gest. 360/970) an ʿAḥudaddaula. Herausgegeben mit Einleitung, kommentierter Übersetzung und Glossar. 1993. ISBN 90 04 09755 4
14. Dhanani, A. *The Physical Theory of Kalām.* Atoms, Space, and Void in Basrian Muʿtazilī Cosmology. 1994. ISBN 90 04 09831 3
15. Abū Maʿšar. *The Abbreviation of the Introduction to Astrology.* Together with the Medieval Latin Translation of Adelard of Bath. Edited and Translated by Ch. Burnett, K. Yamamoto and M. Yano. 1994. ISBN 90 04 09997 2
16. Sābūr Ibn Sahl. *Dispensatorium Parvum (al-Aqrābādhīn al-ṣaghīr).* Analysed, Edited and Annotated by O. Kahl. 1994. ISBN 90 04 10004 0
17. Maróth, M. *Die Araber und die antike Wissenschaftstheorie.* Übersetzung aus dem Ungarischen von Johanna Till und Gábor Kerekes. 1994. ISBN 90 04 10008 3
18. Ibn Abī al-Dunyā. *Morality in the Guise of Dreams.* A Critical Edition of *Kitāb al-Manām*, with Introduction, by Leah Kinberg. 1994. ISBN 90 04 09818 6
19. Kügelgen, A. von. *Averroes und die arabische Moderne.* Ansätze zu einer Neubegründung des Rationalismus im Islam. 1994. ISBN 90 04 09955 7
20. Lameer, J. *Al-Fārābī and Aristotelian Syllogistics.* Greek Theory and Islamic Practice. 1994. ISBN 90 04 09884 4
22. Adang, C. *Muslim Writers on Judaism and the Hebrew Bible.* 1996. ISBN 90 04 10034 2
23. Dallal, A.S. *An Islamic Response to Greek Astronomy. Kitāb Taʿdīl Hayʾat al-Aflāk* of Ṣadr al-Shariʿa. Edited with Translation and Commentary. 1995. ISBN 90 04 09968 9
24. Conrad, L.I. (ed.). The World of Ibn Ṭufayl. *Interdisciplinary Perspectives on* Ḥayy ibn Yaqṣān. 1995. ISBN 90 04 10135 7
25. Hermansen, M.K. (tr.). *The Conclusive Argument from God.* Shāh Walī Allāh of Delhi's *Ḥujjat Allāh al-Bāligha.* 1996. ISBN 90 04 10298 1
26. Abrahamov, B. *Anthropomorphism and Interpretation of the Qurʾān in the Theology of al-Qāsim ibn Ibrāhīm.* Kitāb al-Mustarshid. 1996. ISBN 90 04 10408 9

27. Wild, S. (ed.). *The Qur'an as Text*. 1996. ISBN 90 04 10344 9
28. Riddell, P.G. and T. Street (eds.). *Islam: Essays on Scripture, Thought and Society*. A Festschrift in Honour of Anthony H. Johns. 1997. ISBN 90 04 10692 8
29. Jolivet, J. and R. Rashed (eds.). *Œuvres philosophiques et scientifiques d'al-Kindī*. Volume I. *L'Optique et la Catoptrique*. Edited by R. Rashed. 1997. ISBN 90 04 09781 3
30. Rudolph, U. *Al-Māturīdī und die sunnitische Theologie in Samarkand*. 1997. ISBN 90 04 10023 7
31. Endress, G. and J. A.Aertsen (eds.). *Averroes and the Aristotelian Tradition*. Sources, Constitution and Reception of the Philosophy of Ibn Rushd (1126-1198). Proceedings of the Fourth Symposium Averroicum (Cologne, 1996). ISBN 90 04 11308 8
32. Elmore, G.T. *Islamic Sainthood in the Fullness of Time*. Ibn al-ʿArabī's *Book of the Fabulous Gryphon*. 1999. ISBN 90 04 10991 9
33. Abū Maʿšar. *Kitāb al-milal wa-d-duwal*. Arabic text edited by K. Yamamoto with an English translation by K. Yamamoto and Ch. Burnett. 1999. ISBN 90 04 10725 8
34. Albumasar. *De Magnis Coniunctionibus* (On the Great Coniunctions). Latin text edited by Ch. Burnett and Arabic-Latin, Latin Arabic Glossaries by K. Yamamoto and Ch. Burnett. 1999. ISBN 90 04 11074 7
35. Stroumsa, S. *Freethinkers of Medieval Islam*. Ibn al-Rāwandī, Abū Bakr al-Rāzī, and Their Impact on Islamic Thought. 1999. ISBN 90 04 11374 6
36. King, D.A. *World-Maps for Finding the Direction and Distance to Mecca*. Innovation and Tradition in Islamic Science. 1999. ISBN 90 04 11367 3
37. Bar-Asher, M.M. *Scripture and Exegesis in Early Imāmī Shiism*. 1999. ISBN 90 04 11495 5
38. Sayyid Jalāl al-Dīn Ãshtiyānī et al. *Consciousness and Reality*. Studies in Memory of Toshihiko Izutsu. 2000. ISBN 90 04 11586 2
39. Schmidtke, S. *Theologie, Philosophie und Mystik im zwölfershiitischen Islam des 9./15. Jahrhunderts*. Die Gedankenwelten des Ibn Abī ṛumhūr al-Aḥsāʾī (um 838/1434-35 — nach 906/1501). 2000. ISBN 90 04 11531 5
40. Griffel, F. *Apostasie und Toleranz im Islam*. Die Entwicklung zu al-Ġazālīs Urteil gegen die Philosophie und die Reaktionen der Philosophen. 2000. ISBN 90 04 11566 8
41. Heemskerk, M.T. *Suffering in the Muʿtazilite Theology*. ʿAbd al-Jabbār's Teaching on Pain and Divine Justice. 2000. ISBN 90 04 11726 1
42. Rashed, R. and H. Bellosta. *Ibrāhīm Ibn Sinān. Logique et Géométrie au Xe Siècle*. 2000. ISBN 90 04 11804 7
43. Tamer, G. *Islamische Philosophie und die Krise der Moderne*. Das Verhältnis von Leo Strauss zu Alfarabi, Avicenna und Averroes. 2001. ISBN 90 04 12029 7
44. Genequand, C. *Alexander of Aphrodisias on the Cosmos*. 2001. ISBN 90 04 11963 9
45. Calverley, E.E. and J.W. Pollock. *Nature, Man and God in Medieval Islam*.
    Volume one. 2002. ISBN 90 04 12381 4
    Volume two. 2002. ISBN 90 04 12382 2

46. Swartz, M. *A Medieval Critique of Anthropomorphism.* Ibn al-Jawzī's Kitāb Akhbār aṣ-Ṣifāt. 2002. ISBN 90 04 12376 8
47. Kusuba, T. and D. Pingree. *Arabic Astronomy in Sanskrit.* Al-Birjandī on Tadhkira II, Chapter 11 and its Sanskrit Translation. 2002. ISBN 90 04 12475 6
48. İskenderoğlu, M. *Fakhr al-dīn al-Rāzī and Thomas Aquinas on the Question of the Eternity of the World.* 2002. ISBN 90 04 12480 2
49. Reisman, D.C. *The making of the Avicennan Tradition.* The Transmission, Contents, and Structure of Ibn Sīnā's al-Mubāḥaṯāt (The Discussions). 2002. ISBN 90 04 12504 3
50. Behloul, S.M. *Ibn Ḥazms Evangelienkritik.* Eine methodische Untersuchung. 2002. ISBN 90 04 12527 2
51. Charette, F. *Mathematical Instrumentation in Fourteenth-Century Egypt and Syria.* The Illustrated Treatise of Najm al-Dīn al-Mīṣrī. 2003. ISBN 90 0413015 2
52. Reisman, D.C. (ed.). *Before and After Avicenna.* Proceedings of the First Conference of the Avicenna Study Group. 2003. ISBN 90 04 12978 2
53. Kahl, O. *Sābūr Ibn Sahl. The Small Dispensatory.* 2003. ISBN 90 04 12996 0
54. Burnett, C., J.P. Hogendijk, K. Plofker and M. Yano. *Studies in the History of the Exact Sciences in Honour of David Pingree.* 2004. ISBN 90 04 13202 3
56. McGinnis, J. (ed.). *Interpreting Avicenna: Science and Philosophy in Medieval Islam.* Proceedings of the Second Conference of the Avicenna Study Group. 2004. ISBN 90 04 13960 5
57. Eschraghi, A. *Frühe Šaiḫī- und Bābī-Theologie.* Die Darlegung der Beweise für Muḥammads besonderes Prophetentum (*Ar-Risāla fī Iṯbat an-Nubūwa al-Ḫāṣṣa*). 2004. ISBN 90 04 14034 4
58. Acar, R. *Talking about God and Talking about Creation.* Avicenna's and Thomas Aquinas' positions. 2005. ISBN 90 04 14477 3
59. Akasoy, A. *Philosophie und Mystik in der späten Almohadenzeit.* Die *Sizilianischen Fragen* des Ibn Sabʿīn. 2006. ISBN 90 04 14522 2
60. Schöck, C. *Koranexegese, Grammatik und Logik.* Zum Verhältnis von arabischer und aristotelischer Urteils-, Konsequenz- und Schlußlehre. 2006. ISBN 90 04 14588 5
61. Endress, G. (ed.). *Organizing Knowledge.* Encyclopædic Activities in the Pre-Eighteenth Century Islamic World. 2006. ISBN 978 90 04 14697 6
62. Mourad, S.A. *Early Islam between Myth and History.* Al-Ḥasan al-Baṣrī (d. 110H/728CE) and the Formation of His Legacy in Classical Islamic Scholarship. 2006. ISBN 978 90 04 14829 1
63. Bertolacci, A. *The Reception of Aristotle's* Metaphysics *in Avicenna's* Kitāb al-Šifāʾ. A Milestone of Western Metaphysical Thought. 2006. ISBN 978 90 04 14899 4
64. Shihadeh, A. *The Teleological Ethics of Fakhr al-Dīn al-Rāzī.* 2006. ISBN 978 90 04 14991 5
65. Pourjavady, R. and S. Schmidtke. *A Jewish Philosopher of Baghdad.* ʿIzz al-Dawla Ibn Kammūna (d. 683/1284) and His Writings. 2006. ISBN 978 90 04 15139 0

66. Zakeri, M. (ed. and tr.). *Persian Wisdom in Arabic Garb*. ʿAlī b. ʿUbayda al-Rayḥānī (D. 219/834) and his *Jawāhir al-kilam wa-farāʾid al-ḥikam*. 2007. ISBN 978 90 04 15168 0
67. Wakelnig, E. (Hrsg.). *Feder, Tafel, Mensch*. Al-ʿĀmirīs *Kitāb al-Fuṣūl fī l-Maʿālim al-ilāhīya* und die arabische Proklos-Rezeption im 10. Jh. 2006. ISBN 978 90 04 15255 7
68. Schmidl, P. *Volkstümliche Astronomie im islamischen Mittelalter*. Zur Bestimmung der Gebetszeiten und der Qibla bei al-Aṣbaḥī, Ibn Raḥīq und al-Fārisī. 2007. ISBN 978 90 04 15390 5
69. Belo, C. *Chance and Determinism in Avicenna and Averroes*. 2007. ISBN 978 90 04 15587 9
70. Kahl, O. *The Dispensatory of Ibn at-Tilmīḏ*. Arabic Text, English Translation, Study and Glossaries. 2007. ISBN 978 90 04 15620 3
71. Hildebrandt, T. *Neo-Muʿtazilismus?* Intention und Kontext im modernen arabischen Umgang mit dem rationalistischen Erbe des Islam. 2007. ISBN 978 90 04 15099 7
72. Gleave, R. *Scripturalist Islam*. The History and Doctrines of the Akhbārī Shīʿī School. 2007. ISBN 978 90 04 15728 6
73. Hoover, J. *Ibn Taymiyya's Theodicy of Perpetual Optimism*. ISBN 978 90 04 15847 4
74. Vrolijk, A. and J. Hogendijk (eds.). *O ye Gentlemen: Arabic Studies on Science and Literary Culture*. In Honour of Remke Kruk. 2007. ISBN 978 90 04 15794 1